Control of Permanent Magnet Actuators for Robotics Applications

Other related titles:

You may also like

- PBCE077| Rigatos | Nonlinear Optimal and Flatness-based Control Methods and Applications for Complex Dynamical Systems | 2025
- PBCE127 | Ida | Sensors, Actuators, and Their Interfaces, 2nd Edition: A multidisciplinary introduction | 2020 (3rd Ed under contract)
- PBCE126 | Monteriù | Fault Diagnosis and Fault-Tolerant Control of Robotic and Autonomous Systems | 2020
- PBCE121 | Davoodi | Integrated Fault Diagnosis and Control Design of Linear Complex Systems | 2018

We also publish a wide range of books on the following topics:
Computing and Networks
Control, Robotics and Sensors
Electrical Regulations
Electromagnetics and Radar
Energy Engineering
Healthcare Technologies
History and Management of Technology
IET Codes and Guidance
Materials, Circuits and Devices
Model Forms
Nanomaterials and Nanotechnologies
Optics, Photonics and Lasers
Production, Design and Manufacturing
Security
Telecommunications
Transportation

All books are available in print via https://shop.theiet.org or as eBooks via our Digital Library https://digital-library.theiet.org.

CONTROL, ROBOTICS AND SENSORS SERIES 138

Control of Permanent Magnet Actuators for Robotics Applications

Chao Gong, Yihua Hu, Jinqiu Gao and Wenzhen Li

The Institution of Engineering and Technology

About the IET

This book is published by the Institution of Engineering and Technology (The IET).

We inspire, inform and influence the global engineering community to engineer a better world. As a diverse home across engineering and technology, we share knowledge that helps make better sense of the world, to accelerate innovation and solve the global challenges that matter.

The IET is a not-for-profit organisation. The surplus we make from our books is used to support activities and products for the engineering community and promote the positive role of science, engineering and technology in the world. This includes education resources and outreach, scholarships and awards, events and courses, publications, professional development and mentoring, and advocacy to governments.

To discover more about the IET please visit https://www.theiet.org/

About IET books

The IET publishes books across many engineering and technology disciplines. Our authors and editors offer fresh perspectives from universities and industry. Within our subject areas, we have several book series steered by editorial boards made up of leading subject experts.

We peer review each book at the proposal stage to ensure the quality and relevance of our publications.

Get involved

If you are interested in becoming an author, editor, series advisor, or peer reviewer please visit https://www.theiet.org/publishing/publishing-with-iet-books/ or contact author_support@theiet.org.

Discovering our electronic content

All of our books are available online via the IET's Digital Library. Our Digital Library is the home of technical documents, eBooks, conference publications, real-life case studies and journal articles. To find out more, please visit https://digital-library.theiet.org.

In collaboration with the United Nations and the International Publishers Association, the IET is a Signatory member of the SDG Publishers Compact. The Compact aims to accelerate progress to achieve the Sustainable Development Goals (SDGs) by 2030. Signatories aspire to develop sustainable practices and act as champions of the SDGs during the Decade of Action (2020-2030), publishing books and journals that will help inform, develop, and inspire action in that direction.

In line with our sustainable goals, our UK printing partner has FSC accreditation, which is reducing our environmental impact to the planet. We use a print-on-demand model to further reduce our carbon footprint.

Whilst every reasonable effort has been undertaken by the Publisher and its licensors to acknowledge copyright on material reproduced, if there has been an oversight, please contact the Publisher and we will endeavour to correct this upon a reprint.

Trade mark notice: Product or corporate names referred to within this publication may be trade marks or registered trade marks and are used only for identification and explanation without intent to infringe.

Where an author and/or contributor is identified in this publication by name, such author and/or contributor asserts their moral right under the CPDA to be identified as the author and/or contributor of this work.

British Library Cataloguing in Publication Data

A catalogue record for this product is available from the British Library

ISBN 978-1-83724-033-3 (hardback)
ISBN 978-1-83724-034-0 (PDF)

Typeset in India by MPS Limited

Cover image: Yuichiro Chino/Moment via Getty Images

Contents

About the authors

Dr Chao Gong is currently a professor at the School of Automation, Northwestern Polytechnical University, China. He received his M.Eng. degree in electrical engineering from Northwestern Polytechnical University, Xi'an, China, and his Ph. D. degree in electrical engineering from the University of York, UK. He was a postdoctoral fellow in the Department of Electrical and Computer Engineering, University of Alberta, Canada, between 2022 and 2023. His research interests include robot actuators, motion control, electrical machine design and drives, and power electronics. He holds 10+ granted patents in the area of motor drives. He was awarded the IET Postgraduate Prize and IET Travel Award for his research on *the safety of high-voltage powertrain based electric vehicles* and the KM Stott Prize for excellence in scientific research in 2021.

Dr Yihua Hu is a reader at King's College London, UK. He was previously a research associate with the Power Electronics and Motor Drive Group, University of Strathclyde, a lecturer with the Department of Electrical Engineering and Electronics, University of Liverpool, and Electrical Engineering group head with the Department of Electronics Engineering, University of York, UK. He has authored and co-authored over 160 papers in *IEEE Transactions Journals*. His research interests include robotics, power electronics converters and control, electric vehicles, electric ships and aircrafts, smart energy system, and nondestructive testing technology. He is an associate editor for the *IET Renewable Power Generation*, *IET Intelligent Transport Systems*, *IEEE Transactions on Industrial Electronics*, and *Power Electronics and Drive*s. He is a fellow of the Institution of Engineering and Technology (FIET) and a member of the UK Young Academy. He was a recipient of the Royal Society Industry Fellowship. He holds a Ph.D. degree in power electronics and drives from China University of Mining and Technology, Xuzhou, China.

Dr Jinqiu Gao was born in Shaanxi province, China, on 7 January 1996. She received the B.Eng. degree in electrical engineering from Northwestern Polytechnical University, Xi'an, China, in 2017, where she is currently pursuing a master's degree in electrical engineering at the School of Automation. She is currently a lecturer at Xi'an University of Technology. Her research interests include electrical machine design, drives, and motion control.

Dr Wenzhen Li was born in Xinjiang, China, in April 1991. She received the B.E. degree in electrical engineering from Henan University of Technology, Zhengzhou, China, in 2012, and the M.E. degree in control engineering from Xi'an Polytechnic University, Xi'an, China, in 2015. She received her Ph.D. degree in electrical engineering with Northwestern Polytechnical University, Xi'an. Her research interests include permanent magnet synchronous motor drives and sensorless control of AC motors. She is currently a lecturer with Xi'an Aeronautical Institute.

Chapter 1

Introduction to permanent magnet robot actuators

1.1 Introduction to robotics and actuation systems

Robotics, as a field, has undergone a remarkable transformation, evolving from rudimentary mechanical devices to highly sophisticated systems capable of automating complex tasks. This journey reflects a synergy of advancements in mechanics, electronics, and computing, with actuation technologies playing a pivotal role in enabling robots to interact with and manipulate their environments. The history of robotics is one of ingenuity, innovation, and relentless pursuit of automation, which has shaped the modern world.

Robotic concepts date back to ancient times, when mechanical ingenuity laid the foundation for automated systems. Greek mathematician Hero of Alexandria, in the 1st century CE, designed mechanical devices powered by steam, water, and weights. These automata, while primitive by today's standards, were ingenious in their ability to mimic basic human actions. Similarly, in the Islamic Golden Age, Al-Jazari crafted intricate automata, including humanoid machines and water-powered clocks, showcasing mechanical principles that predated the industrial use of robotics for centuries. The Industrial Revolution marked a pivotal shift, transforming human labor through mechanization. Machines designed during this era, though not robots in the modern sense, introduced gears, levers, and cams to automate repetitive processes. Among the significant inventions of this period was Jacquard's Loom. This programmable textile loom utilized punched cards to control intricate weaving patterns, heralding the concept of machine programmability that underpins contemporary robotics.

In the early 20th century, the concept of robots as autonomous, labor-performing entities gained prominence. Karel Čapek's 1920 play R.U.R. (Rossum's Universal Robots) popularized the term "robot," describing mechanical beings designed to replace human labor. This was followed by the intellectual contributions of Isaac Asimov, whose "Three Laws of Robotics" introduced ethical considerations into the design and deployment of robots. These cultural milestones paralleled technological advancements in control systems and electronics, setting the stage for the development of modern robotics.

The mid-20th century saw a transition from conceptual exploration to practical implementation. George Devol's invention of the programmable robotic arm in

1954, later commercialized as the Unimate, revolutionized manufacturing pro-cesses. The Unimate's ability to perform repetitive tasks with precision paved the way for its adoption in automotive assembly lines, exemplifying the practical utility of robotic automation. Similarly, the Stanford Arm, developed in 1968, demon-strated the potential of joint electric actuators, which allowed robots to perform complex and precise motions, further enhancing industrial automation capabilities.

As robotics entered the late 20th and early 21st centuries, technological con-vergence accelerated its adoption across diverse fields. The integration of compu-ters, sensors, and artificial intelligence, enabled robots to transition from static, preprogrammed systems to dynamic entities capable of interacting with and adapting to their environments. This era witnessed the rise of service robots, medical robotics, and autonomous vehicles, underscoring the versatility of robotic systems in solving real-world challenges.

The development of actuation technologies has been a cornerstone in the evolution of robotics. Early robotic systems predominantly relied on hydraulic and pneumatic actuators, as shown in Figure 1.1. Hydraulic actuators, favored for their ability to generate substantial force, were ideal for applications requiring heavy lifting, such as industrial assembly lines. Pneumatic actuators, on the other hand, offered reliability and simplicity, making them suitable for lightweight tasks like pick-and-place operations. Despite their advantages, these systems faced sig-nificant limitations. Hydraulic systems, for instance, were bulky and required

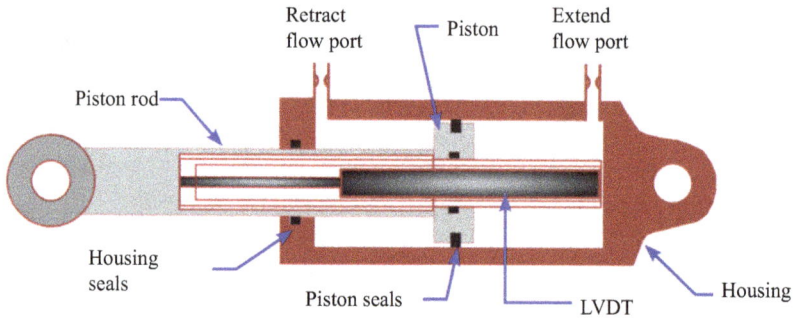

http://www.daerospace.com/HydraulicSystems/ActuatorHydraulicDesc.php

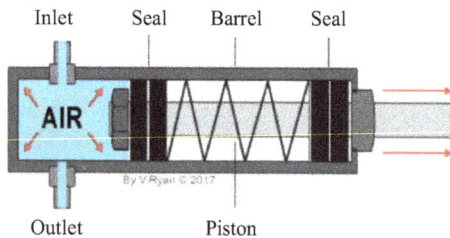

Figure 1.1   Hydraulic and pneumatic actuators

extensive maintenance due to potential fluid leaks. Pneumatic systems, while cost-effective, lacked the precision necessary for advanced robotic applications.

The limitations of hydraulic and pneumatic actuators catalyzed the transition to electric actuators, which now dominate modern robotics. Advances in materials science played a crucial role in this shift. The advent of rare-earth magnets, such as neodymium and samarium cobalt, enabled the creation of compact, high-performance electric actuators. These materials, combined with innovations in digital control systems, facilitated the development of permanent magnet brushless DC motors (PM-BLDCs) and permanent magnet synchronous motors (PMSMs), as shown in Figure 1.2, which are now standard in robotic applications. Electric actuators offer numerous advantages over their hydraulic and pneumatic counter-parts. Their high torque-to-weight ratio makes them ideal for applications requiring compact and lightweight designs, such as drones and robotic arms. Additionally, their low maintenance requirements and ability to integrate seamlessly with digital control systems allow for precise and adaptive motion control. These character-istics have expanded the application of electric actuators across fields, from man-ufacturing and healthcare to exploration and service robotics.

The ongoing evolution of robotic systems and their underlying technologies reflects humanity's quest for efficiency and innovation. As robots become more autonomous and versatile, their reliance on advanced actuation systems will remain a critical factor in their success. The transition from hydraulic and pneumatic sys-tems to electric actuators is not merely a technological advancement but a testa-ment to the adaptability and ingenuity of the field of robotics. This historical journey underscores the importance of continued research and development in actuator design, ensuring that robotic systems meet the demands of an increasingly complex and dynamic world.

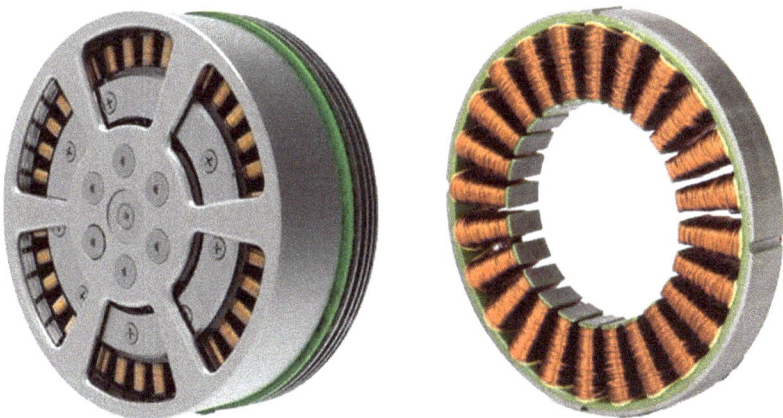

Figure 1.2   Robotic permanent magnet motors

The history of robotics is a narrative of human creativity and problem-solving, demonstrating how mechanical and electronic ingenuity can transform industries and redefine possibilities. From ancient automata to state-of-the-art robotic systems, the field of robotics continues to push boundaries, driven by the relentless pursuit of innovation and the ever-expanding capabilities of actuation technologies.

1.2   Role of actuators in robotics

As shown in Figure 1.3, actuators serve as the "muscles," "joints," and "skeletons" of robotic systems, providing the mechanical motion necessary for robots to perform tasks and interact with their environments. They are the critical components that convert electrical energy into mechanical motion, enabling robots to perform actions such as lifting, rotating, gripping, or walking. The functionality of actuators directly influences a robot's capabilities, including its speed, precision, strength, and efficiency. Without actuators, robots would lack the ability to physically engage with their surroundings, reducing them to passive computational systems.

Actuators come in various forms, each suited to specific applications. Hydraulic actuators, for instance, are known for their ability to generate high force and are commonly used in heavy-duty applications such as construction robots and industrial machinery. They operate by using pressurized fluids to produce linear or rotary motion. Despite their power, hydraulic actuators are bulky and require significant maintenance, limiting their use in applications where size and precision are critical. Pneumatic actuators, which rely on compressed air, are another widely used type. They are favored for their simplicity, reliability, and cost-effectiveness, making them suitable for repetitive tasks like pick-and-place operations in manufacturing. However, pneumatic actuators struggle to achieve the precision and fine

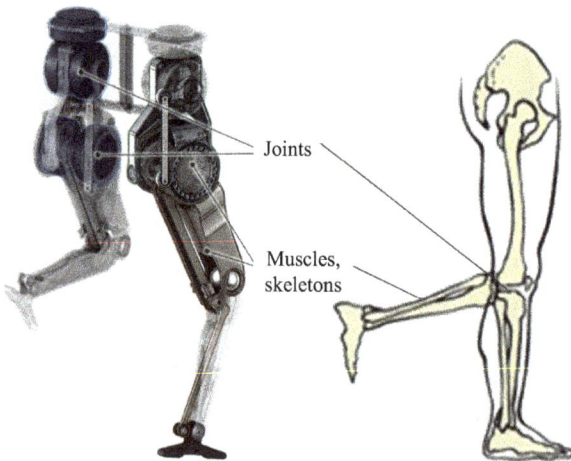

Figure 1.3   Role of actuators in robotics

control required for more intricate robotic applications, as their motion can be less predictable due to the compressibility of air. Electric actuators, including PM-BLDCs and PMSMs, have become the preferred choice for modern robotics. These actuators offer exceptional precision, high torque-to-weight ratios, and energy efficiency. Their compact design makes them ideal for applications where space and weight are limited, such as drones, robotic arms, and surgical robots. Moreover, electric actuators integrate seamlessly with digital control systems, allowing for real-time feedback and adaptive motion control. This ability to deliver precise, programmable motion has driven the widespread adoption of electric actuators in advanced robotic systems.

The growing preference for electric actuators stems from their superior performance and adaptability. Unlike hydraulic and pneumatic systems, electric actuators require minimal maintenance and exhibit high reliability, even in demanding environments. Additionally, advancements in materials, such as rare-earth magnets, and innovations in power electronics have further enhanced their capabilities, ensuring they remain at the forefront of robotic actuation technology. As robotic systems become more complex and diverse in their applications, the role of actuators as the mechanical backbone of robots will continue to be central, enabling new possibilities and pushing the boundaries of what robots can achieve.

1.3   Permanent magnet actuators

Permanent magnet actuators (PMAs) represent a transformative class of electric actuators that leverage the properties of permanent magnets to generate electromagnetic force [1]. These actuators play an integral role in modern robotics, delivering exceptional efficiency, precision, and compactness, which are critical for a wide range of applications. Unlike traditional actuators that often rely on electromagnets to create motion, PMAs utilize permanent magnets made of advanced materials such as neodymium or samarium cobalt. These magnets create a persistent magnetic field that interacts with electric currents to produce mechanical motion. This configuration allows PMAs to achieve unparalleled levels of performance and reliability, solidifying their importance in both industrial and service robotics.

One of the primary applications of PMAs is the use of PMSMs. PMSMs are designed to synchronize the rotor's movement with the magnetic field generated by the stator. This synchronization results in smooth and highly controllable motion, making PMSMs ideal for tasks that require precision and consistency. They are widely used in robotic arms, CNC machines, and advanced manufacturing systems. Another key application is the use of PM-BLDC, which is characterized by its robust design and high operational speeds. Unlike brushed motors, PM-BLDCs eliminate the use of brushes and commutators, which are prone to wear and tear. This design enhances their durability and reduces the need for frequent maintenance, making them an excellent choice for applications such as drones, autonomous vehicles, and medical robots.

The functional performance of PMAs is enabled by their key structural components: motors, controllers, brakes, and reducers (reducers refer to mechanisms used for speed reduction, such as gearboxes). These components work in harmony to deliver precise motion control and high efficiency. Motors are the core element of PMAs, converting electrical energy into mechanical motion. Depending on the application, these motors can be compact, integrated into single units, or modular, allowing for flexibility in system design, as shown in Figure 1.4. Their properties are compared in Table 1.1. PMSMs and BLDCs dominate the motor technology used in PMAs due to their efficiency, precision, and durability. Controllers are vital for managing the operation of PMAs. They regulate the power supplied to the motors, ensuring that they operate within desired parameters. Advanced controllers use feedback systems and sophisticated algorithms to provide precise control over speed, torque, and position. Modern controllers also integrate communication interfaces for seamless integration into complex robotic systems. Reducers serve to modify the output speed and torque of PMAs. By adjusting the gear ratios, reducers enable actuators to deliver the necessary force or speed for specific tasks. Integrated PMAs often combine the motor, controller, and reducer into a single compact unit, enhancing performance and simplifying installation. In contrast, modular designs keep these components separate, allowing for greater customization and easier maintenance.

The performance of PMAs is further enhanced by the integration of sensors. Sensors provide critical data for real-time monitoring and control, enabling

Figure 1.4   Integrated and modular PMAs

Table 1.1   Comparisons of integrated and modular PMAs

Feature	Integrated PMAs	Modular PMAs
Compactness	High	Moderate
Customization	Limited	High
Maintenance	Moderate	Easier
Performance tuning	Pre-optimized	Customizable
Installation	Simplified	Requires additional configuration

actuators to perform complex tasks with high accuracy. Key sensors used in PMAs include:

1. Position sensors: Encoders and resolvers measure the precise position of the motor shaft and also the output of the reducer, ensuring accurate movement and positioning.
2. Torque sensors: These measure the force generated by the actuator, ensuring that torque output matches the requirements of the task.
3. Temperature sensors: Monitoring the temperature of the motor and surrounding components helps prevent overheating and ensures safe operation.
4. Current sensors: These measure the electrical current supplied to the motor, providing insights into energy consumption and identifying potential faults.

The unique advantages of PMAs make them indispensable in robotics. Their high efficiency ensures that the energy consumed is effectively converted into mechanical motion, reducing wastage and optimizing operational costs. The compact and lightweight design of PMAs allows them to be integrated into small-scale robotic systems without compromising on performance. For instance, mobile robots and prosthetic devices benefit greatly from this characteristic. Furthermore, PMAs exhibit a high torque-to-weight ratio, providing the necessary force for demanding applications while maintaining energy efficiency. Unlike traditional brushed actuators, PMAs require minimal maintenance as they lack components that are susceptible to wear, such as brushes. This reliability is crucial in applications where downtime must be minimized, such as in industrial automation or healthcare robotics.

The performance of PMAs can be evaluated through several key parameters. Torque density, which refers to the amount of torque produced per unit volume, is a critical measure of their effectiveness in space-constrained environments. PMAs excel in this aspect, making them ideal for compact robotic systems. Power efficiency, another vital parameter, indicates how effectively the actuator converts electrical energy into mechanical work. High power efficiency reduces operational costs and extends battery life in mobile and portable robotic applications. Dynamic response, which measures the actuator's ability to respond to changes in input, is essential for tasks requiring agility and precision, such as robotic surgery or high-speed pick-and-place operations. Lastly, control precision ensures that the actuator can execute movements with high accuracy, a requirement for applications in aerospace and advanced manufacturing.

To provide a clearer perspective, Table 1.2 compares the characteristics of PMAs with hydraulic and pneumatic actuators.

In summary, PMAs are at the forefront of modern robotics, enabling the development of systems that are efficient, reliable, and capable of handling complex tasks with precision. Their unique advantages over traditional actuator types have solidified their position as a cornerstone of robotic innovation, driving advancements across industries and paving the way for future breakthroughs in automation and artificial intelligence.

Table 1.2   Comparative characteristics of PMAs with hydraulic and pneumatic actuators

Feature	Permanent magnet actuators	Hydraulic actuators	Pneumatic actuators
Efficiency	High	Moderate	Low
Precision	High	Low	Moderate
Maintenance	Low	High	Moderate
Compactness	High	Low	High
Torque-to-Weight Ratio	High	High	Low
Reliability	High	Moderate	Low
Cost-effectiveness	Moderate	High	High

1.4   Permanent magnet synchronous motors in PMAs

The term "permanent magnet" in PMAs primarily stems from the fact that the driving motors used in these actuators are permanent magnet motors, which generate force through permanent magnet materials, regardless of whether they are PMSMs or PM-BLDCs. In practice, while PMAs can be driven by both PMSMs and PM-BLDC motors, PMSMs offer greater potential in terms of control precision. This advantage is primarily due to the use of high-precision position sensors, such as resolvers or encoders, which provide significantly higher accuracy compared to the Hall-effect sensors typically used in BLDC motors. As a result, PMSMs exhibit lower torque ripple and higher precision, making them more suitable for robotic applications that demand high control accuracy and smooth operation. The stator structure of a conventional PMSM is highly similar to that of an induction motor, as illustrated in Figure 1.5.

To minimize eddy current losses and hysteresis losses caused by the magnetic field, the stator core of a PMSM is typically constructed using laminated silicon steel sheets with a thickness of less than 0.5 mm. These sheets are stacked and punched with evenly distributed slots, which accommodate embedded three-phase symmetrical windings. The stator slot structure usually adopts a semi-closed slot design, which helps reduce cogging effects. Among various slot types, the pear-shaped slot is widely used due to its high slot area utilization, extended die lifespan, and minimal bending of slot insulation, reducing the risk of wear and damage. The stator windings are typically made of round copper wires. To minimize stray losses, double-layer short-pitch windings with a star connection are commonly used. In low-power motors, single-layer windings may also be applied, while in special applications, sinusoidal windings are sometimes employed. The key distinction between PMSMs and induction motors lies in the rotor structure, which incorporates permanent magnets. Depending on the positioning of the magnets within the rotor, PMSMs are generally classified into three major types: surface-mounted PMSM, inset PMSM, and interior PMSM (IPMSM) [2].

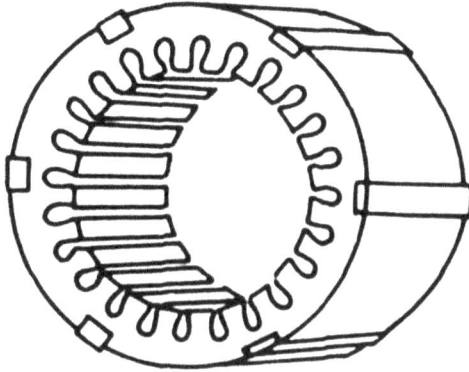

Figure 1.5 Stator structure of a conventional PMSM

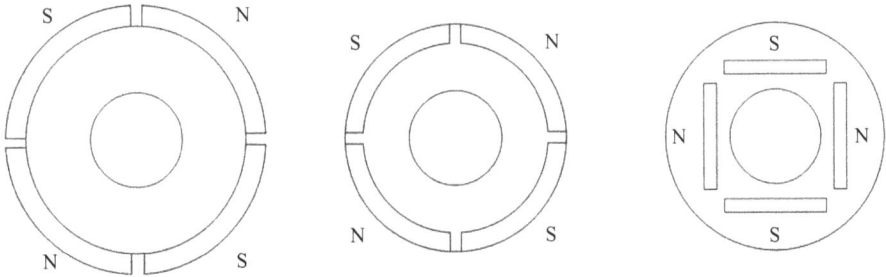

Figure 1.6 Configurations of permanent magnet motors

As illustrated in Figure 1.6, the first two configurations are collectively referred to as external-mounted structures. These designs allow for a smaller rotor diameter and lower inertia, which is particularly advantageous for high dynamic performance. If the permanent magnets are directly bonded to the shaft, the motor exhibits low inductance, further enhancing dynamic characteristics. Due to these advantages, permanent magnet servo motors predominantly adopt external-mounted structures. In contrast, the IPMSM features permanent magnets embedded within the rotor core, where each magnet is enclosed by the iron core. This design provides higher mechanical strength and a reduced air gap in the magnetic circuit, making it more suitable for field-weakening operations compared to external-mounted rotors. From a structural perspective, surface-mounted PMSMs essentially function as salient-pole synchronous motors, as the relative permeability of permanent magnet materials is close to that of air, resulting in almost equal d-axis and q-axis inductances. Meanwhile, both inset PMSMs and IPMSMs are classified as salient-pole synchronous motors, where the q-axis inductance is larger than the d-axis inductance. This inductance difference enables the generation of both electromagnetic torque and reluctance torque. By effectively utilizing reluctance torque, higher operational efficiency can be achieved.

The working principle of a PMSM is based on the interaction between a rotating magnetic field and the rotor-mounted permanent magnets. When a three-phase current is applied to the stator windings, it generates a rotating magnetic field in the motor's air gap. The rotor, equipped with permanent magnets, is then driven to rotate due to the attraction and repulsion between the stator's rotating field and the rotor's magnetic field. Eventually, the rotor reaches a steady state where its rotational speed matches that of the stator's rotating magnetic field, ensuring synchronous operation. The design of PMSMs used in PMAs is a complex and iterative process. Currently, the prevailing design approach follows a sequence that begins with dimensional calculations based on performance requirements, followed by finite element simulation for validation, prototype testing, and iterative optimization. The first step, determining the motor's fundamental dimensions, is crucial as it directly impacts the subsequent workload and the likelihood of achieving a well-optimized design in a single iteration.

To efficiently develop a PMSM that meets the desired specifications, the design process must address several key aspects, including motor dimensions, stator slot configuration, three-phase winding structure, permanent magnet selection and placement, and rotor design. This process requires not only theoretical derivations and numerical calculations but also the pre-selection of empirical coefficients, which are influenced by various factors such as manufacturing processes, material properties, structural constraints, and cooling conditions. These empirical factors play a significant role in ensuring that the designed motor achieves the intended performance characteristics. The fundamental equation governing motor dimensioning is given by (1.1), which establishes the key parameters necessary for determining the motor's primary design specifications.

$$\frac{D^2 l_{ef} n_N}{P'} = \frac{6.1}{\alpha'_p K_{Nm} K_{dp} A B_\delta} = C_A \tag{1.1}$$

where

D, Armature diameter (m);

L_{ef}, Effective length of the armature core (m);

n_N, Rated motor speed (rpm);

P', Calculated power (W);

α'_p, Pole arc coefficient;

K_{Nm}, K_{dp}: Air-gap magnetic field waveform coefficient and winding coefficients, respectively;

C_A, Motor constant.

Once the linear current density (A) and air-gap flux density (B_δ) are selected, the product $D^2 L_{ef}$ (where D is the armature diameter and L_{ef} is the effective length of the armature core) can be determined using (1.1). Motors with the same $D^2 L_{ef}$

can be designed with different aspect ratios, either elongated or compact, with the final geometric proportions determined by the length-to-diameter ratio $\lambda = L_{ef}/D$. The electromagnetic loadings A and B_δ not only affect the material consumption of the motor but also play a crucial role in determining motor performance. To reduce motor size, volume, and material usage, typical design values are chosen as $A = 280$ A/cm and $B_\delta = 6,900$ GS. Additionally, for PMSMs, the pole arc coefficient is set to 0.76, while the air-gap magnetic field waveform coefficient (K_{Nm}) and winding coefficient (K_{dp}) are chosen as 1.05 and 0.96, respectively.

From a performance and thermal perspective, a higher number of stator slots is generally preferable, as it results in a better magnetomotive force (MMF) wave-form, allows for more flexible winding pitch selection, and reduces heat generation per slot, leading to improved thermal management. A higher slot count also reduces harmonic leakage reactance, thereby enhancing starting torque and maximum tor-que. However, an excessive number of slots can lead to thin armature teeth, com-promising mechanical strength. Typically, the number of slots per pole per phase is chosen in the range of 1 to 2.

For motors below 100 kW and 500 V, pear-shaped slots and trapezoidal slots are commonly used. These are semi-closed slots where the slot width at the bottom is greater than at the top, making the tooth section nearly parallel. Such slot shapes are suitable for random-wound windings made of round conductors. Semi-closed slots also help reduce core surface losses and slot ripple losses, thereby improving the power factor.

Once the double-layer, three-phase star-connected winding configuration is determined, additional winding parameters must be designed, including the number of parallel branches, pitch, turns per coil element, number of parallel conductors per turn, and conductor specifications. Multi-branch windings are typically used in high-current, low-voltage motors, and the number of branches is usually set to $a = 1$. Since an uneven magnetic field distribution can cause significant torque ripple, in addition to using skewed stator slots, short-pitch windings can effectively reduce torque pulsations. Here, the slot count is used to define the winding pitch. The number of turns per phase winding is then calculated using (1.2):

$$W'_\Phi = 7.5\alpha'_p \frac{U_N - 2\Delta U}{pn_N\Phi_\delta} \tag{1.2}$$

where
U_N, DC-link voltage (V);
$\triangle U$, Turn-on voltage of transistor (V);
p, Pole pairs;
Φ_δ, Air-gap flux linkage (Wb).

To ensure the desired speed and torque performance of the motor, it is sometimes necessary to adjust the motor dimensions so that W'_Φ (the product of the number of turns and the magnetic flux per pole) is as close to an integer as possible. When this condition is met, the number of conductors per slot is subsequently determined.

The selection of conductor specifications and the number of parallel strands per turn depends on another critical parameter: the current density (J_a). A high conductor current density leads to significant winding heating, requiring high-grade insulation materials to prevent thermal degradation. For motors used in PMAs, F-class insulation materials can be employed, allowing a current density of up to 1,200 A/mm^2. Additionally, the thermal load (AJ_a), which represents the product of the linear current density (A) and the conductor current density (J_a), can be increased to 5,000 A^2/(cm·mm^2), ensuring an optimal balance between thermal performance and electrical efficiency.

The rotor design is determined based on the air-gap dimensions, shaft size, permanent magnet structure, and magnet dimensions. In PMSMs, rotor iron losses are relatively low. If heat dissipation and efficiency requirements are not particularly stringent, a solid magnetic material can be used for the rotor core. However, when high efficiency and thermal performance are critical, the rotor—like the stator—is typically constructed using laminated silicon steel sheets to minimize eddy current losses. Unlike electrically excited motors, where the air-gap flux density is determined by the excitation current, the flux density in a permanent magnet motor is governed by the magnetic circuit structure, permanent magnet material, and core material. As a result, accurately calculating the dimensions of the permanent magnets is highly complex. However, given the high cost of rare-earth permanent magnet materials, accurately estimating the required magnet volume during motor design is crucial to achieving cost-effectiveness. A common approach is to use an estimation method that establishes a relationship between the motor's power rating and the volume of permanent magnets:

$$V_m = 51 \frac{P_N \sigma_0 K_{ad} K_{fd}}{f K_a K_{Nm} C(BH)_{max}} \times 10^6 \tag{1.3}$$

where

P_N, Rated power (W);

σ_0, Leakage coefficient;

K_{ad}, Conversion coefficient for d-axis armature MMF;

K_{fd}, A factor indicating the multiple of the d-axis armature MMF that corresponds to the permanent magnet MMF per pole pair when the motor is in a short-circuit condition;

F, Magnetic field frequency (Hz);

K_a, Voltage coefficient;

B_H, Permanent magnet energy product (J/m^3).

The design of permanent magnet dimensions must also take into account factors such as temperature rise and demagnetization effects caused by armature reaction. When conditions allow, increasing the volume of the permanent magnets and maximizing their thickness in the magnetization direction can help mitigate the risk of demagnetization failures due to heat accumulation, which is particularly critical given the compact size constraints in PMAs.

Additionally, finite element simulation can play a crucial role in verifying design accuracy before prototype fabrication. If the simulation results are sufficiently accurate, design flaws can be identified and corrected early, reducing the need for costly rework and optimizing the motor before physical testing. Currently,

Maxwell, a widely used electromagnetic field finite element analysis (FEA) software, has become an essential design tool in the field of electromagnetics.

Maxwell 2D/3D is an integral component of electromechanical system design and analysis. It is a powerful, accurate, and user-friendly FEA software for two-dimensional electromagnetic field simulations. The software includes modules for analyzing electric fields, static magnetic fields, eddy current fields, transient fields, and thermal fields. It is extensively used for studying motors, sensors, transformers, permanent magnet devices, and actuators, allowing users to evaluate their behavior under static, steady-state, transient, normal, and fault conditions.

Maxwell 2D, while offering different calculation modules for various applications, follows a consistent FEA process, which includes model creation, material assignment, excitation and boundary condition setup, mesh generation, solver configuration, and post-processing of simulation results. The following sections provide a concise overview of the simulation workflow for the designed motor.

Modeling	Maxwell's RMxprt module facilitates automatic modeling and analysis of electric machines, offering a variety of permanent magnet motor configurations with different magnet structures. It also provides options for reasonable winding arrangements, significantly simplifying the motor modeling process. However, the module does not include a rotor with a V-shaped magnet configuration. Therefore, before conducting FEA in Maxwell 2D, it is necessary to design the motor using appropriate CAD software, such as AutoCAD or SolidWorks, and then import the model into Maxwell 2D.

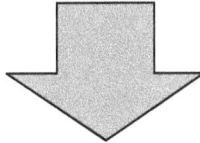

Material assignment	After constructing the motor's geometric model, it is essential to assign appropriate materials to each component. This process involves defining materials for the stator and rotor laminations, conductors, and permanent magnets, as well as specifying the magnetization direction for the magnets. Maxwell includes an internal material library with a variety of predefined materials. Additionally, it allows users to define custom materials and expand the material library as needed. For instance, when using the permanent magnet material NdFe33UH, it is necessary to manually input its specific properties, such as remanence and coercivity, to ensure accurate simulation results. Accurate material assignment is crucial for the precision of the simulation. It is advisable to consult Maxwell's official documentation or relevant tutorials to ensure all material properties are correctly configured.

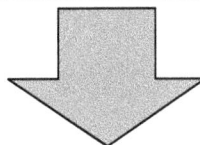

| Boundary conditions and excitation sources | In FEA, defining boundary conditions is essential to inform the software about the specific regions to analyze. When performing a full-model simulation, it is common practice to apply Vector Potential Conditions (Dirichlet boundary conditions) to the outer circumference of the stator and the inner circumference of the rotor. This approach directs the software to focus its analysis on the area between these two boundaries. In static magnetic field analysis, applying current to the windings is necessary for calculating inductance parameters. For transient simulations, it is important to incorporate an external circuit to serve as the motor's excitation source. |

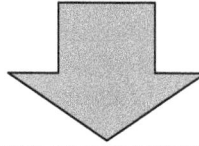

| Mesh generation | In FEA, mesh generation is a critical step that involves discretizing the motor's geometry into smaller elements. This process enables the computer to perform simulations by analyzing each individual mesh element, ultimately providing a comprehensive analysis of the entire motor. By applying appropriate mesh parameters, the simulation can achieve a balance between computational efficiency and result accuracy. A well-designed mesh not only reduces computational time but also enhances the precision of the simulation outcomes.
 For instance, in Ansys Mechanical, users can control local mesh densities, allowing for finer meshes in areas with high-stress gradients while maintaining coarser meshes elsewhere. This targeted approach ensures that critical regions are analyzed with higher accuracy without unnecessarily increasing the overall computational load. Therefore, careful consideration of mesh density and element quality is essential to ensure that the FEA results are both accurate and computationally efficient. |

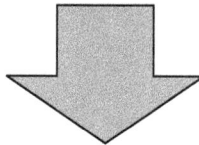

| Static simulation setup | In FEA, multiple solution setups can be configured within a single project to evaluate different operating conditions. For static simulations, the primary focus is on assessing the magnetic flux density distribution within various parts of the motor and ensuring that the design of the air-gap magnetic field meets specified requirements.
 To achieve these objectives, it is essential to configure simulation parameters such as error tolerances, the number of iterations, and simulation time steps. Typically, if the computer's performance is adequate, these parameters can be set to their default values, as illustrated in Figure 1.7. |

General | Convergence | Expression Cache | Solver | Defaults |

Name: Setup1   ☑ Enabled

Adaptive Setup

  Maximum Number of Passe   10

  Percent Erroı   1

Standard

  Refinement Per Pass: 30   %

  Minimum Number of Passes   2

  Minimum Converged Passes   1

Figure 1.7   Static simulation settings

| Transient simulation setup | In transient simulations of electric motors, the primary focus is on analyzing characteristics such as back electromotive force (EMF) waveforms, load capacity, speed range, and losses. To accurately assess these aspects, it is essential to configure various parameters appropriately. Depending on the analysis requirements, voltage or current excitation is applied to the motor windings. The rotor speed is set to reflect the specific operating conditions under investigation. Defining the rotor's moment of inertia is crucial for accurately simulating its dynamic response. Specifying the load characteristics allows for evaluating the motor's performance under different scenarios. Adjusting the total simulation time and the time step size based on the motor's speed ensures a balance between computational efficiency and result accuracy. For instance, higher rotational speeds may require smaller time steps to capture transient phenomena accurately. By meticulously setting these parameters, the transient simulation can provide valuable insights into the motor's dynamic behavior, facilitating optimization, and performance evaluation. |

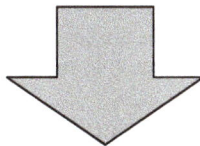

Post-processing of simulation results	After completing the simulation, post-processing involves analyzing and interpreting the data to extract meaningful insights. This process includes tasks such as harmonic analysis of magnetic flux density and calculating the root mean square (RMS) values of voltage and current. For instance, performing a Fast Fourier Transform (FFT) on the magnetic flux density in the air gap can help identify harmonic components that may affect motor performance. Accurate calculation of RMS values for voltage and current is essential for assessing the motor's efficiency and thermal characteristics. By thoroughly analyzing these parameters, engineers can gain a comprehensive understanding of the motor's electromagnetic behavior, identify potential issues, and implement design optimizations to enhance performance.

1.5   Applications of PMAs in robotics

PMAs have revolutionized a wide array of robotics applications, offering unparalleled precision, efficiency, and adaptability. Their integration into various sectors underscores their pivotal role in advancing robotic systems. Below are some key domains where PMAs are extensively applied. PMAs are fundamental components in industrial robotics, particularly in robotic arms, Automated Guided Vehicles (AGVs), collaborative robots (cobots), etc. Robotic arms equipped with PMAs excel in precision-intensive tasks such as assembly, welding, and material handling. The high torque-to-weight ratio and efficiency of PMAs ensure smooth and accurate movements, critical for maintaining product quality and minimizing errors on manufacturing lines. AGVs, which rely on PMA-driven motors, have transformed logistics by offering efficient and reliable transportation of goods in warehouses and production facilities. Their compactness and low energy consumption make PMAs an ideal choice for such mobile robotic systems. Similarly, cobots, designed for close collaboration with human workers, benefit from the high precision and safety provided by PMAs. The seamless integration of sensors and controllers enables cobots to adapt dynamically to changing environments, enhancing workplace productivity. In medical robotics, PMAs are indispensable in surgical robots, prosthetics, and exoskeletons. Surgical robots, such as those used in minimally invasive procedures, require extreme precision and reliability. PMAs provide the necessary control and smooth motion, enabling surgeons to perform delicate operations with enhanced accuracy. Prosthetics powered by PMAs offer users improved mobility and functionality. The lightweight and compact design of PMAs ensures that prosthetic limbs are not only efficient but also comfortable for daily use. Exoskeletons, used in rehabilitation and mobility assistance, leverage the high torque-to-weight ratio of PMAs to support and amplify human movement. These devices provide critical assistance to individuals recovering from injuries or living with mobility impairments.

Humanoid robots represent a specialized and rapidly evolving domain within robotics, leveraging PMAs for a range of critical functionalities. These robots are

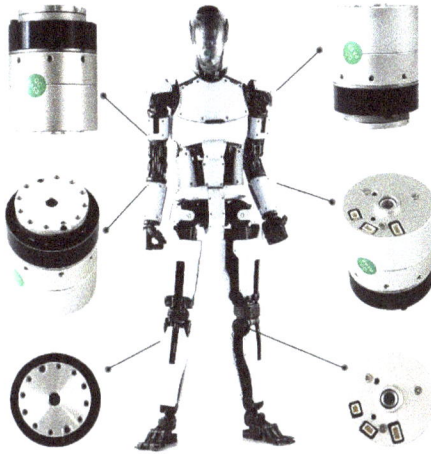

Figure 1.8   PMAs used in humanoid robots

designed to emulate human form and motion, requiring precision, adaptability, and efficiency to replicate complex biomechanical movements. PMAs are integral to achieving these characteristics, enabling humanoid robots to perform tasks that demand high degrees of dexterity and coordination. For instance, Tesla's humanoid robot, Optimus, exemplifies the application of PMAs in humanoid robotics. It is equipped with 50 PMAs strategically distributed across its body to enable natural and efficient movement. These actuators are utilized in the arms, legs, torso, and hands, allowing the robot to execute a wide range of human-like motions. Each PMA contributes to the precise control of joints and ensures smooth transitions between various postures and activities. In humanoid robots, PMAs are predominantly used in joint actuation systems. The high torque-to-weight ratio of PMAs ensures that humanoid robots can perform lifelike movements without compromising stability or efficiency. For instance, PMAs in knee and hip joints allow for smooth walking, running, and climbing motions, closely mimicking human locomotion, as shown in Figure 1.8. The precision and control offered by PMAs are critical for the operation of humanoid robot arms and hands. These components require fine-grained motion control to perform delicate tasks such as object manipulation, tool usage, and even handwriting. The compact size of PMAs facilitates their integration into small, complex structures like robotic fingers. Humanoid robots often rely on advanced PMA-driven actuators for maintaining dynamic balance during motion. Real-time adjustments in torque and speed, enabled by PMAs and their integrated sensors, allow robots to adapt to uneven surfaces and recover from potential instability. The high-power efficiency of PMAs reduces energy consumption, making humanoid robots suitable for extended operation, whether in research, service, or industrial environments. The lightweight and compact nature of PMAs enables the design of agile and proportionate humanoid robots, closely resembling human anatomy. PMAs allow for high-precision movements, essential for tasks requiring a delicate touch or adaptive responses.

PMAs are also at the heart of service and mobile robotics, powering systems designed for household tasks, cleaning, delivery, and exploration. Household robots, such as robotic vacuum cleaners and kitchen assistants, benefit from the compactness and efficiency of PMAs, ensuring reliable and long-lasting performance. Delivery robots, increasingly used for last-mile logistics, rely on PMA-driven actuators for precise navigation and payload management.

As for exploration robotics, PMAs enable systems to operate in challenging environments, such as underwater or rugged terrains. The robustness and energy efficiency of PMAs are essential for ensuring mission success, whether in deep-sea exploration or planetary research. These actuators provide the adaptability needed to overcome obstacles and execute complex maneuvers in unpredictable conditions. Underwater robotics, another frontier of exploration, leverages PMAs for propulsion, manipulation, and data collection. The corrosion resistance and efficiency of PMAs make them suitable for prolonged use in harsh aquatic environments. These systems play a vital role in marine research, resource extraction, and environmental monitoring.

Chapter 2

Fundamentals of PMA control in robotics

2.1 Introduction to control theory in robotics

Control systems are fundamental to permanent magnet actuators (PMAs), and further robotics, serving as the bridge between the mechanical components of a robot and its desired behavior. The primary purpose of these systems is to ensure that robotic systems operate with stability, accuracy, and efficiency. Stability refers to the system's ability to maintain consistent performance under internal or external disturbances. Accuracy involves achieving precise outcomes, such as reaching a specific position or following a trajectory. Efficiency emphasizes the optimal use of energy and resources to maximize performance.

In robotics, the complexity of control systems arises from the need to manage dynamic, nonlinear, and sometimes unpredictable environments. For example, a robotic arm operating in an assembly line must consistently align components with sub-millimeter precision despite potential variations in part placement. Similarly, autonomous mobile robots require real-time control to navigate dynamically changing environments while avoiding obstacles. Robotics applications often demand high-performance control systems that can handle intricate tasks such as manipulation, locomotion, and interaction with humans. Achieving this requires integrating advanced control theories with sophisticated hardware and sensors.

To understand the role of control systems in PMAs, it is essential to consider the unique characteristics of these actuators. PMAs operate based on the interaction between a permanent magnet and electromagnetic fields, providing precise control over motion parameters. Unlike traditional actuators, PMAs exhibit high efficiency, compact design, and the ability to produce high torque and force densities. These features make PMAs ideal for robotic applications where precision and responsiveness are paramount. For example, in robotic surgical systems, PMAs are used to control the fine motion of instruments, enabling precise and minimally invasive procedures. In such applications, control systems must ensure smooth and stable operation while responding to the surgeon's commands in real time. Similarly, in robotic arms used for manufacturing, PMAs enable accurate positioning and force control, ensuring consistent quality and performance.

Advanced control strategies play a critical role in optimizing the performance of PMAs. These strategies include techniques such as adaptive control, which dynamically adjusts control parameters based on real-time feedback, and robust control, which ensures system performance despite uncertainties or variations in operating conditions. By integrating these techniques with PMA systems, roboticists can achieve unparalleled levels of precision and reliability. The importance of control systems extends beyond individual PMAs to the coordination of multiple actuators in robotic systems. For instance, a humanoid robot with multiple degrees of freedom requires synchronized control of its joints to perform complex movements such as walking, grasping, or balancing. Control systems for such applications must manage interactions between actuators, sensors, and the environment, ensuring seamless operation.

In addition to their functional benefits, control systems contribute to the safety and energy efficiency of robotic systems. PMAs equipped with advanced control algorithms can detect and respond to unexpected forces or collisions, preventing damage to the robot or its surroundings. Furthermore, energy-efficient control strategies, such as minimizing current consumption during operation, reduce the environmental impact and operational costs of robotic systems. Another critical aspect of control systems in PMAs is their ability to operate in real time. Real-time control involves processing sensor data, calculating control actions, and updating actuator commands within milliseconds. This capability is essential for applications such as autonomous vehicles or robotic manipulators, where rapid decision-making is crucial for success.

In summary, control systems are the backbone of PMAs in robotics, enabling precise, efficient, and reliable operation. By addressing the challenges of dynamic environments, integrating advanced control strategies, and ensuring real-time performance, control systems unlock the full potential of PMAs, paving the way for innovative robotic applications. The following sections will explore specific control objectives and strategies in greater detail, providing a comprehensive understanding of PMA control in robotics.

2.1.1   Closed-loop control for PMAs

One of the fundamental concepts in control theory is the distinction between open-loop and closed-loop control. Closed-loop control, also known as feedback control, is critical for ensuring that robotic systems achieve their desired behavior despite uncertainties or disturbances. In a closed-loop system, sensors measure the actual state of the system, such as position, speed, or force. This information is compared with the desired state or reference signal. The difference, called the error, is processed by a controller, which adjusts the system's inputs to minimize the error. This feedback mechanism enables the system to adapt dynamically and achieve the desired performance.

For PMAs in robotics, closed-loop control is essential to regulate motion parameters such as position, velocity, and torque. For example, Figure 2.1 illustrates a commonly employed three-loop control scheme for a robotic joint driven by a PMA. This hierarchical structure typically comprises an outer position loop, an intermediate speed loop, and an inner current loop, each designed to achieve specific objectives. In the case of a robotic joint powered by a PMA, the desired joint angle serves as the reference signal, representing the system's target state. Position

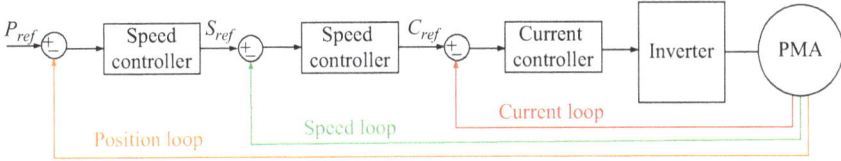

Figure 2.1   Three closed-loop control scheme for a robot joint PMA

sensors continuously measure the actual joint angle, providing real-time feedback to the control system. The controller compares the desired angle with the actual angle, calculating the error signal. Based on this error, the position controller generates a command signal, typically a desired velocity, which is forwarded to the speed control loop. The speed control loop processes the velocity error, refining the control signal into a desired current or torque command. Finally, the inner current loop ensures precise current regulation in the actuator, directly influencing the electromagnetic torque generated by the PMA. This layered control structure not only achieves high-precision motion but also enhances the system's adaptability to external factors, such as varying loads, dynamic conditions, and unexpected disturbances. Moreover, the inner current loop offers a fast response and high bandwidth, essential for compensating for rapid changes in load or electromagnetic dynamics. The intermediate speed loop ensures smooth motion and accurate trajectory tracking, while the outer position loop guarantees precise endpoint control.

The closed-loop control scheme, as applied to PMAs, demonstrates the critical role of feedback in enabling robust and adaptable operations. It can be noticed that robotic systems rely on several key control objectives to ensure optimal performance. These objectives guide the design and implementation of control systems for PMAs. Each objective addresses a specific aspect of robotic functionality, ensuring that the system performs reliably and efficiently.

1. Position Control: Position control involves regulating a robot's position with high precision. For example, in robotic surgery, position control ensures that the surgical tool moves to the exact location required. In the context of PMAs, position control uses feedback from position sensors to adjust the actuator's input and achieve the target position. Advanced algorithms, such as proportional-integral-derivative (PID) control, are often employed to minimize position error and ensure stability. These methods are particularly effective in applications requiring high precision, such as semiconductor manufacturing or medical robotics.

2. Speed Control: Speed control or velocity control, is essential for tasks requiring consistent motion. For example, conveyor belt systems in industrial robots demand precise speed regulation to synchronize with other components. Speed control for PMAs involves maintaining a desired angular or linear velocity, regardless of changes in load or external forces. This is achieved through feedback mechanisms that monitor the actual speed and adjust the motor input dynamically. Advanced speed control strategies may also

incorporate adaptive control techniques to handle variations in system dynamics or external disturbances.

3. Torque Control: Torque control is critical for applications involving force interaction, such as robotic grippers or haptic devices. By controlling the torque output of PMAs, robotic systems can apply appropriate forces to manipulate objects or interact with humans safely. For instance, in exoskeletons or assistive robotic devices, torque control ensures that the actuators provide the required force without causing discomfort or injury to the user. Techniques such as model-based control or feedforward control are often employed to achieve precise torque regulation.

4. Trajectory Tracking: Trajectory tracking ensures that a robot follows a predefined path over time. This objective is vital for applications such as autonomous vehicles, robotic arms, and drones. For PMAs, trajectory tracking involves coordinating position, speed, and acceleration to achieve smooth and precise motion along the desired trajectory. Advanced algorithms, such as model predictive control (MPC), are commonly used to optimize trajectory tracking performance while considering constraints on actuator capabilities and system dynamics. Closed-loop control extends beyond the basic objectives of position, speed, torque, and trajectory tracking. In modern robotics, these systems are integral to achieving complex behaviors and ensuring robustness under diverse operating conditions.

- Adaptive Control in PMAs: Adaptive control is a technique that allows the control system to adjust its parameters in real time to accommodate changes in system dynamics or environmental conditions. For PMAs, adaptive control is particularly useful in applications involving variable loads or operating environments. For example, an industrial robot arm equipped with adaptive control can handle different payloads seamlessly, maintaining consistent performance without requiring manual re-tuning of the control system.

- Robust Control for PMAs: Robust control focuses on ensuring system performance despite uncertainties or disturbances. This approach is essential for PMAs used in environments with unpredictable conditions, such as outdoor robotics or underwater exploration. By accounting for potential variations in system parameters, robust control strategies provide a high degree of reliability and safety.

- Energy Optimization in PMA Control Systems: Energy efficiency is a critical consideration in robotic systems, particularly for battery-powered applications such as drones or mobile robots. Closed-loop control systems can be designed to minimize energy consumption by optimizing actuator inputs. Techniques such as energy-aware control or regenerative braking are increasingly used to enhance the efficiency of PMAs in robotics.

2.1.2   Brief introduction to vector control for PMAs

Vector control, also known as field-oriented control (FOC), is a sophisticated control strategy widely used in PMAs and other types of motors. This technique enables precise control of motor parameters by decoupling torque and flux components in the motor's magnetic field. The essence of vector control lies in transforming the three-phase permanent magnet synchronous motor (PMSM) currents into a two-axis coordinate system: the direct axis (*d*-axis) and the quadrature axis (*q*-axis). The *d*-axis represents the flux-producing component, while the *q*-axis corresponds to the torque-producing component. By independently controlling these components, vector control achieves high-performance operation, even under dynamic conditions.

In robotics, vector control is particularly advantageous for PMAs because it provides:

- Enhanced Precision: Fine control of torque and speed enables high accuracy in robotic applications.
- Improved Efficiency: Optimal current distribution reduces energy consumption.
- Dynamic Response: Rapid adjustments to changes in load or operating conditions.

An example of vector control in action is a robotic joint driven by a PMA. The control system uses real-time feedback to calculate the *d*-axis and *q*-axis currents required to achieve the desired torque and position. By modulating the motor inputs accordingly, the system ensures smooth and efficient operation.

The decoupling of torque and flux components is the cornerstone of vector control. Traditional control strategies often struggle to manage the interdependence between these components in a three-phase system. By employing mathematical transformations, vector control isolates these components, allowing for precise regulation. This decoupling is achieved through Park and Clarke transformations, which convert the three-phase stator currents into a two-dimensional rotating reference frame. In this frame, the *d*-axis aligns with the rotor's magnetic field, and the *q*-axis remains orthogonal. This transformation simplifies the control process and provides a clear framework for adjusting torque and flux independently.

In PMAs, maintaining optimal flux is crucial for minimizing energy losses and ensuring efficient operation. Overfluxing can lead to saturation and increased heat generation while underfluxing can reduce torque production. Vector control addresses these challenges by actively monitoring and adjusting the *d*-axis current to maintain the desired flux level.

Vector control's ability to decouple torque and flux provides several benefits for robotic systems using PMAs. First, robotic systems often require precise control over position and velocity, especially in applications such as assembly lines, surgical robotics, and pick-and-place tasks. Vector control ensures that PMAs deliver consistent and accurate motion by maintaining a tight grip on torque production. Second, in robotics, smoothness of motion is essential to avoid jerky movements that could damage components or disrupt tasks. Vector control's real-time adjustments to motor inputs minimize oscillations and vibrations, ensuring fluid operation. Third, energy efficiency is a critical factor in robotic applications, particularly for battery-powered systems such as drones or mobile robots. Vector control optimizes current usage, reducing energy consumption without compromising performance. This capability is especially valuable for autonomous robots operating in remote or challenging environments. Fourth, robots often operate in environments where load conditions can change unpredictably. For example, a robotic arm lifting objects of varying weights must adapt its torque output accordingly. Vector control's rapid response to feedback signals allows it to adjust to these changes dynamically, maintaining consistent performance.

The implementation of vector control involves several steps, each requiring precise calculations and real-time processing capabilities. These steps include:

1. Sensing and Feedback: Sensors measure key parameters, such as rotor position, current, and voltage. These measurements are essential for calculating the d-axis and q-axis components.
2. Transformations: Using Clarke and Park transformations, the three-phase currents are converted into a two-axis reference frame. This transformation simplifies the control process by decoupling torque and flux components.
3. Controller Design: Proportional-integral (PI) controllers are commonly used to regulate the d-axis and q-axis currents. These controllers adjust the motor inputs to minimize errors and achieve the desired torque and flux levels.
4. Pulse Width Modulation (PWM): The calculated motor inputs are converted into voltage signals using PWM techniques. These signals drive the PMA, ensuring precise execution of the control commands.

Despite its advantages, vector control presents several challenges, particularly in the context of PMAs:

1. Complexity: Vector control requires sophisticated algorithms and high-speed processors to perform real-time calculations. The computational burden increases with the complexity of the robotic system, necessitating robust hardware and software solutions.
2. Sensor Dependency: Accurate measurements of rotor position and current are critical for vector control. Sensor errors or delays can degrade performance, making sensor calibration and maintenance essential.
3. Nonlinearities and Saturation: Nonlinear behaviors, such as magnetic saturation, can complicate the control process. Advanced compensation techniques are often needed to address these issues and ensure reliable operation.

2.2   Field-oriented control theory

2.2.1   Working principles

The fundamental concept of vector control technology lies in coordinate transformation: converting the three-phase currents of an AC motor into the rotating dq-axis reference frame. In this frame, the direct-axis (d-axis) and quadrature-axis (q-axis) currents become decoupled and can be controlled independently. Essentially, vector control transforms the control of an AC motor to resemble that of a DC motor, where the d-axis and q-axis currents are equivalent to the field current and torque-producing current of a separately excited DC motor, respectively. Field-oriented vector control technology has reached a high level of maturity, significantly enhancing the control performance of AC motors. Today, it has been widely adopted across various industrial applications due to its superior dynamic response and precision control capabilities.

2.2.1.1 Coordinate transformation

When performing a simplified analysis of the mathematical model of a PMA, it is essential to employ several coordinate systems to transform stator current, voltage, and flux linkage components between these systems. To achieve vector control of a PMA, transformations such as the Clarke (3/2) transformation, Park (2/2) transformation, inverse Park (iPark) transformation, and (2/3) transformation are utilized. The fundamental premise of these coordinate transformations is the invariance of the rotating magnetomotive force (MMF). The following sections introduce the principles behind these coordinate transformations, providing the theoretical foundation necessary for the effective implementation of PMSM vector control.

The 3/2 transformation is a coordinate transformation that converts a three-phase stationary abc coordinate system into a two-phase stationary $\alpha\beta$ coordinate system. As shown in Figure 2.2, suppose each phase winding has N_2 turns in the transformed $\alpha\beta$ coordinate system and N_3 turns in the abc coordinate system. It is important to note that the MMF remains unchanged before and after the transformation. This transformation is used to simplify the analysis of the motor's electrical quantities and is one of the key steps in PMA control, allowing for easier manipulation and decoupling of the phase components.

Then, it can be derived that:

$$\begin{cases} N_2 i_\alpha = N_3 i_a \cos 0 + N_3 i_b \cos \dfrac{2\pi}{3} + N_3 i_c \cos \dfrac{4\pi}{3} \\[2mm] N_2 i_\beta = N_3 i_b \sin \dfrac{2\pi}{3} + N_3 i_c \sin \dfrac{4\pi}{3} \end{cases} \tag{2.1}$$

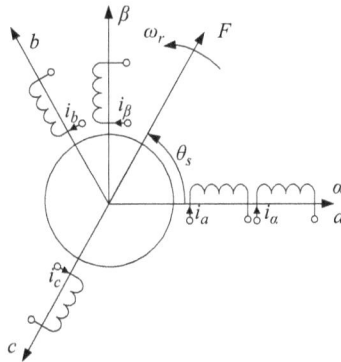

Figure 2.2   *Relation of stationary abc coordinate system and stationary $\alpha\beta$ coordinate*

where i_a, i_b, and i_c are the three-phase currents. i_α, i_β are the $\alpha\beta$-axis currents. Because $N_2/N_3 = \sqrt{3}/\sqrt{2}$, i_α, i_β can be calculated by:

$$\begin{bmatrix} i_\alpha \\ i_\beta \end{bmatrix} = \sqrt{\frac{2}{3}} \begin{bmatrix} 1 & -\frac{1}{2} & -\frac{1}{2} \\ 0 & \frac{\sqrt{3}}{2} & -\frac{\sqrt{3}}{2} \end{bmatrix} \begin{bmatrix} i_a \\ i_b \\ i_c \end{bmatrix} \tag{2.2}$$

Equation (2.2) represents the 3/2 transformation matrix, which is commonly referred to as the Clarke transformation matrix. Further, the inverse Clarke transformation matrix can be obtained:

$$\begin{bmatrix} i_a \\ i_b \\ i_c \end{bmatrix} = \sqrt{\frac{2}{3}} \begin{bmatrix} 1 & 0 \\ -\frac{1}{2} & \frac{\sqrt{3}}{2} \\ -\frac{1}{2} & -\frac{\sqrt{3}}{2} \end{bmatrix} \begin{bmatrix} i_\alpha \\ i_\beta \end{bmatrix} \tag{2.3}$$

The Park transformation is a coordinate transformation that converts the two-phase stationary $\alpha\beta$ coordinate system into the two-phase rotating dq coordinate system. As illustrated in Figure 2.3, the $\alpha\beta$-axis currents i_α, i_β, and the dq-axis current i_d, i_q generate stator MMF vectors (F) of equal magnitude, rotating at the synchronous speed ω_r.

Based on the principle of MMF equivalence theory, i_α, i_β and i_d, i_q can be written as:

$$\begin{cases} i_\alpha = i_d \cos \theta_r - i_q \sin \theta_r \\ i_\beta = i_d \sin \theta_r + i_q \cos \theta_r \end{cases} \tag{2.4}$$

Equation (2.4) can be further expressed as a matrix version:

$$\begin{bmatrix} i_\alpha \\ i_\beta \end{bmatrix} = \begin{bmatrix} \cos \theta_r & -\sin \theta_r \\ \sin \theta_r & \cos \theta_r \end{bmatrix} \begin{bmatrix} i_d \\ i_q \end{bmatrix} \tag{2.5}$$

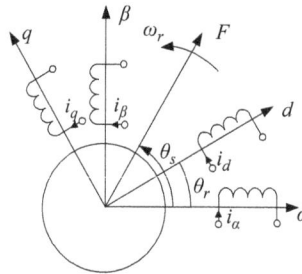

Figure 2.3   Relation of stationary $\alpha\beta$ coordinate and rotating dq coordinate

Then, the Park transformation operation is:

$$
\begin{bmatrix} i_d \\ i_q \end{bmatrix} = \begin{bmatrix} \cos \theta_r & \sin \theta_r \\ -\sin \theta_r & \cos \theta_r \end{bmatrix} \begin{bmatrix} i_\alpha \\ i_\beta \end{bmatrix}
\tag{2.6}
$$

Based on (2.3) and (2.6), the relation between i_d, i_q and i_a, i_b, i_c can be obtained:

$$
\begin{bmatrix} i_d \\ i_q \end{bmatrix} = \sqrt{\frac{2}{3}} \begin{bmatrix} \cos \theta_r & \cos\left(\theta_r - \dfrac{2\pi}{3}\right) & \cos\left(\theta_r - \dfrac{4\pi}{3}\right) \\ -\sin \theta_r & -\sin\left(\theta_r - \dfrac{2\pi}{3}\right) & -\sin\left(\theta_r - \dfrac{4\pi}{3}\right) \end{bmatrix} \begin{bmatrix} i_a \\ i_b \\ i_c \end{bmatrix}
\tag{2.7}
$$

$$
\begin{bmatrix} i_a \\ i_b \\ i_c \end{bmatrix} = \sqrt{\frac{2}{3}} \begin{bmatrix} \cos \theta_r & -\sin \theta_r \\ \cos\left(\theta_r - \dfrac{2\pi}{3}\right) & -\sin\left(\theta_r - \dfrac{2\pi}{3}\right) \\ \cos\left(\theta_r - \dfrac{4\pi}{3}\right) & -\sin\left(\theta_r - \dfrac{4\pi}{3}\right) \end{bmatrix} \begin{bmatrix} i_d \\ i_q \end{bmatrix}
\tag{2.8}
$$

2.2.1.2 Modeling of PMSM used in PMA

Due to the complex electromagnetic interactions between the rotor permanent magnets and the stator windings in permanent magnet motors, as well as the influence of nonlinear factors such as magnetic circuit saturation, it is often challenging to establish an accurate mathematical model. To facilitate control and analysis, the mathematical model of the permanent magnet motor is typically simplified based on the following assumptions:

1. The magnetic permeability of the iron core is considered infinite, and both eddy current losses and hysteresis losses are neglected.
2. The influence of stator winding slots on the air-gap magnetic field distribution is ignored.
3. There are no damping windings on the rotor.
4. The permanent magnets generate a sinusoidal magnetic field distribution along the circumferential direction of the air gap.

Figure 2.4 shows the motor structure, where ω_r represents the rotor's rotational direction. The excitation magnetic field generated by the permanent magnets is denoted as ψ_f, which rotates synchronously with the rotor. The phase of this magnetic field (flux) in the abc coordinate system can be represented by the electrical angle θ_r. In the abc coordinate system, the stator voltage vector equations are expressed as follows:

$$
u_s = R_s i_s + \frac{d}{dt}(L_s i_s) + \frac{d}{dt}\left(\psi_f e^{j\theta_r}\right)
\tag{2.9}
$$

where i_s represents the stator current. u_s is the stator voltage, R_s is the resistance and L_s is the inductance.

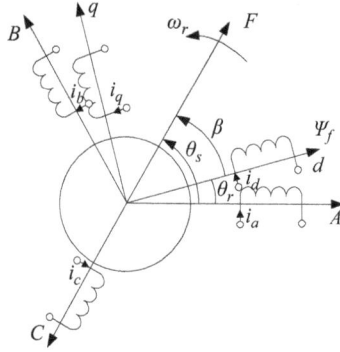

Figure 2.4   Motor structure

In the abc coordinate system, the electrical properties are:

$$
\begin{bmatrix} u_a \\ u_b \\ u_c \end{bmatrix} = \begin{bmatrix} R_a & 0 & 0 \\ 0 & R_b & 0 \\ 0 & 0 & R_c \end{bmatrix} \begin{bmatrix} i_a \\ i_b \\ i_c \end{bmatrix} + \frac{d}{dt} \begin{bmatrix} \psi_a \\ \psi_b \\ \psi_c \end{bmatrix}
\tag{2.10}
$$

where the magnetic flux can be expressed as:

$$
\begin{bmatrix} \psi_a \\ \psi_b \\ \psi_c \end{bmatrix} = L_s \begin{bmatrix} \cos 0 & \cos \dfrac{2\pi}{3} & \cos \dfrac{4\pi}{3} \\ \cos \dfrac{4\pi}{3} & \cos 0 & \cos \dfrac{2\pi}{3} \\ \cos \dfrac{2\pi}{3} & \cos \dfrac{4\pi}{3} & \cos 0 \end{bmatrix} \begin{bmatrix} i_a \\ i_b \\ i_c \end{bmatrix} + \psi_f \begin{bmatrix} \cos \theta \\ \cos\left(\theta - \dfrac{2\pi}{3}\right) \\ \cos\left(\theta - \dfrac{4\pi}{3}\right) \end{bmatrix}
\tag{2.11}
$$

where u_a, u_b, and u_c are the voltages of the stator three-phase windings. i_a, i_b, and i_c are the stator currents. ψ_a, ψ_b, and ψ_c are the magnetic flux linkages. ψ_f is the magnetic flux linkage of the rotor permanent magnet. R_s is the stator winding resistance. θ denotes the rotor position angle.

The mathematical model in the two-phase rotating *dq* coordinate system can be obtained from the *dq*-axis model shown in Figure 3.3 and the transformation formula from the abc coordinate system to the *dq* coordinate system derived earlier. Its voltage equation is:

$$
\begin{cases} u_d = R_s i_d + \dfrac{d\psi_d}{dt} - \omega_r \psi_q \\ u_q = R_s i_q + \dfrac{d\psi_q}{dt} - \omega_r \psi_d \end{cases}
\tag{2.12}
$$

where the magnetic flux linkage equation is:

$$
\begin{cases} \psi_d = L_d i_d + \psi_f \\ \psi_q = L_q i_q \end{cases}
\tag{2.13}
$$

where u_d and u_q are the voltage components in the dq-axis system; i_d and i_q are the stator current components in the dq-axis system; ψ_d and ψ_q are the stator magnetic flux linkage vector components; ω_r is the rotor angular velocity; L_d and L_q are the direct-axis and quadrature-axis synchronous inductances.

Substituting (2.13) into (2.12), we obtain:

$$\begin{cases} u_d = R_s i_d + L_d \dfrac{di_d}{dt} - \omega_r L_q i_q \\ u_q = R_s i_q + L_q \dfrac{di_q}{dt} + \omega_r L_d i_d + \omega_r \psi_f \end{cases} \tag{2.14}$$

Figure 2.5 is the dynamic equivalent circuit of the PMSM based on (2.14).

When the motor is operating in a steady state, the differential terms in (2.14) become 0, and it can be rewritten as:

$$\begin{cases} u_d = R_s i_d - \omega_r L_q i_q \\ u_q = R_s i_q + \omega_r L_d i_d + \omega_r \psi_f \end{cases} \tag{2.15}$$

The electromagnetic torque of the PMSM can be expressed as:

$$T_e = p_n \psi_s \times i_s \tag{2.16}$$

where p_n is the number of pole pairs.

Substituting $\psi_s = \psi_d + j\psi_q$ and $i_s = i_d + ji_q$ into (2.16) gives the torque equation of the PMSM in the dq-axis system:

$$T_e = p_n \left[\psi_f i_q + (L_d - L_q) i_d i_q \right] \tag{2.17}$$

Additionally, $i_d = i_s \cos\beta$, $i_q = i_s \sin\beta$, and β is the torque angle, as shown in Figure 2.4. Thus, the torque equation of the PMSM can be obtained as:

$$T_e = p_n \left[\psi_f i_q \sin\beta + \frac{1}{2} (L_d - L_q) i_s^2 \sin 2\beta \right] \tag{2.18}$$

Equation (2.18) consists of two parts: the first term is the permanent magnet torque, which is generated by the interaction between the rotor permanent magnet field and the stator air-gap magnetic field; the second term is the reluctance torque caused by the saliency effect, which arises from the difference in the direct and quadrature axis inductances due to the rotor asymmetry.

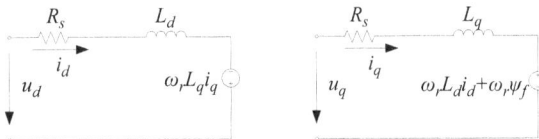

Figure 2.5 Dynamic equivalent circuit of the PMSM

Ignoring static and Coulomb friction in the mechanical system, and considering only viscous friction, the motion equation of the PMSM can be obtained as:

$$T_e = J\frac{d\omega}{dt} + F\omega + T_l \qquad (2.19)$$

In the equation, J is the moment of inertia of the electrical spindle rotor, including the load; F is the viscous friction coefficient; T_l is the load torque; ω is the rotor mechanical rotational speed; and $\omega = \omega_r / p_n$.

2.2.2 Principle of vector control for permanent magnet synchronous motor

Figure 2.6 is the schematic diagram of a typical vector control system for PMSMs, which independently controls the current components i_d and i_q of the PMSM. The actual i_d and i_q components are made to track the given i^*_d and i^*_q values, enabling independent control of the magnetic flux and torque of the PMSM.

In the dq-axis system, using the control method of a DC motor, control signals are obtained and then transformed from the dq coordinate system to the abc coordinate system to control the PMSM. Vector control effectively controls the motor current i_d, with the control system being relatively simple while achieving superior control performance.

In Figure 2.6, the system's speed loop and current loop form the outer loop and inner loop, respectively. First, the encoder pulses are read to calculate the motor speed n and rotor position θ. The difference between the actual speed and the reference speed is processed by a speed PI controller to obtain the reference values i^*_q for the stator current torque components. The difference between the reference

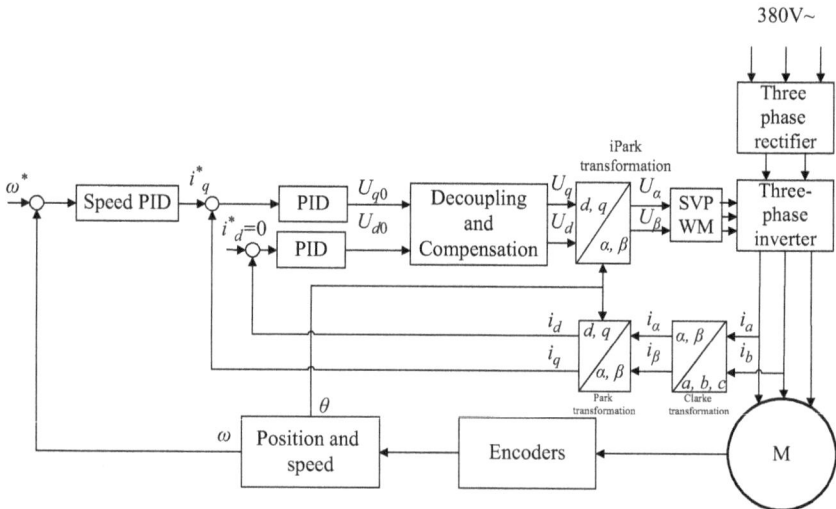

Figure 2.6   Implementation diagram of the vector control system

value i^*_q and the actual value i_q is processed by the q-axis current loop PI controller to determine the quadrature axis voltage control signal U_{q0}. The difference between the reference value i^*_d of the excitation current component and the actual value i_d is processed by the d-axis current loop PID controller to determine the direct axis voltage control signal U_{d0}.

Then, based on the dq-axis mathematical model of the PMSM, the dq-axis currents are decoupled, and a voltage compensation controller is designed. After the voltage compensation regulator, the dq-axis voltage regulation signals are obtained. The voltage's torque component and excitation component are processed by the iPark transformation to obtain the voltage components in the two-phase stationary reference frame. Finally, the Space Vector Pulse Width Modulation (SVPWM) technique is used to control the inverter, which outputs voltage directly to the motor, enabling the motor to run normally.

2.2.3   PID controller

P stands for proportional amplification, I represents integration, and D stands for differentiation. PID control is one of the most widely used methods in system control, as it can effectively ensure and improve the system's dynamic performance. Due to their simple algorithm and high control accuracy, PID controllers are commonly used in modern motor control systems, with some systems requiring two or even multiple PID controllers.

PID control works by manipulating the error, meaning a feedback loop is essential. First, the error $e(t)$ is calculated as the difference between the system's setpoint $r(t)$ and the actual feedback value $y(t)$. The proportional term (P) adjusts the error $e(t)$ by amplifying or reducing it, the integral term (I) accumulates the error over time using integration, and the derivative term (D) processes the rate of change of the error. After processing the error $e(t)$, the results are summed to obtain the control signal $u(t)$, which is used to control the system. Its block diagram is shown in Figure 2.7.

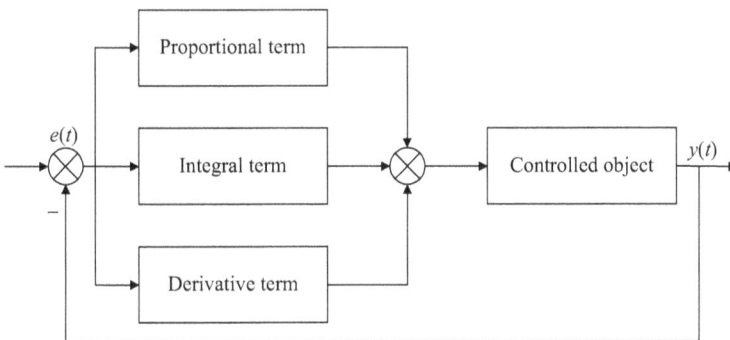

Figure 2.7   Block diagram of the PID controller

The general form of a continuous PID controller is:

$$u(t) = K_p e(t) + K_I \int_0^{T_1} e(\tau)d\tau + K_D \frac{de(t)}{dt} \tag{2.20}$$

where K_P is the proportional gain; K_I is the integral gain; K_D is the derivative gain; and T_1 is the integral time.

The transfer function of the continuous PID controller is:

$$G(s) = K_p + \frac{K_I}{s} + K_D s \tag{2.21}$$

With the development of high-speed microprocessors, analog PID controllers are increasingly being replaced by digital PID controllers, which are implemented through programming on digital processors. This requires engineers to discretize the continuous PID using appropriate algorithms. When the sampling period T is sufficiently small, the integral and differential expressions of the error $e(t)$ at the k-th sampling period are given by:

$$\frac{de(t)}{dt} = \frac{e(kT) - e[(k-1)T]}{T} \tag{2.22}$$

$$\int_0^{kT} e(t)dt \approx T \sum_{i=0}^{k} e(iT) \tag{2.23}$$

Substituting (2.22) and (2.23) into (2.20) gives the discrete PID expression:

$$u(kT) = K_p e(kT) + K_I T \sum_{i=0}^{k} e(iT) + K_D \frac{e(kT) - e[(k-1)T]}{T} \tag{2.24}$$

In the equation, T is the sampling period; $e(kT)$ and $e[(k-1)T]$ are the errors at the k-th and $(k-1)$-th sampling periods, respectively.

Therefore, the pulse transfer function of the discrete PID controller is:

$$G(z) = K_p + \frac{K_I z}{z - 1} + K_D(1 - z^{-1}) \tag{2.25}$$

Under the conditions determined by the control algorithm, the control effectiveness of the PID controller primarily depends on the proportional gain, integral gain, and derivative gain. The following text provides a detailed introduction to the tuning effect of each PID parameter and its impact on the system:

1. Proportional Control is the simplest form of control, where the output is directly proportional to the input error signal. As soon as a deviation occurs in the system, the proportional adjustment immediately acts to reduce the error. A larger proportional gain can accelerate the adjustment and reduce the error, but if the proportional gain is too large, it can reduce the system's stability and even cause instability. When only proportional control is used, there will be a steady-state error in the system output.

2. Integral Control involves the controller output being proportional to the integral of the input error signal. In an automatic control system, if there is a steady-state error after the system reaches a steady state, the system is considered to have a steady-state error, or it is referred to as a "system with steady-state error." To eliminate steady-state error, an "integral term" must be introduced in the controller. The integral term depends on the accumulated error over time, and as time increases, the integral term grows. Thus, even if the error is small, the integral term will continue to increase over time, pushing the controller's output higher and reducing the steady-state error until it reaches zero. Therefore, a PI controller can eliminate steady-state error once the system reaches a steady state. The strength of the integral action depends on the integral time constant—smaller time constants result in stronger integral action, while larger time constants weaken it. Adding integral control can reduce the system's stability and slow down dynamic response. Integral action is often combined with the other two types of control to form a PI or PID controller.

3. Derivative Control reflects the rate of change of the system's error signal, providing a predictive function. It can foresee the trend of the error and thus take preemptive control actions. The derivative action works before the error becomes significant, effectively eliminating it. This improves the system's dynamic performance. When the derivative time is properly selected, it can reduce overshoot and shorten the settling time. However, derivative action can amplify noise interference, so overly strong derivative control may reduce the system's ability to resist disturbances. Additionally, derivative control outputs zero when there is no change in the input.

The three PID coefficients have a significant impact on the control performance, but they also complicate the tuning process. The traditional parameter adjustment process generally follows the sequence of adjusting the proportional term first, then the integral term, and finally the derivative term. However, in more complex control systems, such as the dual-loop control system of a permanent magnet motor, this process can still be very time-consuming and labor-intensive. Even when an appropriate set of PID parameters for steady-state conditions is found, the system's dynamic performance may still be limited. Achieving automatic PID adjustment based on the real-time parameters of the controlled object has always been a goal pursued by engineers.

2.3   Space vector pulse width modulation

The application of SPWM control in motor control aims to approximate the inverter's output voltage to a sinusoidal wave, without focusing on the waveform of the output current. However, an AC motor requires a rotating magnetic field that forms a circular shape in the motor space to drive the motor more efficiently. In the control system, the switching operation of the inverter is controlled to form a circular rotating magnetic field. This is known as the flux linkage tracking control technology, also referred to as Voltage SVPWM technology [3].

2.3.1   *The concept of space vector pulse width modulation control*

Equation (2.26) defines the stator voltage space vector based on the stator three-phase voltage, where the introduced rotating vector factor represents the spatial electrical angle.

$$u_s = 2\left[u_a(t) + u_b(t)e^{j\frac{2\pi}{3}} + u_c(t)e^{j\frac{4\pi}{3}}\right]/3 \tag{2.26}$$

When the motor is running, if the three-phase windings are supplied with three-phase symmetrical sinusoidal voltages, the three-phase voltage vectors can be combined into a voltage space vector. In (2.26), u_s represents a rotating space vector. At this time, the stator of the motor generates a circular space-rotating magnetic field with a constant amplitude and a constant rotational speed. Based on this theory, it is clear that if the voltage supplied by the inverter to the motor can ensure the formation of a circular space-rotating magnetic field with a constant amplitude and controllable speed in the motor's windings, the motor speed can then be controlled. The central idea of SVPWM is to control the duty cycle of the inverter's switching devices, rapidly alternating between different voltage vectors, so that the stator magnetic flux linkage closely follows a circular trajectory, thereby constructing an approximate circular voltage vector circle.

Figure 2.8 represents the stator voltage vectors of the PMSM in the three-phase stationary reference frame. The three axes in the diagram represent the directions of the voltage vectors for the three-phase windings. In the stator's three-phase windings, each winding is supplied with a voltage that varies sinusoidally with a constant amplitude and fixed frequency, and the direction of each voltage vector always lies along the axis of its respective winding. The voltage vectors of the three-phase windings A, B, and C can be combined into a voltage space vector u_s, which rotates at a constant speed with an angular velocity equal to the power supply's angular frequency ω.

When symmetric three-phase sinusoidal voltages are applied to the three-phase windings of a PMSM, the three-phase voltage balance equations can be written. By

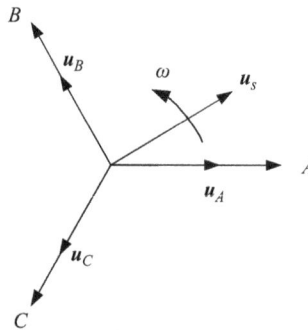

Figure 2.8   Voltage space vector

summing the three-phase voltage balance equations, the composite space voltage vector equation can be obtained as:

$$u_s = R_s i_s + \frac{d\psi_s}{dt} \tag{2.27}$$

When the motor speed is relatively high, the stator resistance voltage drop $R_s i_s$ in (2.27) can be neglected. Therefore, the relationship between the voltage space vector and the flux linkage space vector can be expressed as:

$$u_s \approx \frac{d\psi_s}{dt} \tag{2.28}$$

or

$$\psi_s \approx \int u_s dt \tag{2.29}$$

Additionally, when the motor is running stably, the trajectory of the tip of the stator flux linkage space vector forms a circle, which is referred to as the flux linkage circle. Thus, (2.29) can be expressed as (2.30):

$$\psi_s = \psi_m e^{j\omega t} \tag{2.30}$$

ψ_m, the magnitude of the stator flux vector.
By substituting (2.30) into (2.28), we can obtain (2.31):

$$u_s = \frac{d(\psi_m e^{j\omega t})}{dt} = j\omega \psi_m e^{j\omega t} = \omega \psi_m e^{j(\omega t + \pi/2)} \tag{2.31}$$

From (2.31), it can be seen that when the magnitude of the flux linkage space vector ψ_m is fixed, the magnitude of the voltage space vector u_s is proportional to ω, and its direction is orthogonal to the flux linkage space vector ψ_s. In other words, the direction of u_s always lies along the tangent to the flux linkage circle, as shown in Figure 2.9. When ψ_s completes one full rotation, the stator's three-phase

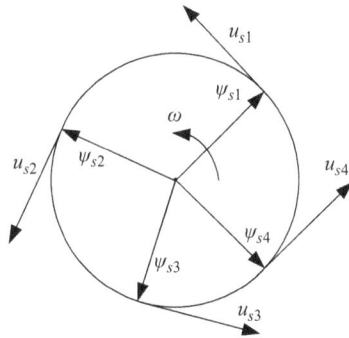

Figure 2.9   Motion trajectories of the rotating magnetic field and voltage space vector

voltage space vector u_s also rotates once along the tangent direction of the flux linkage circle, and the trajectory of BB's motion coincides with the flux linkage circle, as shown in Figure 2.10. Therefore, the motion trajectory of the rotating magnetic field of the PMSM can be analyzed by studying the motion trajectory of the voltage space vector u_s in the form of a circle.

Figure 2.11 shows a typical voltage-source PWM inverter circuit. From the circuit topology in Figure 2.9, it can be seen that the six power electronic devices have a total of eight switching modes. The switching function is defined as:

$$S_K = \begin{cases} 1 & \text{One of the controllable devices in the upper bridge arm of the inverter conducts} \\ 0 & \text{One of the controllable devices in the lower bridge arm of the inverter conducts} \end{cases} \quad (k = a, b, c)$$

The value of function S_a, S_b, S_c determines the switching mode, for example, when $(S_a, S_b, S_c) = (1, 0, 0)$, it means that the upper bridge arm T_1 of the phase A and the lower bridge arms T_4 and T_6 of the phase B and C are on, and the equivalent circuit diagram is shown in Figure 2.12. According to the above provisions, the eight switching states are 000, 010, 100, 101, and 111, of which and these two states indicate that the inverter output voltage to the motor is zero, so these two states are called "0" state or "0" vector.

When $(S_a, S_b, S_c) = (1,0,0)$, From Figure 2.12, it can be calculated that at this time $U_{AB} = U_{dc}$ $U_{BC} = 0$ $U_{CA} = -U_{dc}$, from the relationship between line voltage and phase voltage, $U_{AN} = \dfrac{2}{3}U_{dc}$ $U_{BN} = -\dfrac{1}{3}U_{dc}$ $U_{BN} = -\frac{1}{3}U_{dc}$, and

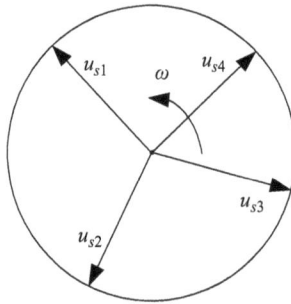

Figure 2.10   Voltage vector circular trajectory

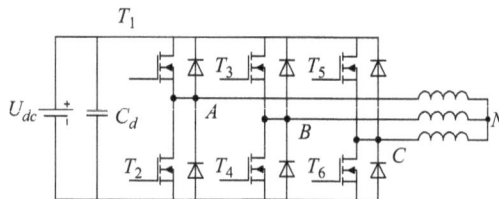

Figure 2.11   Three-phase voltage source inverter circuit

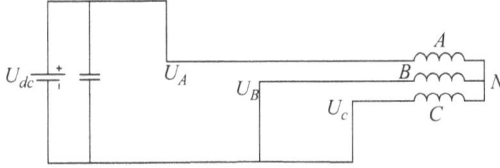

Figure 2.12   Equivalent circuit diagram of 100 states

Table 2.1   Correspondence of space voltage vectors with phase voltages and line voltages

$S_aS_bS_c$	Line voltages			Phase voltages		
	U_{AB}	U_{BC}	U_{CA}	U_{AN}	U_{BN}	U_{CN}
000	0	0	0	0	0	0
100	U_{dc}	0	$-U_{dc}$	$\frac{2}{3}U_{dc}$	$-\frac{1}{3}U_{dc}$	$-\frac{1}{3}U_{dc}$
110	0	U_{dc}	$-U_{dc}$	$\frac{1}{3}U_{dc}$	$-\frac{1}{3}U_{dc}$	$-\frac{2}{3}U_{dc}$
010	$-U_{dc}$	U_{dc}	0	$-\frac{1}{3}U_{dc}$	$\frac{2}{3}U_{dc}$	$-\frac{1}{3}U_{dc}$
011	$-U_{dc}$	0	U_{dc}	$-\frac{2}{3}U_{dc}$	$\frac{1}{3}U_{dc}$	$\frac{1}{3}U_{dc}$
001	0	$-U_{dc}$	U_{dc}	$-\frac{1}{3}U_{dc}$	$-\frac{1}{3}U_{dc}$	$\frac{2}{3}U_{dc}$
101	U_{dc}	$-U_{dc}$	0	$\frac{1}{3}U_{dc}$	$-\frac{2}{3}U_{dc}$	$\frac{1}{3}U_{dc}$
111	0	0	0	0	0	0

the same can be deduced from the other seven states of the motor line voltage and phase voltage, you can get the relationship between the line voltage vector $[U_{AB} \quad U_{BC} \quad U_{CA}]^T$ and the switching state vector $[S_a \quad S_b \quad S_c]^T$ as shown in (2.32):

$$\begin{bmatrix} U_{AB} \\ U_{BC} \\ U_{CA} \end{bmatrix} = U_{dc} \begin{bmatrix} 1 & -1 & 0 \\ 0 & 1 & -1 \\ -1 & 0 & 1 \end{bmatrix} \begin{bmatrix} S_a \\ S_b \\ S_c \end{bmatrix} \tag{2.32}$$

The relationship between the phase voltage vector $[U_{AN} \quad U_{BN} \quad U_{CN}]^T$ and the switching state vector $[S_a \quad S_b \quad S_c]^T$ is shown in (2.33):

$$\begin{bmatrix} U_{AN} \\ U_{BN} \\ U_{CN} \end{bmatrix} = \frac{1}{3}U_{dc} \begin{bmatrix} 2 & -1 & -1 \\ -1 & 2 & -1 \\ -1 & -1 & 2 \end{bmatrix} \begin{bmatrix} S_a \\ S_b \\ S_c \end{bmatrix} \tag{2.33}$$

Accordingly, the magnitude of the line and phase voltages at the output of the inverter for the eight switching states indicated in Table 2.1 can be obtained. The reason why the rotating magnetic field produced by a six-tap staircase wave inverter is positively hexagonal is that the switching operating state of the inverter

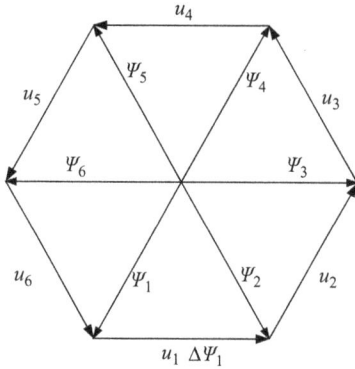

*Figure 2.13   Relationship between voltage space vectors and magnetic chain
vectors*

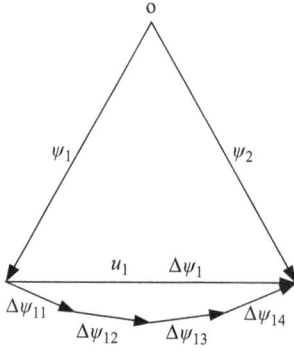

*Figure 2.14   Incremental trajectory of the magnetic chain when approaching a
circle*

is switched only six times in a cycle, forming six active voltage vectors as represented in Figure 2.13, which does not allow obtaining a circular rotating magnetic field. It is thus conceivable to increase the number of switching operations to obtain a more polygonal magnetic chain trajectory to more closely approximate the circular magnetic field. One approach is to still use a six-tap stepped-wave inverter with eight fundamental voltage vectors, using a linear combination of voltage vectors, as shown in Figure 2.14, to synthesize the two adjacent effective voltage vectors to obtain the desired output voltage vector and magnetic chain increment, which is the basic idea of the control.

Figure 2.15 shows the voltage space vector map divided into six sectors by six effective vectors. When chain increment $\Delta\Psi$ and output voltage space vector u_s fall within a sector, the two effective fundamental voltage space vectors adjacent to it are used to form a synthesis u_s.

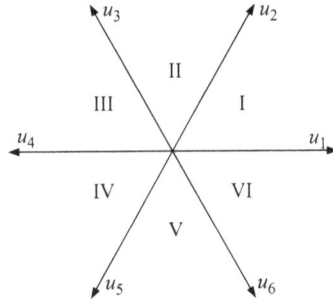

Figure 2.15   Six sectors of the fundamental voltage space vector

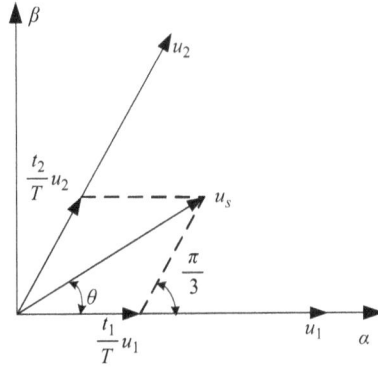

Figure 2.16   Synthesis of desired voltage vectors

From Figure 2.16, it can be seen that when u_s is in the first sector, it can be linearly combined by the voltage space vectors u_1 and u_2 into, and in the phase change cycle T of the inverter, the time to be in the state of the voltage vector u_1 is t_1, and the time to be in the state of the voltage vector u_2 is t_2, so there is:

$$u_s = (t_1/T)u_1 + (t_2/T)u_2 = u_s \cos \theta + j u_s \sin \theta \tag{2.34}$$

Substituting u_1 and u_2 from Table 2.1 into (2.34) yields (2.29):

$$
\begin{aligned}
u_s &= \frac{t_1}{T} U_{dc} + \frac{t_2}{T} U_{dc} e^{j\pi/3} = U_{dc}\left(\frac{t_1}{T} + \frac{t_2}{T} e^{j\pi/3}\right) \\
&= U_{dc}\left[\frac{t_1}{T} + \frac{t_2}{T}\left(\cos \frac{\pi}{3} + j \sin \frac{\pi}{3}\right)\right] = U_{dc}\left[\frac{t_1}{T} + \frac{t_2}{T}\left(\frac{1}{2} + j\frac{\sqrt{3}}{2}\right)\right] \\
&= U_{dc}\left[\left(\frac{t_1}{T} + \frac{t_2}{2T}\right) + j\frac{\sqrt{3}t_2}{2}\right]
\end{aligned}
\tag{2.35}
$$

Let the real and imaginary parts of (2.34) and (2.35) be equal, then there is:

$$\begin{cases} u_s \cos \theta = \left(\dfrac{t_1}{T} + \dfrac{t_2}{2T}\right) U_{dc} \\ u_s \sin \theta = \dfrac{\sqrt{3}t_2}{2T} U_{dc} \end{cases} \tag{2.36}$$

Then the time occupied by the space voltage vectors u_1 and u_2 corresponding to u_S is shown in (2.37):

$$\begin{cases} t_1 = \left[\dfrac{u_s \cos \theta}{U_{dc}} - \dfrac{1}{\sqrt{3}}\dfrac{u_s \sin \theta}{U_{dc}}\right] T \\ t_2 = \dfrac{2}{\sqrt{3}}\dfrac{u_s \sin \theta}{U_{dc}} T \end{cases} \tag{2.37}$$

The phase change period T in (2.37) is not necessarily equal to $t_1 + t_2$, the size of which is determined by the required frequency of the rotating magnetic field, and the remaining time $T - (t_1 + t_2)$ is filled by two zero vectors u_0 and u_7. And the time occupied by the two zero vectors 6 and 7 is half each to reduce the switching times of the switching tubes, as shown in (2.38). Since the six sectors are symmetrical, the above calculation method is applicable to each sector.

$$t_0 = t_7 = \frac{1}{2}[T - (t_1 + t_2)] \geq 0 \tag{2.38}$$

The results analyzed above show that the amplitude and sector of the desired voltage space vector can be obtained from the two selected adjacent fundamental voltage vectors and their action times, and then the zero vector action time is calculated according to (2.38). However, the order of the action times of the two fundamental voltage vectors and the zero vector is not decided, so there are different solutions for the implementation. In the selection of the general implementation scheme, the selection of the zero states is very flexible, usually, there are two principles for the selection of the sequence of the fundamental voltage vector and the zero vector, one is to minimize the number of switching times of the switching tubes to reduce the switching loss, and the other is to try to make the PWM waveforms of the inverter output symmetrical to reduce the harmonic components. The following is the first sector as an example to introduce two common implementation methods.

2.3.1.1 "Zero vector concentration" implementation methods

In accordance with the principle of symmetry of the output waveform, the action times t_1 and t_2 of the two fundamental voltage vectors u_1 and u_2 are equally divided and placed at the beginning and the end of this switching cycle, and the zero vector action time is placed in the middle. Figure 2.15 shows the implementation of SVPWM zero vector concentration.

From Figure 2.17, it can be seen that during the switching cycle T of generating the desired voltage vector, the switching state of phase A of the inverter does not change, while the switching states of phases B and C each change twice, and each time only one

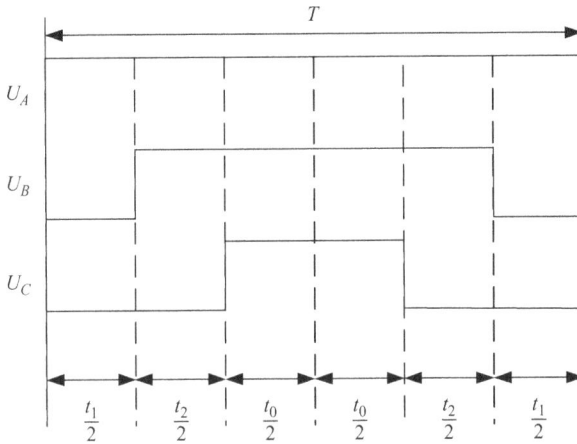

Figure 2.17 Implementation of "Zero Vector Concentration"

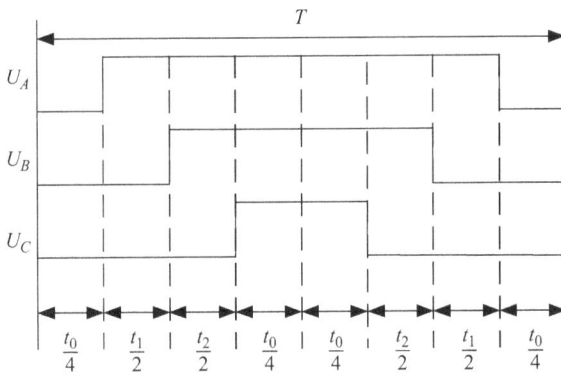

Figure 2.18 The implementation method of "Zero Vector Dispersal"

phase switches its switching state. It can be seen that the "zero vector concentration" method can reduce the number of switching cycles and the switching loss.

2.3.1.2 "Zero-vector dispersion" implementation approach

Figure 2.18 shows the implementation of "zero-vector dispersion." It can be seen that the zero vector is distributed at the beginning, end, and middle of the switching period T. The action times t_1, t_2 of u_1, u_2 are equally distributed in the middle action of the zero vector.

This method is characterized by placing zero vectors at the start and end of each switching cycle. During each transition of the fundamental voltage vector, only one inverter leg (phase) changes its switching state. However, within one switching cycle, each of the three inverter legs switches twice. As a result, the switching loss of the "zero-vector dispersion" method is slightly higher than that of other methods. Nonetheless, in practical applications, this method is widely adopted because it offers a smoother motor output torque and improves the overall performance of the SVPWM scheme.

2.3.2   Implementation of space vector pulse width modulation control algorithm

2.3.2.1   Judgment of the sector in which the voltage space vector is located

Firstly, it is necessary to determine the sector in which the output voltage vector U_{out} is located. When the phase angle of U_{out} is known, the sector can be determined by its phase angle; when U_{out} is given in the form of components U_α, U_β in the $\alpha\beta$ coordinate system, the positive and negative and the range of the values of U can be used to determine the conditions for the determination of the sector in which the spatial voltage vector is located (Table 2.2).

The following definitions can be made:

$$U_1 = U_\beta$$

$$U_2 = \frac{\sqrt{3}}{2}U_\alpha - \frac{U_\beta}{2} \tag{2.39}$$

$$U_3 = -\frac{\sqrt{3}}{2}U_\alpha - \frac{U_\beta}{2}$$

where $\mathrm{sign}(x)$ is the sign function. If $x > 0$, $\mathrm{sign}(x) = 1$; if $x < 0$, $\mathrm{sign}(x) = 0$. The sector number can be determined by looking up in Table 2.3.

Table 2.2   Judgment conditions for sectors where space voltage vectors are located

Sector	Judgmental condition	Equivalence judgmental condition		
		A	B	C
I	$U_\alpha > 0, U_\beta > 0,$ $U_\beta/U_\alpha < \sqrt{3}$	1	1	0
II	$U_\beta > 0, U_\beta/|U_\alpha| > \sqrt{3}$	1	0	0
III	$U_\alpha < 0, U_\beta > 0,$ $U_\beta/|U_\alpha| > -\sqrt{3}$	1	0	1
IV	$U_\alpha < 0, U_\beta < 0,$ $U_\beta/U_\alpha < \sqrt{3}$	0	0	1
V	$U_\beta < 0, U_\beta/|U_\alpha| < -\sqrt{3}$	0	1	1
VI	$U_\alpha > 0, U_\beta < 0,$ $U_\beta/U_\alpha > -\sqrt{3}$	0	1	0

2.3.2.2 Basic voltage vector action time

Completing the judgment of the sector where U_{out} is located, calculating the basic vector action time according to the method of (2.37) and then substituting (2.39) into the calculation result, we can get the adjacent basic voltage vector working time in the six sectors indicated in Table 2.4.

2.3.2.3 Calculate the value of the comparator

After determining the sector where U_{out} is located, the voltage vector action time is calculated, and according to the principle of PWM operation, the value of each corresponding comparator can be obtained:

$$\begin{cases} t_{aon} = (T - t_1 - t_2)/2 \\ t_{bon} = t_{aon} + t_1 \\ t_{con} = t_{bon} + t_2 \end{cases} \tag{2.40}$$

In (2.40), T_{aon}, T_{bon}, and T_{con} are the corresponding on-time variables, and Table 2.5 shows the values of the comparators for different sectors.

Table 2.3 Judgment condition for the sector where the space voltage vector is located

N	3	1	5	4	6	2
Sector number	I	II	III	IV	V	VI

Table 2.4 Basic voltage vector operating time for each sector

Sector	Time of action of two neighboring fundamental voltage vectors	
I	$t_1 = \frac{2}{\sqrt{3}} \frac{U_2}{U_{dc}} T$	$t_2 = \frac{2}{\sqrt{3}} \frac{U_1}{U_{dc}} T$
II	$t_2 = -\frac{2}{\sqrt{3}} \frac{U_3}{U_{dc}} T$	$t_3 = -\frac{2}{\sqrt{3}} \frac{U_2}{U_{dc}} T$
III	$t_3 = \frac{2}{\sqrt{3}} \frac{U_1}{U_{dc}} T$	$t_4 = \frac{2}{\sqrt{3}} \frac{U_3}{U_{dc}} T$
IV	$t_4 = \frac{2}{\sqrt{3}} \frac{U_2}{U_{dc}} T$	$t_5 = -\frac{2}{\sqrt{3}} \frac{U_1}{U_{dc}} T$
V	$t_5 = \frac{2}{\sqrt{3}} \frac{U_3}{U_{dc}} T$	$t_6 = \frac{2}{\sqrt{3}} \frac{U_2}{U_{dc}} T$
VI	$t_6 = -\frac{2}{\sqrt{3}} \frac{U_1}{U_{dc}} T$	$t_1 = -\frac{2}{\sqrt{3}} \frac{U_3}{U_{dc}} T$

Table 2.5 Compare register values across sectors

Sector number	I	II	III	IV	V	VI
T_a	t_{bon}	t_{aon}	t_{aon}	t_{con}	t_{con}	t_{bon}
T_b	t_{aon}	t_{con}	t_{bon}	t_{bon}	t_{aon}	t_{con}
T_c	t_{con}	t_{bon}	t_{con}	t_{aon}	t_{bon}	t_{aon}

2.4 PMSM dual closed-loop control system model

The PID/PI controller is the basis for vector control implementation. Simple and accurate parameter design method of digital PID/PI controller for PMSM has high engineering value. The traditional method of PID/PI parameter tuning is to simplify the motor control system according to the classical control theory, derive the PI parameter analytical calculation formula, and combine the numerical analysis calculation method, the dominant pole configuration method, and the phase margin analysis method to analyze the system in the s-domain, etc. This type of method is relatively simple, but it requires that the PI controller be regarded as a continuous module, however, the commonly used PI controller is a digital module programmed and implemented within a digital processor, so better results can be achieved by analyzing the PMSM dual closed-loop control system in the discrete domain, mainly due to the following reasons: (1) the digital PID/PI controller parameters have a close relationship with the sampling period of the control system, which is not reflected by the method of rectifying the PID/PI parameters in the continuous domain; (2) the additional phase angle hysteresis caused by the zero-order keeper (ZOH) is negligible only when the sampling frequency is sufficiently high relative to the system operating frequency, while the operating frequency of high-speed motors does not differ much from the sampling frequency, especially the current loop. This chapter introduces a digital PI parameter tuning method for the closed-loop control system of a PMSM, which adopts a discrete design method of parameter tuning in the w' domain, and the phase angle hysteresis caused by the sampling time and the ZOH is directly embodied in the design process, and the specific research contents are as follows: first, the mathematical models of the system segments are established and the block diagrams of discrete model of the closed-loop system are obtained; then the discrete model block diagram of the closed-loop system is obtained; then, the time domain performance indexes are converted into open-loop frequency domain eigen quantities by using engineering experience, and the PI controller parameters are obtained by analytical calculation in the w' domain. The calibration process adopts the method of the current inner loop followed by the speed outer loop, and the simulation and experiments show that the results obtained by this design method meet the requirements of performance indexes, and the study shows that this method can provide guidance for the calibration of the parameters of the digital PI controller of PMSM [4].

2.4.1 PMSM dual closed-loop control system model

2.4.1.1 PMSM model

Neglecting magnetic saturation of the core and disregarding eddy currents and magnetic losses, the voltage equation of the PMSM in the *dq* synchronous rotating coordinate system is given by:

$$\left.\begin{array}{l} v_q = i_q R_s + L_q \dfrac{di_q}{dt} + \omega_e L_d i_d + \psi_f \omega_e, \\[2mm] v_d = i_d R_s + L_d \dfrac{di_d}{dt} - \omega_e L_q i_q \circ \end{array}\right\} \tag{2.41}$$

where v_q and v_d are the q and d axis equivalent voltages respectively, i_q and i_d are the q and d axis currents, L_q and L_d are the q and d axis synchronous inductances, R_s is the winding resistance, Ψ_f is the magnetic chain of permanent magnets, and ω_e is the electrical angular velocity of the motor.

$$T_e = \frac{3}{2}p\left[(L_d - L_q)i_d i_q + \psi_f i_q\right] \tag{2.42}$$

where T_e is the output electromagnetic torque and p is the number of pole pairs.

When the weak magnetic control of the motor is not considered, the control method of $i_d = 0$ is commonly used in engineering, and there is $L_d = L_q$ for the hidden pole PMSM, and the motor model is thus simplified. Then the equation of motion of the PMSM is:

$$\frac{d\omega}{dt} = \frac{T_e}{J} - \frac{T_L}{J} - \frac{B}{J}\omega = \frac{K_t}{J}i_q - \frac{T_L}{J} - \frac{B}{J}\omega \tag{2.43}$$

where T_L is the load torque, B is the friction coefficient, J is the rotational inertia of the shaft, ω is the mechanical angular velocity, and $K_t = 3p\Psi_f/2$, is the torque constant.

After using the state feedback compensation method to achieve the model decoupling of the motor, the Laplace transform of (2.41)–(2.43) is used to obtain the motor model in the s-domain (Figure 2.19).

2.4.1.2 Inverter model

The basic model of the voltage inverter can be equated to a first-order inertial link:

$$G_{inv}(s) = \frac{k_{inv}}{sT + 1} \tag{2.44}$$

where T is the inverter switching period, k_{inv} is the inverter voltage amplification, and $k_{inv} = 1$ when controlled by the space vector modulation (SVPWM) method.

In addition, the dead time and switching delay can be equated to a pure delay link e^{-sT_d}, which can be approximated as a first-order inertial link since T_d is very small:

$$G_d(s) = e^{-sT_d} \approx \frac{1}{sT_d + 1} \tag{2.45}$$

The transfer function of the inverter is:

$$G_{inv\_d}(s) = G_{inv}(s) \cdot G_d(s) = \frac{1}{(sT + 1)(sT_d + 1)} \tag{2.46}$$

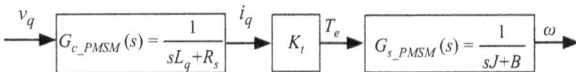

Figure 2.19   Model of PMSM in the s-domain

2.4.1.3   Filter modeling

After sampling the current, it needs to be filtered in a certain way before it can be used, otherwise, the motor control effect will deteriorate due to the inaccuracy of the measurement, and the hardware is generally used to implement the second-order active low-pass filtering, and the transfer function is:

$$G_{c\_f}(s) = \frac{A \cdot \omega_0^2}{s^2 + \frac{\omega_0}{Q}s + \omega_0^2} \qquad (2.47)$$

where A is the gain of the filter, with a value of 1 when the filter has no amplification; ω_0 is the cut-off frequency; and Q is the quality factor.

The speed is taken as it is, and the filter link is realized by programming, which is equivalent to a proportional link with a gain of 1:

$$G_{s\_f}(z) = 1 \qquad (2.48)$$

2.4.1.4   Digital PI controller models

The transfer function of the PI controller in the continuous domain is:

$$G_{c\_PI}(s) = K_{cp} + \frac{K_{ci}}{S} \qquad (2.49)$$

Using the post-differential method to discretize the integral and differential terms, the widely used positional PI in engineering can be obtained, and the digital PI pulse transfer functions for the current loop and speed loop are respectively:

$$G_{c\_PI}(Z) = K_{cp} + \frac{T_c K_{ci}}{1 - z^{-1}} = K'_{cp} + \frac{K'_{ci}}{1 - Z^{-1}} \qquad (2.50)$$

$$G_{s\_PI}(Z) = K_{sp} + \frac{T_s K_{si}}{1 - z^{-1}} = K'_{sp} + \frac{K'_{si}}{1 - Z^{-1}} \qquad (2.51)$$

where K_{cp}, K_{ci} and K_{sp}, K_{si} are the proportional and integral coefficients of the continuous PI controller, K'_{cp}, K'_{ci} and K'_{sp}, K'_{si} are the proportional and integral coefficients of the discrete PI controller, T_c, T_s are the regulation periods of the current loop and speed loop, respectively. Equations (2.50) and (2.51) also show that the PI parameters are very much related to the regulation period, and the parameters designed in the continuous domain cannot be directly applied in practice.

2.4.1.5   Discrete modeling of control systems

Based on the discrete system modeling approach to configuring the position of the sampler and the ZOH, Figure 2.20 shows the block diagram of the discrete model of the PMSM dual closed-loop control system including the ZOH.

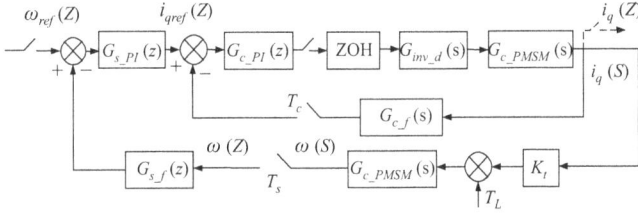

Figure 2.20 Discrete model diagram of PMSM control system

2.4.2 Digital PI controller parameter setting

2.4.2.1 Relationship between open-loop frequency-domain eigen quantities and time-domain metrics for higher-order systems

When analyzing and designing a system in the frequency domain, the open-loop frequency-domain characteristics (phase angle margin γ, amplitude margin h, open-loop gain K, and shear frequency ω_c) are often used as the basis. However, the time-domain specifications of a closed-loop system, i.e., the regulation time t_s (the time for the unit step response to remain within $\pm 2\%$ of its final value), the overshoot $\sigma\%$, and the steady state error e_{ss} are usually given in the project, and therefore, obtaining the relationship between the frequency-domain characteristics and the time-domain specifications can provide a clear direction for the frequency-domain design. For systems higher than the third order, accurately deriving the relationship between ω_c, γ, t_s, and $\sigma\%$ is not only very difficult but also lacks practical value. In control engineering, this relationship is typically described using empirical knowledge gained from engineering practice. As shown in Figure 2.21 and Figure 2.22, we can observe that a larger γ corresponds to a smaller $\sigma\%$. When γ is fixed, an increase in t_s is associated with a decrease in ω_c. The parameter h does not have a definite relationship with the time-domain indicators, but it is generally used as a constraint. However, the general requirement $h \geq 10$ dB, can be used as a constraint to judge the reasonableness of the designed parameters. In engineering, it is hoped that the value of phase angle margin ranges from $30°$ to $70°$.

2.4.2.2 Current loop analysis

From Figure 2.20, the open-loop transfer function from $i_{qref}(z)$ to $i_q(z)$ is:

$$G_c(z) = G_{c\_PI}(z)\mathscr{L}\big(G_{ZOH}(s)G_{inv\_d}(s)G_{c\_PMSM}(s)G_{c\_f}(s)\big) \tag{2.52}$$

The calculation process and the calculation results of (2.12) are complicated, which increases the time cost of the current loop analysis and is not necessary for

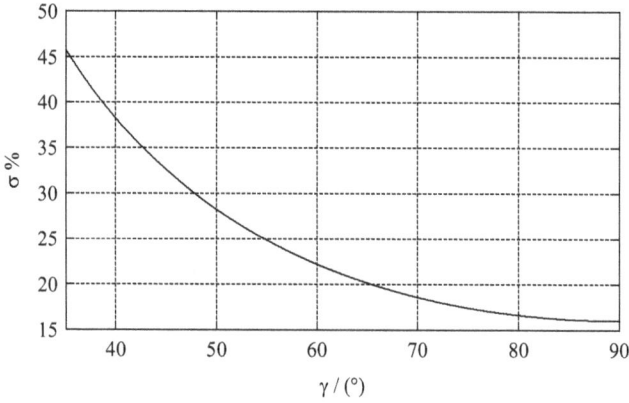

Figure 2.21   Relationship between overshoot σ% and phase margin γ of high-order system

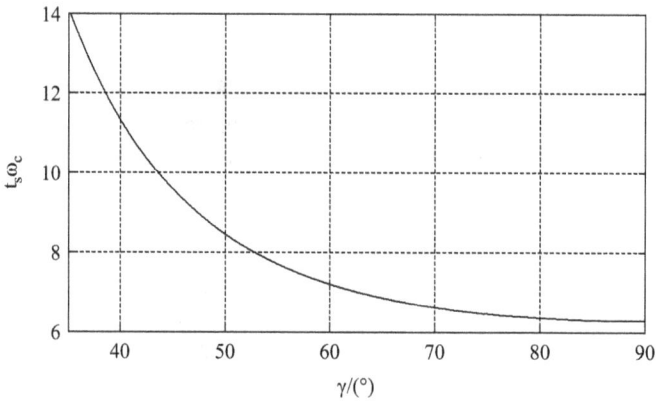

Figure 2.22   Relationship between $t_s \omega_c$ and phase margin γ of high-order system

engineering applications. Through further analysis of the current loop, it is found that the model of the current loop can be simplified to a certain extent, which in turn reduces the complexity of the design of the PI parameters of the current loop: (1) Since the current filter filters only high-frequency harmonics and has a very small effect on the signals near the operating frequency, it can be equated to a proportional link with a gain of 1 in the design of the system; (2) The time constants T and T_d of the inverter model satisfy the requirements of $T_d \ll T \ll L_q$, whose delay effect can be neglected, and is also equivalent to a proportional link with a

Table 2.6   *The current-loop PI controller parameters and amplitude margin corresponding to different value of γ*

$\gamma_c/^\circ$	K'_{cp}	K'_{ci}	h_c/dB
54	29.10	6.96	11.7
57	30.10	5.96	11.5
60	31.08	4.96	11.4
63	31.98	3.95	11.3
66	32.80	2.93	11.2
70	33.76	1.55	11.2

gain of 1. Combined with the motor parameters in Table 2.6, the open-loop pulse transfer function of the current loop can be simplified as:

$$G_c(z) = G_{c\_PI}(z)\mathscr{L}(G_{ZOH}(s)G_{c\_PMSM}(s))$$

$$= \frac{\left[\left(K'_{cp} + K'_{ci}\right)z - K'_{cp}\right]\left(e^{\frac{R_sT_c}{L_q}} - 1\right)}{\frac{R_sT_c}{L_q}}$$

$$= 0.016\frac{\left(K'_{cp} + K'_{ci}\right)z - K'_{cp}}{(z - 0.9953)(z - 1)} \tag{2.53}$$

Before designing the PI controller parameters in the w' domain, the given time-domain indicators are converted to frequency-domain indicators according to Figures 2.21 and 2.22. As the current regulation is the inner loop of the system, in addition to meeting the stability conditions, but also needs to take into account the dynamic performance, that is, the regulation time should be shorter and allow a certain degree of overshooting of the step response. In this chapter, the current loop regulation is required to be completed within 10–100 control cycles, $10T_c \le t_{cs} \le 100T_c$ (2–20 ms), the overshooting amount meets $\sigma_c\% \le 25\%$, the static speed error coefficient $K_{cv} \ge 10$, and the phase angle margin of the current loop is satisfied by combining with Figure 2.21:

$$54^\circ \le \gamma_c \le 70^\circ \tag{2.54}$$

When this phase angle margin condition is satisfied, it is known according to Figure 2.22:

$$6.6. \le t_{cs}\omega_{cc} \le 7.8 \tag{2.55}$$

Substituting $10T_c \le t_{cs} \le 100T_c$ into (2.55) yields a range of values for the current loop shear frequency:

$$\frac{0.066}{T_c} \leq \omega_{cc} \leq \frac{7.8}{T_c} \tag{2.56}$$

The range of values of ω_{cc} can be further narrowed down: without considering the weak magnetic control, when the motor speed reaches the maximum (rated value), the operating frequency also reaches the maximum, and the current loop shear frequency needs to be larger than this value. According to the motor parameters in Table 2.8, the maximum operating frequency of the current is $\omega_{cmax} = \frac{2\pi n_N p}{60} = 2{,}093.3$ rad/s. The range of values for the current loop shear frequency is narrowed to $ax(0.066/T_c, \omega_{cmax}) \leq \omega_{cc} \leq 0.75/T_c$, so $2{,}093.3 \leq \omega_{cc} < 3{,}900$ rad/s.

The current loop PI controller parameters are two-dimensional functions of:

$$\left. \begin{array}{l} \left(K'_{cp} + K'_{ci} \right) = f(\gamma_c, \omega_{cc}), \\ 54° \leq \gamma_c \leq 70°, \\ 2{,}093.3 \leq \omega_{cc} \leq 3{,}900\circ \end{array} \right\} \tag{2.57}$$

To simplify the analysis and design process, the "constant shear frequency method" is used to reduce the problem to a one-dimensional function of γ_c:

$$\left. \begin{array}{l} \left(K'_{cp} + K'_{ci} \right) = f(\gamma_c, \omega_{cc0}), \\ 54° \leq \gamma_c \leq 70°\circ \end{array} \right\} \tag{2.58}$$

The constant shear frequency of the current loop is taken as $\omega_{cc0} = 2{,}800$ rad/s. The frequencies ω analyzed previously are the real frequencies, and the corresponding virtual frequencies v in the w' domain are calculated by (2.59).

$$v = \frac{2}{T_{ex}} \tan \frac{\omega T_{ex}}{2} \tag{2.59}$$

where T_{ex} is the regulation period, ω_{cc0} corresponds to the virtual shear frequency $v_{cc0} = 2{,}877$ rad/s.

2.4.2.3 Current loop design and simulation

Transforming (2.53) to the w' plane yields:

$$G_c(w') = -0.004 \frac{(w' - 10{,}000)[(2K'_{cp} + K'_{ci})w' + 10{,}000K'_{ci}]}{w'(w' + 23.56)} \tag{2.60}$$

The shear frequency is v_{cc0}, then:

$$\left. \begin{array}{l} |G_c(jv_{cc0})| = 5 \times 10^{-6} \sqrt{2{,}877^2 (2K'_{cp} + K'_{ci})^{2'} + 10{,}000^2 K'_{cp}{}^2} = 1 \\ \gamma_c = 180° + \angle G_c(jv_{cc0}) = -15.5° + \arctan \left(\frac{2K'_{cp} + K'_{ci}}{3.5K'_{ci}} \right)\circ \end{array} \right\} \tag{2.61}$$

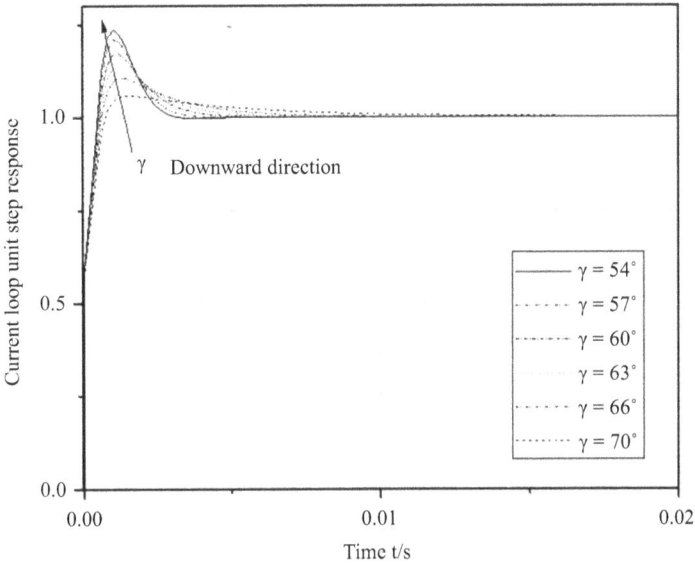

Figure 2.23   The current-loop unit step response under different PI controller parameters

When the phase margin γ_c of the current loop is set to different values, the corresponding PI controller parameters for the current loop can be calculated using (2.59). To obtain suitable controller parameters, the relationship between the current loop characteristic γ_c and the parameters is studied as follows: Table 2.8 shows the calculated PI parameters and gain margin h_c for different values of γ_c. Figure 2.23 presents the simulation curves of the unit step response of the current loop for different PI parameters.

The data in Table 2.6 shows that for every 3° increase in phase margin, the proportional gain K'_{cp} increases by approximately 1, while the integral gain K'_{ci} decreases by 1. Additionally, the current loop gain margin h_c is inversely related to the phase margin γ_c, but all six sets of gain margin values are greater than 10 dB, which meets the engineering requirements. Figure 2.24 visually indicates that as the phase margin increases from 54° to 70°, the overshoot of the current loop shows a significant change, decreasing from 23% to 5.5%, while the settling time remains below $100T_c = 20$ ms.

2.4.2.4   Speed loop analysis

Since the current loop adjustment process is much faster than the speed loop, it is generally treated in the literature as either a pure proportional element or a first-order inertia element. The PI controller parameters obtained through the equivalent method have little effect on the system's steady-state results, but the dynamic performance of the system differs from expectations. Therefore, in this chapter, when designing the speed loop, the results of the current loop design are directly

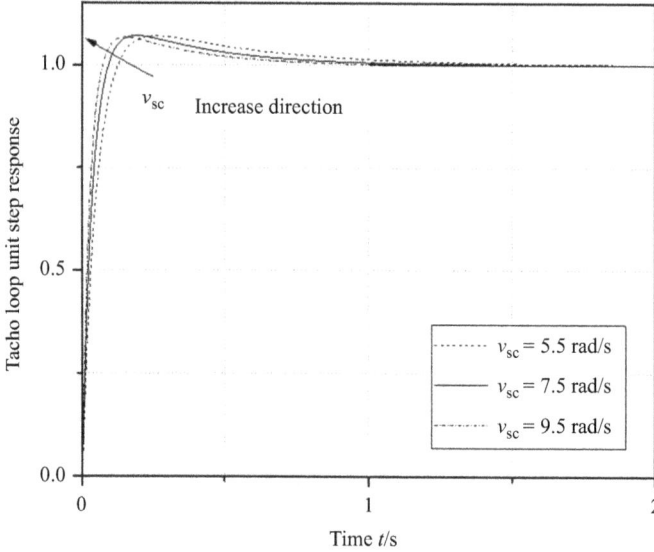

Figure 2.24   The speed-loop unit step response under different PI controller parameters

utilized. To minimize the overshoot in speed regulation, the phase margin of the current loop is kept sufficient, and the current loop PI controller parameters are selected as $K'_{cp} = 33.76$ and $K'_{ci} = 1.55$. This can, to some extent, reduce the design pressure of the speed loop. According to Figure 2.7, the open-loop transfer function from $\omega_{ref}(z)$ to $\omega(z)$ of the speed loop is:

$$G_s(z) = G_{s\_PI}(z)G'_c(z)G_{s\_f}(z)\mathscr{L}(K_tG_{s\_PMSM}(s))$$

$$= 0.565\frac{K_tz(z - 0.9519)\left[(K'_{sp} + K'_{si})z - K'_{sp}\right]}{j(z - 0.9519)(z - 0.4781)(z - 1)\left(z - e^{\frac{B}{J}T_s}\right)}$$

$$= 3.48\frac{z(z - 0.9519)\left[(K'_{sp} + K'_{si})z - K'_{sp}\right]}{(z - 0.9519)(z - 0.4781)(z - 1)^2} \qquad (2.62)$$

where, $G'_c(z)$ is the closed-loop transfer function of the current loop.

The frequency domain characteristics of the speed loop can be determined based on the given time-domain specifications. In this chapter, the system is first required to have a fast response, with a settling time of $0.65\ s \leq t_{ss} \leq 1.2\ s$. Secondly, the speed overshoot should satisfy $\sigma_s\% \leq 7.5\%$. Considering that the speed overshoot decreases with an increase in phase margin, and in contrast to the current loop design, the speed loop PI parameters are tuned using the "constant phase margin method," which directly selects a larger phase margin value for the speed loop:

$$\gamma_{s0} = 70° \tag{2.63}$$

Under this phase margin condition, based on Figure 2.22, it is known that:

$$t_{ss}\omega_{sc} = 6.6 \tag{2.64}$$

Substituting $0.65\ s \le t_{ss} \le 1.2\ s$ into (2.64) gives:

$$5.4 \le \omega_{sc} \le 10\ \text{rad/s} \tag{2.65}$$

According to (2.59), the shear frequency range in the w' domain for the speed loop is $5.4 \le v_{sc} \le 10\ \text{rad/s}$. From the above analysis, it can be observed that the strict limitation on the adjustment time results in a very small range of variation for the shear frequency ω_{sc}.

Transforming (2.63) into the w' plane gives:

$$G_s(w') = -0.5898 \frac{(w'^2 - 500^2)(w' + 11.22)[(2K'_{sp} + K'_{si})w' + 500K'_{si}]}{w'^2(w' + 12.32)(w' + 176.54)} \tag{2.66}$$

Let the speed loop cutting frequency be v_{sc}, and the phase margin be $70°$, then:

$$\left.\begin{array}{l} |G_s(jv_{sc})| = 1, \\ \gamma_{s0} = 180° + \angle G_s(jv_{sc}) = 70°\circ \end{array}\right\} \tag{2.67}$$

The speed loop PI controller parameters are a one-dimensional function of the cutting frequency. To obtain appropriate current loop PI controller parameters, the relationship between the speed loop characteristics and v_{sc} is studied as follows. Table 2.7 presents the calculated speed loop PI parameters and gain margin h_s for different values of v_{sc}. Figure 2.24 shows the unit step response curves of the speed loop with three sets of different PI parameters.

Table 2.7 shows that for each increase of 1 in the speed loop cut-off frequency, the proportional coefficient K'_{sp} increases by approximately 0.0006, and the integral coefficient K'_{si} increases by 0.00001. Since the change in the cut-off frequency is small, the variations in the PI controller parameters are not significant. The speed

Table 2.7 *The speed-loop PI controller parameters and amplitude margin corresponding to different values of v_{sc}*

$v_{sc}/\text{rad/s}$	$K'_{sp} \times 10^{-3}$	$K'_{si} \times 10^{-5}$	h_s/dB
5.5	3.3	2.7	48.2
6.5	3.9	3.7	46.8
7.5	4.4	4.9	45.6
8.5	4.8	6.0	44.6
9.5	5.6	7.6	43.5

loop gain margin is inversely proportional to the cut-off frequency, with all 5 sets of gain margin values (h_s) being greater than 40, meeting the engineering requirements. As shown in Figure 2.24, when v_{sc} gradually increases, the speed loop overshoot remains unchanged at about 5%, but the adjustment time tends to shorten, dropping from 1.2 s to 0.7 s.

The experiment is designed to verify that the PI controller parameters can ensure the system's dynamic and steady-state performance, thereby proving that the PI controller parameter design and analysis method proposed in this chapter is effective. This chapter analyzes a PMSM, with its specific parameters shown in Table 2.8.

According to the design, the current loop PI controller parameters in the experiment are set as $K'_{cp} = 33.76$ and $K'_{ci} = 1.55$. To shorten the adjustment time, the speed loop PI controller parameters are set as $K'_{sp} = 0.0056$ and $K'_{si} = 0.000076$. The designed PI parameters are verified on a PMSM control system platform with the digital signal processor (DSP) TMS320F2812 as the main control chip. The inverter uses the intelligent power module (IPM) PM75RLA12, the motor position detection uses an optical encoder, the load is applied and speed is recorded by a 30 kW dynamometer from Magtrol, and the current waveform is recorded by the WT1800 from Yokogawa.

The set speed is 1,000 rpm. Figures 2.25 and 2.26 show the speed response curves under no-load and constant load ($T_L = 17.5$ N·m) conditions, respectively. From the figures, it can be seen that the speed adjustment times are 1.76 s and 1.92 s, respectively. Additionally, the overshoot in no-load speed and load speed are both relatively small, at 2.3% and 1.8%, respectively, indicating good dynamic performance of the system. The motor's steady-state speed is stable, suggesting that the designed PI controller parameters provide good steady-state performance. Figures 2.27 and 2.28 show the dq-axis current curves and three-phase current waveforms under load steady-state conditions, respectively.

Table 2.8   The parameters of permanent magnet
motor

Parameters	Value
Rated power/kW	7.5
Rated torque T_L/N·m	17.5
Direct axis inductance L_d/H	0.01096
Quadrature axis inductance L_q/H	0.01244
Pole pair number p	5
Moment of inertia J/(kg·m^2)	0.285
Friction coefficient B	0.0003
Rotor flux linkage amplitude Ψ_f/Wb	0.2339
Winding resistance R_s/Ω	0.29
Sampling frequency/Hz	5,000
Current loop regulation period T_c/s	0.0002
Speed loop regulation period T_s/s	0.004

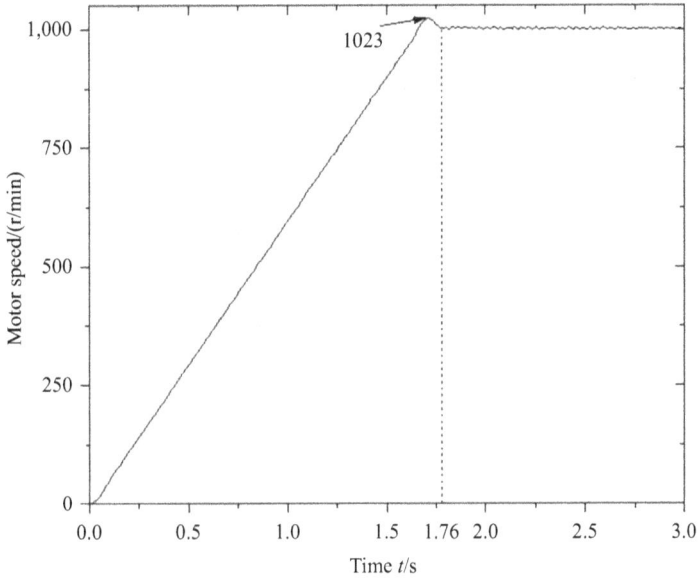

Figure 2.25   The speed response under no-load state

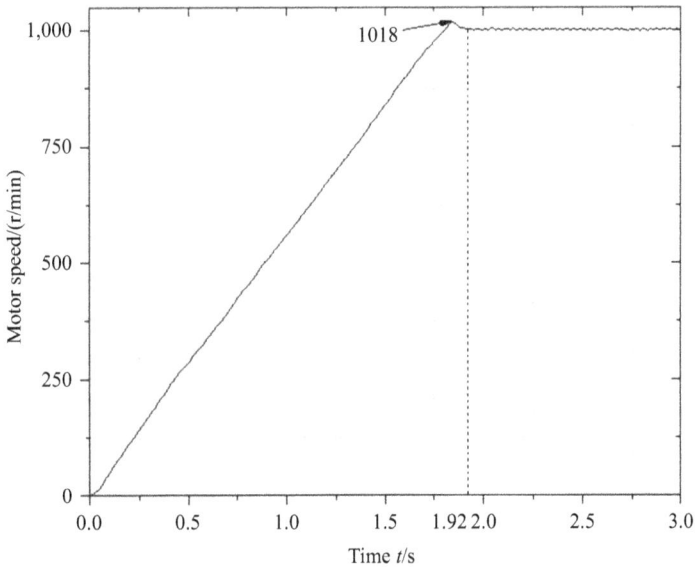

Figure 2.26   The speed response under load state

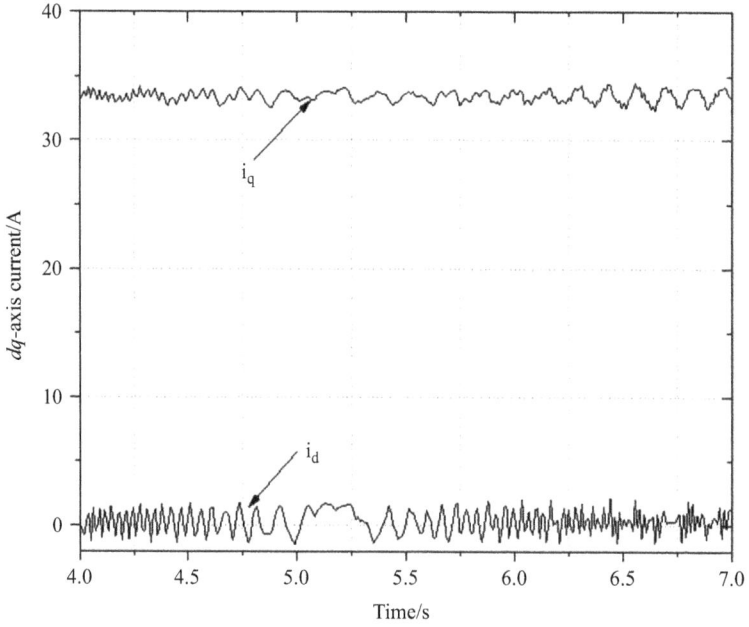

Figure 2.27   The dq-axis current

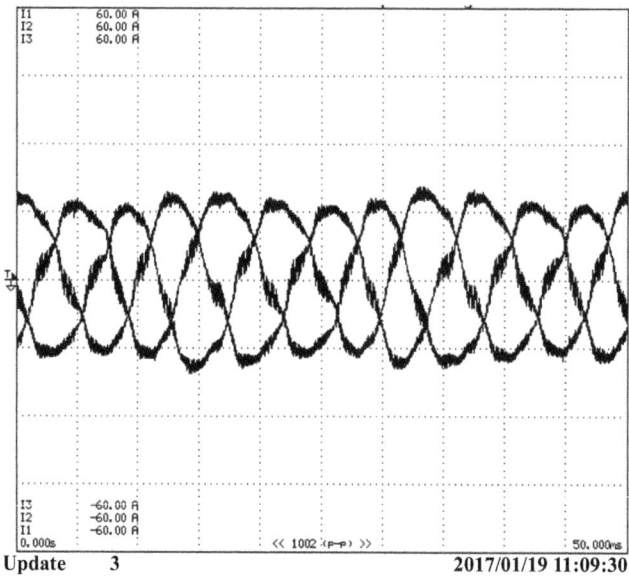

Figure 2.28   Three-phase current after filtering under load state

This section designs the PI controller parameters for the PMSM speed and current loops in the w'-domain using the frequency method. The discretization design method considers the impact of sampling frequency on parameters and the phase lag caused by the zero-order hold (ZOH), making the designed parameters more accurate. For analysis in the w'-domain, engineering experience formulas are directly used to convert time-domain indicators into open-loop frequency domain characteristics, which reduces the complexity of system design. Simulation and experimental results show that the system, corrected using the frequency method, exhibits good steady-state and dynamic performance, meeting engineering requirements.

2.5   Digital implementation of control system

2.5.1   *Introduction to embedded systems*

Embedded systems play an indispensable role in modern control systems, especially in implementing high-performance control algorithms. An embedded system is a specialized computing system that integrates hardware and software to perform specific tasks. In the field of robotics, embedded systems provide real-time control, data processing, and communication capabilities for PMAs.

An embedded system typically consists of the following main components.

Microcontroller (MCU) or Microprocessor: The microcontroller is the core of the embedded system, responsible for processing sensor inputs, running control algorithms, and generating output signals to drive actuators. For example, typical MCUs like the ARM Cortex series are widely used due to their strong computational power and low power consumption. Additionally, Texas Instruments (TI) DSPs play an important role in embedded systems, offering efficient signal processing capabilities suitable for complex control algorithms. TI's C2000 series DSPs are widely used in PMAs due to their high performance and motor control optimization features.

Sensors and Signal Conditioning: Embedded systems rely on sensors to gather system status information such as position, speed, and current. Signal conditioning circuits ensure that the signals from the sensors can be correctly read and processed by the microcontroller. Hall sensors and optical encoders are common sensors in PMA control, used to measure current and position information, respectively.

Communication Interfaces: Embedded systems need to communicate with other devices, such as the main control computer or other sensors. Common communication protocols include CAN, UART, SPI, and I2C, which ensure fast and reliable data transfer. For applications with high real-time requirements, such as multi-axis robot collaborative control, real-time industrial bus protocols like EtherCAT are increasingly popular.

Power Management: The design of embedded systems typically considers power consumption, especially in battery-powered robotic applications. Power management modules optimize energy distribution and extend the system's operational time. Texas Instruments and NXP provide efficient power management ICs that support low-power design requirements.

Real-time operating system (RTOS): For complex control systems that need to handle multiple tasks, an RTOS provides task scheduling, priority management, and time determinism. For example, FreeRTOS is a popular open-source RTOS that enables parallel processing of control tasks. Additionally, TI's SYS/BIOS and NXP's MQX RTOS are commonly used commercial RTOSs in embedded control systems.

The design of embedded systems emphasizes performance, reliability, and compactness. They are typically designed as highly integrated devices capable of performing complex control tasks within resource-limited environments.

2.5.1.1   Overview of embedded chips from different manufacturers

Texas Instruments (TI): TI's C2000 series DSPs are specifically designed for motor and power sets of peripherals such as PWM modules and high-precision analog-to-digital converters (ADCs), making them ideal control applications. They provide powerful floating-point processing capabilities and a real-time library (RL) for vector control of PMAs. TI also offers the code composer studio (CCS) integrated development environment, which simplifies the development of embedded systems.

NXP: NXP's Kinetis series MCUs and i.MX processors perform excellently in embedded systems. The Kinetis series, based on the ARM Cortex-M core, features low power consumption, making them suitable for portable robotic applications. The i.MX series focuses on multimedia and communication capabilities, supporting robots that require advanced human-machine interaction.

STMicroelectronics (ST): ST's STM32 series MCUs are widely popular for their high performance and versatility. Integrated timers, ADCs, and communication interfaces make them an ideal choice for embedded control. Additionally, STM32 supports a wide range of development tools and libraries, such as STM32CubeMX and HAL libraries.

Infineon: Infineon's XMC series MCUs are optimized for industrial control and motor drive applications. They provide efficient PWM generation and power electronics support, making them well-suited for complex robotic applications.

Xilinx: Xilinx's Zynq series SoCs combine ARM processors and FPGAs, enabling highly customizable embedded control solutions. The Zynq SoCs excel in robotic applications requiring ultra-low latency and high computational power.

In the PMA control system, embedded systems ensure the precise operation of the actuators by processing feedback signals in real time and running control algorithms. For example, in the implementation of vector control, the embedded system needs to calculate the d-axis and q-axis currents in real time and generate the corresponding PWM signals to drive the motor. The high computational capability and real-time performance of embedded systems enable them to quickly respond to external disturbances and dynamic changes, thereby maintaining the system's stability and efficiency.

In addition, embedded systems also support various advanced functions, such as:

1.   State Monitoring and Diagnostics: Embedded systems can monitor the operating status of PMAs and detect potential faults. Through data analysis, the

system can issue warnings before issues occur, preventing downtime and losses.

2. Remote Updates and Debugging: Embedded systems support remote firmware updates and code debugging via network connectivity, making system maintenance more convenient.

3. Machine Learning Inference: With the development of embedded AI, some embedded systems are now capable of running simple machine learning models on the device to optimize control strategies and improve efficiency.

2.5.2   PMA digital control hardware system

The hardware circuit system designed in this section is centered around the high-performance DSP TMS320F2812 from TI. Using CCS programming software enables the development of a vector control program based on fuzzy adaptive PID. This serves as an excellent platform for the research of new control algorithms for PMSMs.

2.5.2.1   The overall scheme of the control system

The overall scheme of the control system is shown in Figure 2.29. The control panel is the human-machine interaction module, which receives control information via a matrix keyboard. The control panel can also display system information from the main control board on the LCD screen, making it easier for the experimenter to observe. The core of the main control board is the DSP TMS320F2812, which also includes peripheral circuits for the DSP, PWM output circuits, and signal detection circuits. The software is developed by programming the DSP through CCS. The power module receives the PWM signals sent by the main control board, switches the power transistors on and off, and the output is directly connected to the motor, providing the driving voltage to the motor.

The circuit hardware structure diagram studied in this chapter is shown in Figure 2.30. The control circuit is mainly responsible for collecting and processing the signals required to control the motor and generate control signals. This includes the main control chip DSP support circuit, voltage and current detection circuit, rotor position detection circuit, and PWM signal output circuit. The power circuit is directly connected to the PMSM and provides the driving voltage.

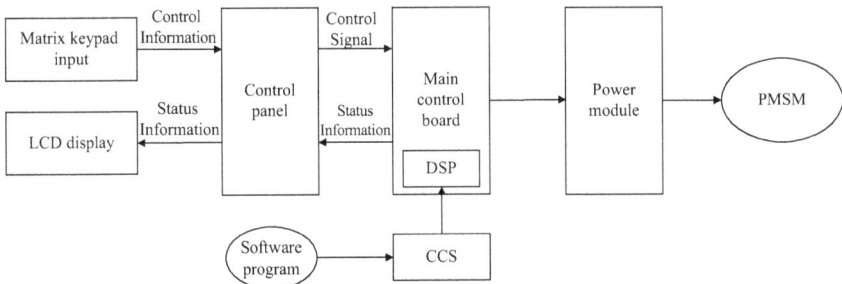

Figure 2.29   Overall design scheme of the control system

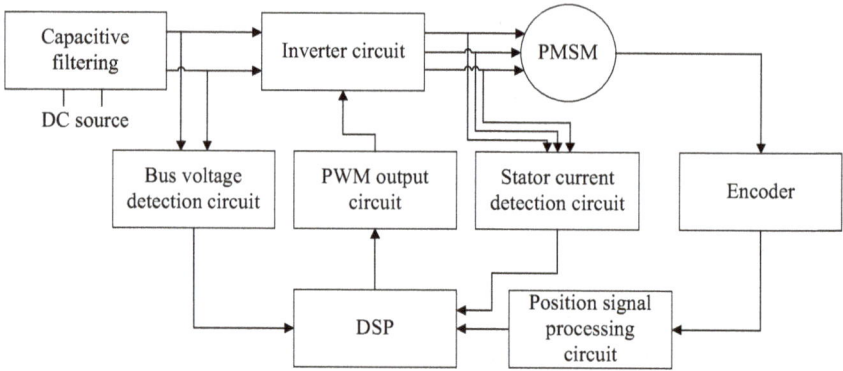

Figure 2.30   Hardware system architecture diagram

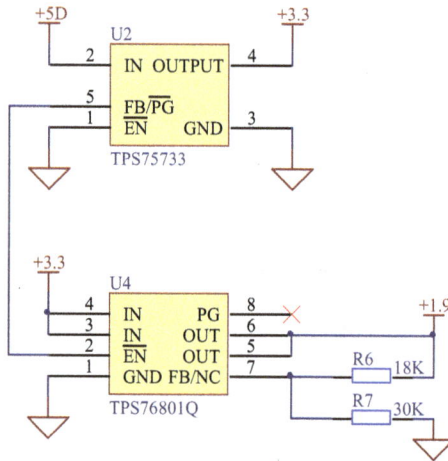

Figure 2.31   DSP power supply circuit diagram

2.5.2.2   DSP support circuit

The TMS320F2812 requires support from the power supply circuit, JTAG interface circuit, clock circuit, and reset circuit for normal operation. During operation, it receives the motor bus voltage, stator three-phase current, and rotor position sampling signals, and outputs six PWM control signals.

2.5.2.3   Power supply circuit

The TMS320F2812 DSP requires a 1.8 V core power supply and a 3.3 V input/output (IO) pin power supply for normal operation. The power supply circuit for the DSP is shown in Figure 2.31. This circuit provides the necessary power to the DSP according to the above requirements.

2.5.2.4   JTAG interface circuit

The JTAG interface is the connection between the DSP and the emulator. It is an 8-pin interface, and its circuit is shown in Figure 2.32.

2.5.2.5   Clock circuit

The clock circuit is responsible for providing the timing pulses to the DSP, with its core component being a 30 MHz passive crystal oscillator. The clock circuit is shown in Figure 2.33.

2.5.2.6   Reset circuit

The reset circuit is used to implement the reset function for the DSP. By pulling the DSP's RST pin to a low level, the DSP will be reset. The circuit design is shown in Figure 2.34.

2.5.2.7   Voltage and current detection circuit

The bus voltage of the controller is collected by a Hall voltage sensor, and its sampling circuit is shown in Figure 2.35.

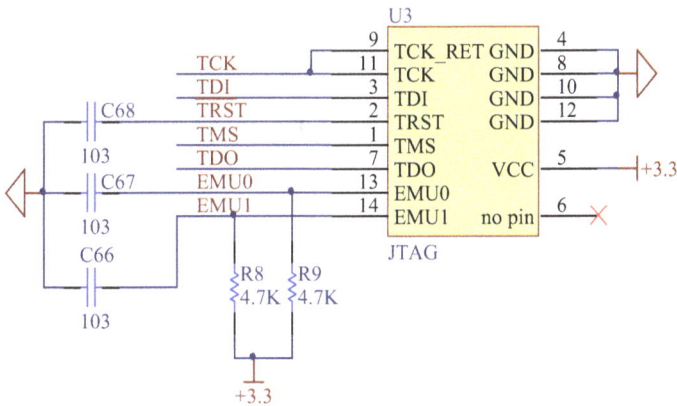

Figure 2.32   JTAG interface circuit diagram

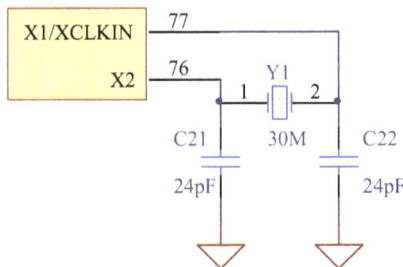

Figure 2.33   Clock circuit diagram

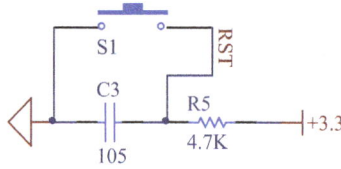

Figure 2.34   Reset circuit diagram

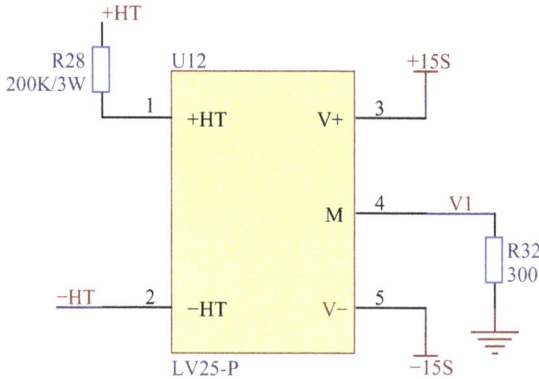

Figure 2.35   Voltage sensor sampling circuit diagram

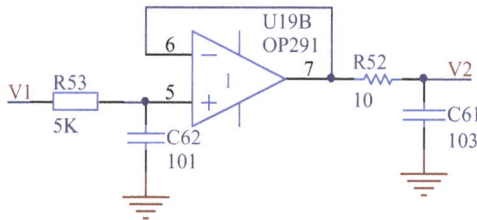

Figure 2.36   Filter circuit

Since the output signal of the Hall voltage sensor is always positive and does not exceed 3.3 V, no biasing is required. The signal only needs to undergo filtering and voltage-following processing before being directly input into the DSP's AD sampling port. The sampling circuit is shown in Figure 2.36.

Given that the motor's rated current is 12.5 A, the peak-to-peak current is 35.4 A. To prevent unexpected situations, a Hall current sensor with a measurement range of 0–76 A is selected. After calculation, when the motor's three-phase current varies between −20 A and 20 A, the voltage signal sampled across the sampling resistor will range from −1 V to 1 V. Therefore, biasing and filtering are required before the signal can be input into the DSP.

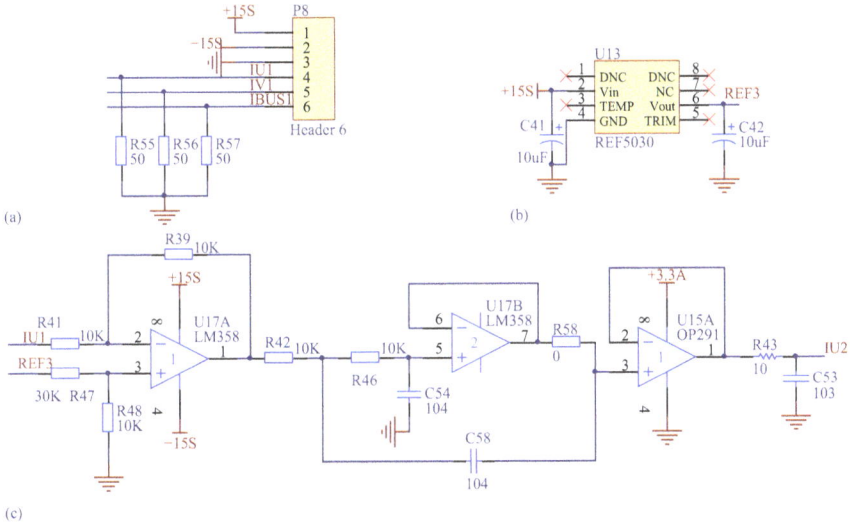

Figure 2.37   Current sampling circuit: (a) converting current to voltage, (b) reference voltage generation, and (c) signal processing circuits

A three-stage operational amplifier circuit processes this signal. The first stage handles the biasing, the second stage filters the signal, and the third stage performs voltage following. Then, the signal can be input into the DSP. The current sampling circuit is shown in Figure 2.37, with (a), (b), and (c) indicating the sampling circuit, the 3 V reference voltage circuit, and the signal conditioning circuit, respectively.

2.5.2.8   Rotor position detection circuit

A 1,024-line incremental optical encoder is chosen as the rotor position sensor for the motor control system in this chapter. This encoder requires a 5 V DC power supply, which can be directly provided by the main control circuit. When it operates normally, it generates three pairs of differential signals: A+, A−, B+, B−, and Z+, Z−. These differential signals are then converted into A, B, and Z single-ended 5 V pulse outputs using a differential-to-single-ended conversion circuit. The conversion circuit is implemented using the MAX3093E chip, as shown in the circuit diagram in Figure 2.38.

The three 5 V single-ended pulse signals output by the conversion circuit need to pass through an optocoupler isolation circuit before being sent to the DSP's QEP port. After passing through the optocoupler isolation circuit, the 5 V single-ended pulse signals are also converted into 3.3 V signals that can be received by the DSP. As shown in Figure 2.39, the optocoupler isolation circuit for the A channel is provided, and the optocoupler isolation circuits for the B and Z channels are the same as the A channel.

After the A, B, and Z signals pass through the optocoupler isolation circuit, they are input to the QEP port of the DSP. The rotor position angle, direction, and speed of the motor are then calculated through computations.

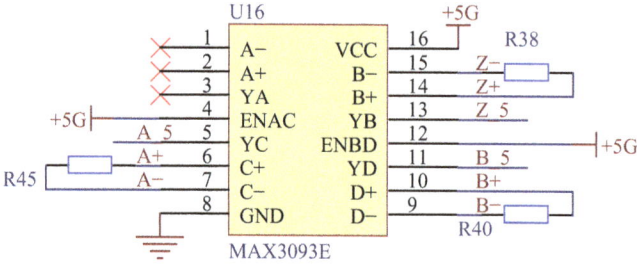

Figure 2.38   Differential signal conversion circuit

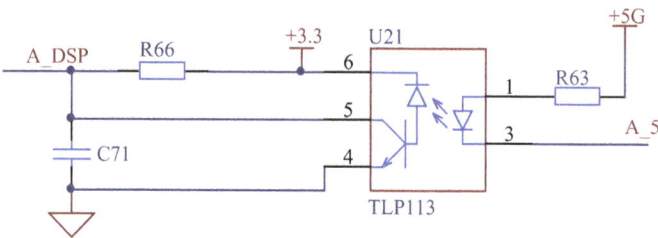

Figure 2.39   Optocoupler isolation circuit for the A channel

2.5.2.9   PWM wave output circuit

After the DSP processes the collected voltage, current, rotor position, and speed signals, it will generate six PWM signals. These six PWM signals are complementary in pairs and are used to control the switching of the power switches in the three-phase inverter. However, these six PWM signals output by the DSP are not only weak in signal strength, making it impossible to directly control the power switches, but they also lack dead time, which could cause a short circuit due to both upper and lower switches in the bridge arm turning on simultaneously. Therefore, these six PWM signals need to be processed by AND and NOT gates to form complementary signals in pairs, dead-time circuits need to be added to prevent shoot-through, and the ULN2003 chip is used to drive and amplify the signals before they can be fed into the three-phase inverter to control the switching of the power switches. The six PWM signal processing circuit is shown in Figure 2.40.

2.5.3   PMA digital implementation software system

Using TI products, the software program needs to be written and debugged in C language within the CCS3.3 development environment. It mainly includes the main program section, main interrupt program section, SVPWM program section, AD conversion program section, and rotor position detection program section.

2.5.3.1   Main function

The main function of the main program is to complete system initialization, interrupt vector configuration, processing of input control signals, and fault handling. The flowchart of its operation is shown in Figure 2.41.

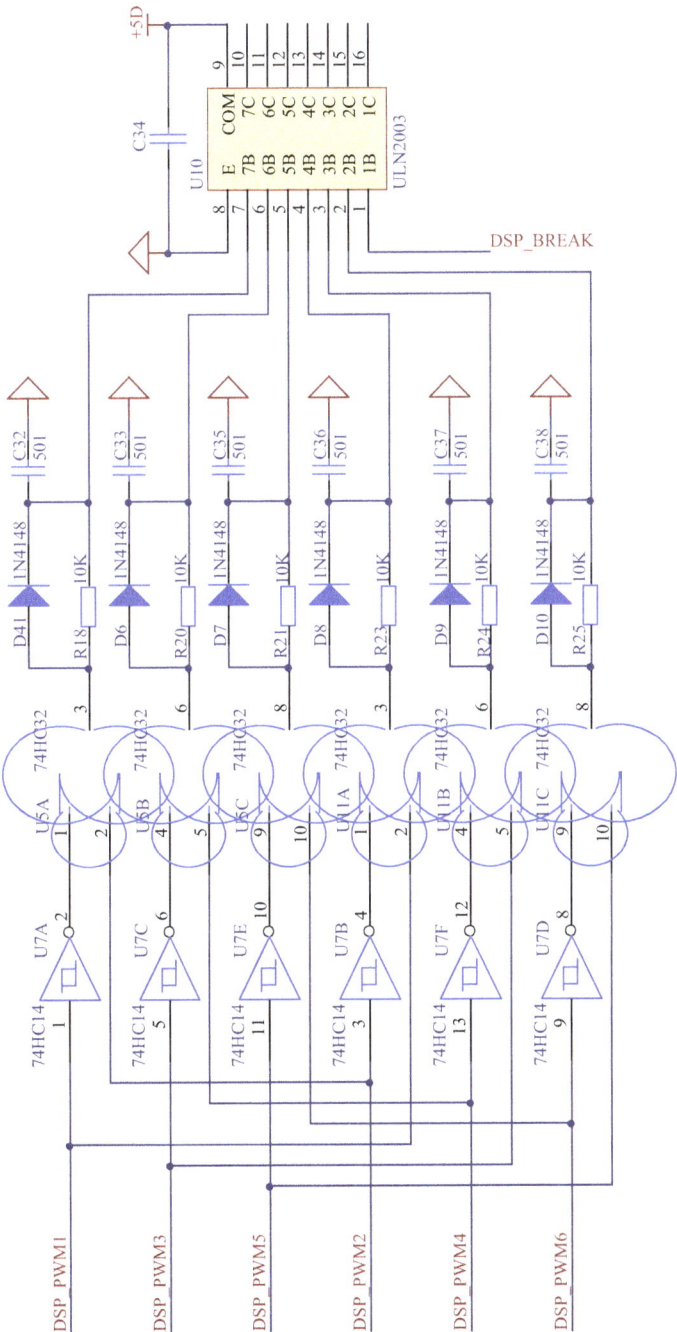

Figure 2.40   PWM signal processing circuit diagram

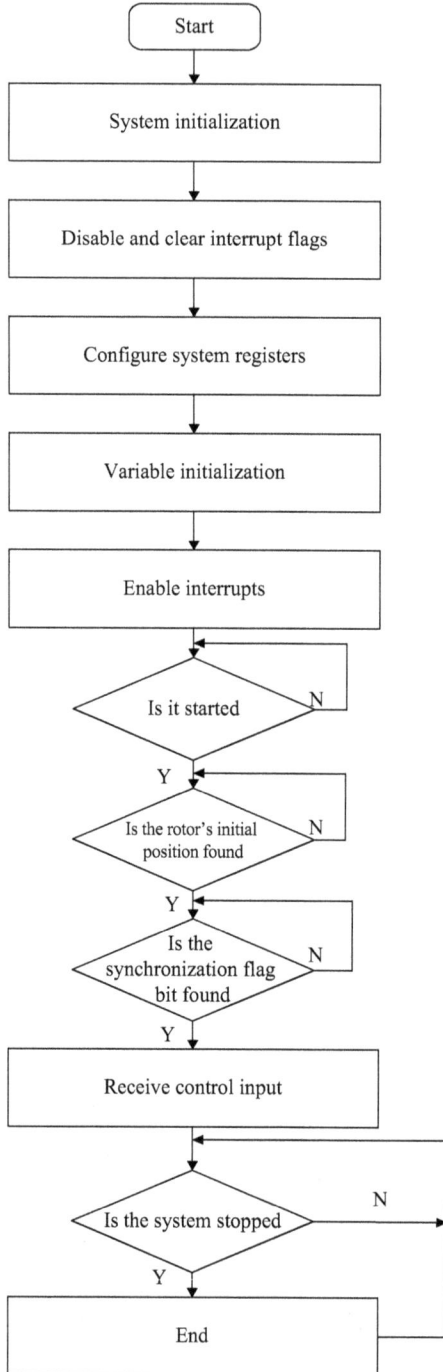

Figure 2.41   Flowchart of the main program execution

2.5.3.2 Main interrupt program section

The main interrupt program is a timer interrupt. Its primary function is to read the motor rotor position signals, process the collected motor status signals, and complete the vector control of the motor. The flowchart of its operation is shown in Figure 2.42.

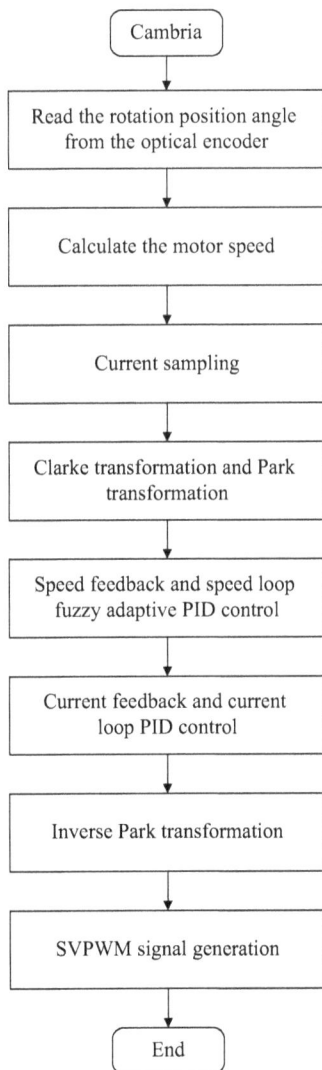

Figure 2.42   Flowchart of the main interrupt program

2.5.3.3   SVPWM program section

The main function of the SVPWM program is to generate the PWM waves for controlling the voltage space vector. Its primary tasks include calculating the sector number, the time duration of the basic space vectors, and the comparison values. The flowchart of its operation is shown in Figure 2.43.

2.5.3.4   AD conversion program section

The main function of the AD conversion program is to collect and process the sampled signals of the motor bus voltage and stator current. The flowchart of its operation is shown in Figure 2.44.

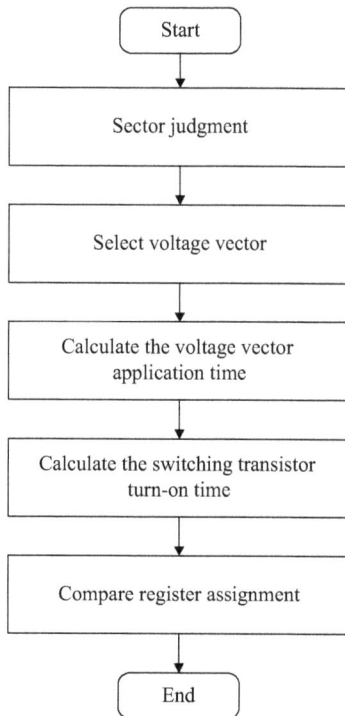

Figure 2.43   Operating flowchart of the SVPWM program

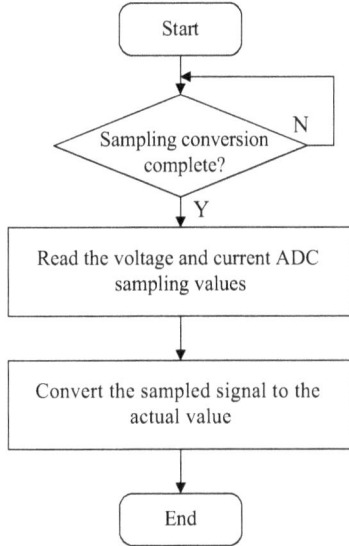

Figure 2.44   Flowchart of AD conversion program

Chapter 3
High-performance control strategies

In Chapter 2, traditional vector control methods that predominantly rely on proportional-integral-derivative (PID)/PI controllers are discussed. While these methods have been widely used due to their simplicity and effectiveness, they also exhibit certain limitations. These include slower dynamic responses, sensitivity to system parameter variations, and limited disturbance rejection capabilities. With advancements in control theory and technology, numerous high-performance control strategies have been developed to overcome these drawbacks. These strategies include adaptive PI/PID controllers, infinite control set (ICS) model predictive controllers, finite control set (FCS) model predictive controllers, and artificial intelligence (AI)-based control algorithms. By addressing the inherent limitations of PID/PI controllers, these advanced approaches offer significant potential for applications in the control of Permanent Magnet Actuators (PMAs) in robotics.

3.1 Performance requirements for PMAs

PMAs are pivotal in robotic systems due to their unique capabilities, compact design, and adaptability. As robotics continues to evolve, with applications ranging from industrial automation to healthcare and autonomous systems, the demand for advanced actuators that can deliver superior performance has surged. PMAs meet these needs by providing precise control, high energy efficiency, and reliability under varying operational conditions. However, achieving optimal performance with PMAs requires meeting a set of stringent performance criteria tailored to the demands of modern robotic systems.

3.1.1 High steady-state performance

High steady-state performance is a critical attribute of PMAs in robotic systems, playing a pivotal role in ensuring precise, reliable, and efficient operation. Steady-state performance refers to the ability of an actuator to maintain a desired operational state, such as position, speed, or torque, with minimal deviations or errors after transient effects have dissipated. This performance criterion is defined by low fluctuations, negligible steady-state error, and robust stability under constant or near-constant conditions. It is essential for applications where accuracy and consistency are paramount, such as in robotic arms used for precision manufacturing, surgical robotics, and autonomous vehicles. In these applications, even small variations in motor speed or torque can lead to errors and inconsistencies in the process, resulting in reduced product quality and decreased efficiency. Moreover, high steady-state performance helps to reduce the wear and tear on the motors, as they are subjected to less stress and strain. More specifically, in manufacturing, it ensures precise movements and positioning in assembly lines, reducing defects and improving productivity. In healthcare, steady-state performance is critical for surgical robots and rehabilitation devices, where even minor errors can have significant consequences. Autonomous vehicles rely on this performance to maintain consistent speed and torque, ensuring smooth and efficient operation. In service robots, low noise and high stability contribute to user comfort and system reliability, enhancing their integration into everyday environments.

Achieving high steady-state performance requires addressing several interconnected factors. One of the most important is the design of the control system. Control algorithms, such as the widely used PI/PID control, are fundamental in minimizing steady-state errors. The proportional component of PI/PID control ensures a quick response to deviations by adjusting the output proportionally to the error. The integral component eliminates steady-state errors by integrating past errors over time and applying corrective action, while the derivative component improves stability by counteracting rapid changes in error. Fine-tuning these parameters is crucial to balance responsiveness and stability, with methods such as Ziegler-Nichols tuning or heuristic optimization often employed for parameter optimization. These techniques enable the PMA system to achieve precise and stable performance under various operating conditions.

As illustrated in Chapter 2, traditional PI/PID methods for improving steady-state performance have laid the foundation for the development of modern PMA systems. Techniques such as pole placement, root locus analysis, and frequency response methods have been used to design controllers that meet stringent performance requirements. State-space control approaches, which represent the system in terms of its state variables, enable the optimization of both transient and steady-state responses. These methods have been instrumental in enabling PMAs to achieve high levels of precision and reliability.

Feedback mechanisms also play a crucial role in maintaining high steady-state performance. Accurate and real-time feedback allows the system to continuously monitor its output and make necessary adjustments. Position sensors, such as encoders or resolvers, provide precise information about the actuator's position, ensuring that it remains within acceptable bounds. Speed sensors, including tachometers, monitor rotational speed, while current sensors measure the torque-generating current to ensure efficient operation. Integrating these sensors into a closed-loop control system enables continuous and precise adjustments to the actuator's output, thereby reducing errors and enhancing overall stability.

The optimization of system parameters is another critical factor influencing steady-state performance. Mechanical parameters such as friction and electrical parameters such as resistance must be carefully managed to ensure the system operates smoothly. Reducing mechanical friction through the use of high-quality bearings or appropriate lubrication enhances precision and minimizes steady-state error. Similarly, optimizing the electromagnetic design of the PMA, including the magnetic circuit and winding configuration, reduces energy losses and improves efficiency. Effective thermal management is equally important, as it prevents overheating, which could lead to performance degradation or system instability.

In addition to traditional control methods, advanced control strategies have been developed to further enhance the steady-state performance of PMAs. For example, sliding mode controllers (SMC) and adaptive controllers have been developed for PMAs in applications where specific mathematical models, like those used in PI controllers, are not available. In these cases, Lyapunov functions are employed to analyze system stability and design optimal controller parameters. While intelligent algorithms often lack well-established stability analysis theories, empirical evidence indicates their stability in real-world scenarios, such as artificial neural networks (ANN) trained with reinforcement learning or other machine learning techniques. The stability of intelligent systems largely depends on the availability of sufficient training data, but challenges such as comprehensive data collection, high-quality data, addressing data imbalance, and ensuring data privacy remain significant hurdles. To effectively mitigate tracking errors in PMA systems, either differential-free controllers or adaptive controllers integrated with specialized techniques can be utilized. On one hand, PI-based and PID-based strategies are widely applied as differential-free control methods in PMAs due to their ability to accurately track speed, current, and position references. These strategies are highly favored in industrial applications where precise tracking of actuator performance is essential. On the other hand, tracking error constraint techniques have been directly incorporated into adaptive control processes to ensure improved tracking performance and stability. Additionally, fuzzy logic control methods have been employed to achieve precise tracking in PMA systems, leveraging expert knowledge to minimize tracking errors and enhance steady-state performance. The integration of fuzzy logic control highlights the use of modern control strategies to address challenges in steady-state performance.

Speed and torque fluctuations are common issues in PMA systems and can adversely affect steady-state performance. To address these challenges, various techniques have been proposed. For instance, Field-Oriented Control (FOC)-based multi-level inverters have been implemented in electrical cascaded configurations, enabling precise control of speed and torque by decoupling and controlling them separately. This approach reduces fluctuations, leading to smoother operation and enhanced performance. Frequency domain analysis methods have also been employed to optimize system bandwidth, effectively minimizing fluctuations caused by high-frequency disturbances in traditional FOC methods. Further, enhancements in direct torque control (DTC), such as optimizing the modulation process and increasing control frequencies, have successfully mitigated speed and torque fluctuations. Additionally, model predictive control (MPC) schemes incorporating multiple-vector-based strategies have been utilized to replace zero vectors with quasi-optimal voltage vectors. This optimization improves the utilization of available voltage vectors, reducing fluctuations and enhancing the overall performance of PMA systems. Collectively, these optimization techniques play a critical role in mitigating fluctuations and improving steady-state performance across various control methods.

In conclusion, high steady-state performance is a cornerstone of effective PMA operation in robotic systems. By focusing on optimized control algorithms, robust feedback mechanisms, advanced control strategies, and meticulous system design, engineers can achieve the precision, stability, and reliability required in demanding applications. As robotics technology continues to advance, the importance of steady-state performance in PMAs will only grow, driving innovation in control methods and actuator design.

3.1.2   High dynamic performance

Dynamic performance is a critical attribute for PMA systems in robotics, as it directly influences their ability to respond swiftly and accurately to command inputs under varying operational conditions. High dynamic performance ensures the system can achieve rapid acceleration, deceleration, and precise tracking of desired trajectories or control objectives. This is vital in applications where timing, precision, and adaptability are paramount, such as in robotic arms, mobile robots, and surgical devices. Dynamic performance of a PMA system is often characterized by fast response time, and the ability of the actuator to respond quickly to control commands and minimize delay. It is also defined by precise trajectory tracking, which involves maintaining minimal deviation from desired position, speed, or torque trajectories. Other critical aspects include controlling overshoot and ensuring a short settling time in transient states, as well as maintaining system stability and performance even during abrupt variations in load or command inputs. Factors influencing dynamic performance include the actuator's physical parameters, control algorithms, and the system's ability to handle disturbances.

Achieving high dynamic performance in PMA systems involves optimizing multiple aspects, including the control strategy, actuator design, and system architecture. Enhanced control strategies play a vital role, with advanced algorithms such as MPC or adaptive fuzzy PI/PID controllers significantly improving transient response. Feedforward control is another important technique that anticipates changes to mitigate lag. Designing actuators with high electromagnetic and mechanical bandwidth ensures rapid response to input changes, while high-resolution encoders and current sensors provide accurate and timely feedback. Minimizing communication and computational delays in digital control implementations is also critical for achieving high dynamic performance. High-speed motor drives capable of fast switching and precise current regulation, coupled with optimized hardware designs, further enhance system responsiveness.

Traditional methods have long been employed to enhance the dynamic performance of PMAs. FOC remains a cornerstone in PMA control, leveraging decoupled control of torque and flux to achieve precise and rapid responses. By controlling the direct (d-axis) and quadrature (q-axis) currents independently, FOC enables fast dynamic adjustments and high-performance operation. PI/PID controllers are also widely used for their simplicity and effectiveness. Proper tuning of PI/PID parameters can significantly enhance response speed and minimize overshoot, with traditional tuning methods like the Ziegler-Nichols approach or frequency response analysis often employed. Feedforward control is another traditional method that allows the system to anticipate changes, thereby reducing lag and improving transient response. Enhancing the stiffness of the control system, particularly in position-controlled applications, mitigates the impact of load variations and external disturbances, thereby improving stability and response time. Addressing system nonlinearities, such as magnetic saturation or cogging torque, through compensation algorithms ensures consistent performance across the operating range.

While traditional methods provide a robust foundation for improving dynamic performance, they have inherent limitations. Many traditional techniques rely on precise system modeling, which may not account for parameter variations or nonlinear effects. Achieving high dynamic performance can sometimes compromise other objectives, such as energy efficiency or noise reduction. Traditional controllers may also struggle to maintain performance under significant disturbances or unmodeled dynamics.

Modern approaches are increasingly being integrated with traditional methods to address these limitations and further enhance dynamic performance. Adaptive controllers dynamically adjust their parameters in real time to accommodate changes in system dynamics, ensuring consistent performance under varying conditions. MPC optimizes control actions over a prediction horizon, considering system constraints and performance criteria, and provides superior dynamic response compared to traditional PI/PID controllers. Machine learning techniques, such as neural networks, are being used to model complex actuator dynamics and design controllers capable of handling nonlinearities and uncertainties. Advances in digital control hardware and algorithms enable faster computation and higher control loop frequencies, reducing delay and improving response time. Combining data from multiple sensors through fusion algorithms improves the accuracy and reliability of feedback, thereby enhancing dynamic performance.

When implementing strategies to achieve high dynamic performance, accurate identification of system parameters is essential for effective control design. Controllers should be tested for robustness against parameter variations, disturbances, and noise. Control strategies must be tailored to the capabilities and limitations of the actuator and drive hardware. Dynamic performance improvements should be balanced with energy consumption to ensure overall system efficiency. For instance, implementing a combination of FOC and feedforward control has significantly improved the dynamic response of robotic arm joints, enabling precise and rapid trajectory tracking. Adaptive PID control has enhanced the dynamic performance of PMA-based drive systems in mobile robots, allowing them to navigate complex terrains with high responsiveness. In surgical robots, high-speed digital control and sensor fusion have been used to achieve the precise and rapid control required for minimally invasive surgical applications.

High dynamic performance is a cornerstone of PMA control in robotics, ensuring systems can meet the stringent demands of modern applications. Traditional methods, such as FOC and PID control, provide a solid foundation, while modern approaches, including adaptive control and MPC, address their limitations and push the boundaries of performance. By integrating these strategies with robust design and implementation practices, PMA systems can achieve the high dynamic performance required for advanced robotics.

3.1.3   Strong robustness

Robustness refers to the ability of a PMA system to maintain stable performance despite uncertainties or variations in system parameters (internal factors). In the context of PMAs in robotics, robustness is crucial for ensuring the reliable and efficient operation of robotic systems under real-world conditions. Robustness in practical applications is essential as PMAs are subjected to varying operating conditions, such as temperature fluctuations affecting the resistances and flux linkage of motors, component aging causing wear in mechanical components and deviations in electrical parameters, and load variations resulting in changes in torque demand and system dynamics. These uncertainties introduce performance challenges such as degraded efficiency, reduced precision, increased wear, and system failures. Robustness is a system's ability to mitigate these issues by compensating for parameter uncertainties and disturbances, ensuring high steady-state performance, high dynamic performance, and resilience to parameter variations.

Robustness performance in PMA systems can be achieved through a combination of robust control strategies and parameter adaptation. The first step involves modeling uncertainties by identifying and quantifying uncertainties in both electrical and mechanical parameters. Mathematical models that incorporate these uncertainties are developed for simulation and analysis. Parameter identification follows, using online parameter estimation techniques to monitor variations in resistance, inductance, flux linkage, inertia, and damping. Adaptive mechanisms are then utilized to update control parameters in real time based on identified changes. Robust control design involves creating control laws that are insensitive to parameter variations, ensuring that the system can reject disturbances and maintain performance objectives. Extensive simulations and hardware-in-the-loop testing validate the robustness of the control strategy under various operating conditions.

Several traditional control methods have been employed to improve the robustness of PMA systems. PI/PID control with robust tuning is one such method, where conventional PI/PID tuning, though simple, often struggles with parameter variations. Robust PID tuning enhances this by using advanced methods such as Ziegler-Nichols rules, the Cohen-Coon method, or heuristic optimization, often combined with gain-scheduling techniques for dynamic adjustment. FOC decouples the control of torque and flux, enabling precise and robust operation, with robustness further improved by integrating parameter adaptation mechanisms such as online inductance estimation into the FOC framework. SMC is another approach known for its inherent robustness against parameter uncertainties and disturbances, working by driving the system states to a predefined sliding surface, where they remain despite uncertainties. H-Infinity Control focuses on minimizing the worst-case gain from disturbances to system outputs, making it particularly useful for systems with well-defined uncertainty bounds. Robust state feedback control incorporates state observers to estimate and compensate for unmeasured disturbances, with linear quadratic regulator designs featuring robust weighting matrices being commonly employed.

To improve robustness, uncertainties in both electrical and mechanical parameters must be addressed. Electrical parameters, such as inductance, resistance, and flux linkage, are prone to variations due to temperature and magnetic effects. Real-time parameter estimation algorithms, adaptive control laws, and temperature compensation techniques are effective solutions. Mechanical parameters, including inertia, damping, and transmission compliance, vary with load and system dynamics. Addressing these involves incorporating dynamic observers for load and compliance estimation, implementing torque feedforward and feedback compensation, and using adaptive control techniques to handle mechanical variations.

Robust control for PMA systems is an active area of research, with trends focusing on advanced robust control algorithms, such as combining traditional techniques with AI-based adaptive control and hybrid approaches integrating sliding mode and H-infinity methods. Data-driven approaches are also gaining traction, utilizing machine learning to predict parameter variations and optimize control laws. Innovations in hardware, such as the development of sensors and actuators with higher robustness to environmental factors and the integration of robust embedded systems for real-time control, are complementing these advancements. Additionally, system-level robustness is being enhanced by improving coordination and communication among multiple actuators in robotic systems to handle distributed uncertainties.

Robustness is a critical performance metric for PMA systems in robotics, enabling them to maintain high performance despite uncertainties. By addressing both electrical and mechanical parameter variations and employing advanced control techniques, robust systems can ensure reliable, efficient, and precise operation in diverse applications. Traditional methods, while effective to some extent, are continually being augmented with modern advancements to meet the growing complexity of robotic systems. New approaches emphasize the integration of adaptive algorithms, predictive modeling, and real-time compensation mechanisms to handle dynamic and unpredictable conditions. Such advancements pave the way for future robotic systems to achieve unparalleled levels of robustness and performance in challenging environments.

3.1.4   High anti-disturbance capacity

High anti-disturbance capacity is a critical aspect of the performance of PMAs in robotics applications, particularly in systems where precision, stability, and efficiency are paramount. Unlike robustness, which focuses on a system's ability to maintain functionality under parameter variations and modeling inaccuracies, anti-disturbance measures a system's capability to respond to and reject external disturbances. These disturbances may include load variations, voltage fluctuations, mechanical irregularities such as uneven bearings, and external noise from the environment or electrical systems. Achieving high anti-disturbance performance is essential for ensuring the reliability and accuracy of robotic systems driven by PMAs. To understand anti-disturbance performance, it is important to consider the types of external disturbances that PMA systems encounter. Load changes, for example, are common in robotics due to dynamic interactions with the environment or variations in the task being performed. These changes can lead to sudden torque fluctuations, affecting the actuator's performance and stability. Voltage changes, on the other hand, may arise from grid instability, power supply issues, or fluctuations in shared power systems, leading to variations in the input voltage supplied to the PMA drive. Mechanical factors, such as uneven bearings, introduce periodic disturbances that can cause vibrations and noise, compromising the actuator's smooth operation and potentially reducing its lifespan. Lastly, environmental noise, including electromagnetic interference and mechanical vibrations from surrounding machinery, can further degrade system performance if not adequately addressed.

To achieve high anti-disturbance capacity in PMA systems, various methods have been developed and implemented over the years. These methods can broadly be categorized into control design techniques, parameter optimization approaches, and structural enhancements. Control design techniques focus on improving the system's ability to detect and counteract disturbances through advanced control algorithms. Parameter optimization involves tuning the system parameters to enhance its resilience to disturbances. Structural enhancements aim to reduce the physical impact of disturbances through improved hardware design.

One of the most common traditional methods to enhance anti-disturbance performance in PMA systems is the use of PI/PID controllers. PI/PID controllers are simple yet effective tools for mitigating disturbances. By carefully tuning the proportional, integral, and derivative gains, the controller can react swiftly to disturbances, minimize steady-state errors, and dampen oscillations. The integral action of the PI/PID controller is particularly effective in addressing load disturbances by eliminating steady-state errors, while the derivative action helps in reducing the impact of high-frequency noise. However, PI/PID controllers have limitations, especially in systems with highly dynamic or nonlinear characteristics, as their fixed parameters may not provide optimal performance under varying conditions.

FOC methods have also been widely employed to improve the anti-disturbance capacity of PMAs. FOC decouples the motor control problem into independent control of torque and flux, enabling precise regulation of these variables even in the presence of disturbances. This decoupling ensures that changes in one parameter do not adversely affect the other, thereby improving the system's overall stability. Furthermore, FOC allows the use of feedback mechanisms to monitor and adjust the motor's performance in real time, enhancing its ability to reject external disturbances.

Disturbance observer-based control has emerged as a powerful technique for enhancing anti-disturbance performance. Disturbance observer-based control involves the use of an observer to estimate the disturbances affecting the system. By feeding this disturbance estimate into the control loop, the system can actively compensate for the effects of the disturbance, thereby maintaining its desired performance. This approach is particularly effective in handling disturbances that are difficult to measure directly, such as torque ripples caused by uneven bearings or voltage fluctuations from the power supply. The success of disturbance observer-based control relies heavily on the accuracy of the disturbance estimation and the speed with which the control system can respond to the disturbance. SMC is another traditional method that has been applied to enhance the anti-disturbance capacity of PMA systems. SMC is characterized by its ability to maintain system performance even in the presence of large uncertainties and disturbances. By driving the system state to a predefined sliding surface and maintaining it there, SMC ensures robustness to both internal parameter variations and external disturbances. However, one of the challenges of SMC is the phenomenon known as chattering, which can lead to undesirable high-frequency oscillations. To mitigate this issue, modifications to the basic SMC algorithm, such as the use of boundary layer techniques or higher-order sliding modes, have been proposed.

Parameter adaptation methods also play a crucial role in achieving high anti-disturbance performance. Adaptive control techniques dynamically adjust the control parameters based on the system's operating conditions, ensuring optimal performance under varying disturbances. For instance, model reference adaptive control and self-tuning regulators have been used to enhance the disturbance rejection capability of PMA systems. These methods rely on the identification of system parameters in real time and the adjustment of control gains accordingly, providing a flexible and effective solution for handling disturbances.

Modern approaches to enhancing anti-disturbance capacity leverage advanced computational techniques and algorithms. For example, AI-based control methods, such as neural networks and fuzzy logic controllers, have shown promise in addressing the limitations of traditional methods. Neural networks can learn the complex relationships between input disturbances and system responses, enabling them to predict and compensate for disturbances more effectively. Fuzzy logic controllers, on the other hand, use rule-based systems to handle uncertainties and disturbances in a more intuitive manner. These AI-based methods can be particularly beneficial in systems with nonlinear dynamics or where disturbances are highly variable and difficult to model accurately.

Another modern technique is the use of MPC, which optimizes the control inputs by predicting the system's future behavior over a finite time horizon. MPC considers the constraints and disturbances affecting the system and calculates the optimal control actions to minimize their impact. FCS MPC and ICS MPC are two variations of this approach that have been applied to PMA systems. These methods offer a high degree of flexibility and precision in disturbance rejection, making them suitable for high-performance robotic applications.

Hardware improvements also contribute to enhanced anti-disturbance performance. For instance, the use of advanced materials and manufacturing techniques can reduce mechanical irregularities and improve the precision of actuator components. Enhanced shielding and filtering techniques can mitigate the effects of electromagnetic interference, while improved bearing designs can minimize mechanical disturbances. Additionally, the integration of sensors and monitoring devices allows for real-time detection and analysis of disturbances, enabling more effective implementation of control strategies.

The importance of hybrid approaches in combining these techniques cannot be overstated. For example, integrating AI-based control methods with traditional PI/PID and FOC frameworks can result in systems that leverage the strengths of each approach. AI algorithms can be used to dynamically adjust PI/PID parameters or enhance FOC-based feedback mechanisms, ensuring robust performance under varying disturbances. Similarly, combining disturbance observer-based control with MPC allows for precise disturbance estimation and optimal control actions, creating a highly resilient system capable of handling complex and variable environments.

3.1.5   High efficiency

Maximizing system efficiency in PMA systems is a critical requirement, especially in applications involving multi-motor drives where energy consumption and operating costs are of paramount concern. High efficiency ensures that the actuators operate with minimal energy wastage, which is particularly significant in robotics applications where extended operational time, reduced thermal stress, and lower maintenance costs are prioritized. Performance related to high efficiency encompasses various aspects, including effective current and torque distribution, smooth operation, and optimized switching frequency in the control systems. To achieve high efficiency in PMA systems, it is necessary to address both electrical and mechanical losses. Electrical losses, primarily comprising copper and iron losses, can be minimized through advanced control strategies that optimize the current and voltage supplied to the actuators. Mechanical losses, such as friction and windage, require precise mechanical design and operation under optimal load conditions. By integrating electrical and mechanical considerations, a holistic approach to efficiency optimization can be implemented.

Traditional methods for achieving high efficiency in PMA systems rely on established techniques such as FOC and maximum torque per ampere strategies. FOC, with its ability to decouple torque and flux control, provides precise control over the actuator's performance. In FOC, the d-axis and q-axis currents are controlled separately, allowing for the adjustment of magnetic flux and torque independently. This decoupling is fundamental to maintaining efficiency as it enables the actuator to operate at its optimal flux level under varying load conditions.

Maximum torque per ampere strategies further enhance efficiency by ensuring that the required torque is produced with the minimum current magnitude. This is achieved by optimizing the current vector in the d-q plane, which reduces copper losses and improves the overall energy efficiency of the system. The maximum torque per ampere control algorithm identifies the optimal current ratio between the d-axis and q-axis, thereby maximizing the torque output for a given current input. This strategy is particularly beneficial in applications where high torque is required at low speeds, as it minimizes energy consumption while maintaining performance.

Another traditional method involves optimizing the switching frequency of the inverter. The switching frequency impacts both the efficiency and the noise characteristics of the PMA system. Lower switching frequencies reduce switching losses in the inverter, but they can lead to increased current ripple, which affects the smooth operation of the actuator. Conversely, higher switching frequencies improve the current waveform quality but result in higher switching losses. Balancing these trade-offs is essential for achieving high efficiency. Advanced pulse-width modulation techniques, such as space vector pulse-width modulation, are commonly used to optimize switching frequency and minimize harmonic distortion.

Smooth operation of the PMA system also plays a crucial role in achieving high efficiency. Mechanical vibrations and torque ripple not only reduce efficiency but also increase wear and tear on the mechanical components. Traditional approaches to minimizing torque ripple include the use of harmonic compensation techniques and precise rotor position estimation. Harmonic compensation involves injecting specific harmonic currents into the stator windings to counteract the effects of cogging torque and other sources of ripple. Accurate rotor position estimation, achieved through advanced sensors or sensorless algorithms, ensures that the control system maintains precise synchronization with the rotor, thereby reducing energy losses due to misalignment.

In multi-motor drive systems, the challenge of achieving high efficiency is compounded by the need to coordinate the operation of multiple actuators. Current and torque distribution strategies are critical in such scenarios. These strategies involve dynamically allocating current and torque demands among the motors based on their individual operating conditions and efficiencies. For example, in a robotic arm with multiple joints, the distribution of torque among the actuators must consider factors such as load distribution, actuator efficiency curves, and thermal limits. By prioritizing the use of the most efficient actuators and redistributing load to balance thermal stress, the overall system efficiency can be significantly improved.

Adaptive control strategies, including adaptive PI/PID controllers, also contribute to efficiency optimization. These controllers adjust their parameters in real time based on the operating conditions of the PMA system. For example, an adaptive PID controller can modify its gain values to maintain optimal performance as the load or speed changes. This adaptability ensures that the system operates at its highest efficiency across a wide range of conditions.

The integration of AI and machine learning techniques has opened new avenues for efficiency optimization in PMA systems. AI-based algorithms can analyze large amounts of operational data to identify patterns and optimize control parameters. For instance, reinforcement learning algorithms can be used to develop control policies that maximize efficiency by learning from the system's performance over time. Additionally, neural networks can be employed to model the nonlinear characteristics of the PMA system and predict optimal operating points.

Emerging technologies, such as energy harvesting and regenerative braking, further enhance the efficiency of PMA systems. Energy harvesting techniques capture ambient energy, such as vibrations or heat, and convert it into electrical energy that can be used to power the system. Regenerative braking, commonly used in electric vehicles and robotic systems, recovers kinetic energy during deceleration and stores it for future use. These technologies not only improve efficiency but also contribute to the sustainability of PMA systems.

In conclusion, achieving high efficiency in PMA systems requires a comprehensive approach that combines traditional methods, such as FOC, maximum torque per ampere, and harmonic compensation, with advanced strategies like adaptive control, and AI-based optimization. By addressing both electrical and mechanical losses, optimizing current and torque distribution, and leveraging emerging technologies, it is possible to maximize the efficiency of PMA systems. This not only reduces energy consumption and operating costs but also enhances the performance and reliability of robotic applications.

3.1.6   Low noise and vibration

In the context of PMAs, low noise and vibration performance is a critical aspect of achieving high-performance operation, particularly in applications where comfort, precision, and durability are paramount. Noise and vibration issues, if unaddressed, can significantly compromise the overall system's efficiency and reliability. The combined effects of noise and vibration generated by multiple motors operating simultaneously can amplify each other, resulting in heightened levels of acoustic emissions and mechanical disturbances. These phenomena are not merely nuisances; they carry tangible consequences. Excessive noise can lead to discomfort for operators and occupants, creating an unfavorable environment in collaborative robotic settings or industrial operations. Moreover, persistent vibration can adversely impact system stability, reduce performance, and accelerate mechanical wear and tear, thereby increasing maintenance requirements and shortening the lifespan of the robotic system.

The sources of noise and vibration in PMAs are multifaceted and stem from a combination of electromagnetic, mechanical, and structural factors. Electromagnetic noise, for instance, arises from the interaction of magnetic fields within the motor, including harmonics caused by non-idealities in the magnetic circuit. Mechanical sources include imbalances in the rotor, misalignments, and irregularities in the bearings. Structural components, such as the housing and mounting, may further amplify vibrations, leading to resonances that exacerbate the issue. Addressing these sources requires a comprehensive understanding of their origins and the development of targeted mitigation strategies.

Achieving low noise and vibration performance in PMAs involves a combination of design optimizations, advanced control strategies, and precise implementation.

From a design perspective, the use of high-quality materials and precision manufacturing processes is essential to minimize inherent imbalances and irregularities. Advanced rotor balancing techniques and the selection of low-friction bearings contribute to reducing mechanical sources of noise and vibration. Furthermore, the structural design of the motor housing and its mounting system can be optimized to dampen vibrations and avoid resonant frequencies that might amplify disturbances.

From a control standpoint, sophisticated algorithms play a crucial role in mitigating noise and vibration. Traditional control methods, such as FOC, can be adapted to address these challenges by incorporating additional layers of compensation. For instance, torque ripple minimization techniques can be integrated into the control loop to reduce periodic disturbances that contribute to noise and vibration. Additionally, current harmonics can be actively suppressed using harmonic compensation methods, ensuring smoother motor operation. One of the fundamental approaches to reducing noise and vibration in PMAs involves the careful design of the stator winding configuration and the use of advanced materials in the rotor and stator cores. Skewed stator slots, for example, are a widely used design technique to minimize cogging torque, a primary source of vibration in permanent magnet motors. By angling the stator slots relative to the rotor axis, the cogging effect is distributed over a wider range, reducing the intensity of the torque ripple and, consequently, the associated noise and vibration.

Traditional methods for achieving low noise and vibration performance in PMAs have relied on mechanical and structural optimizations. For example, vibration isolation systems are commonly employed to decouple the motor from its mounting structure, thereby reducing the transmission of vibrations to the surrounding environment. These systems often use elastomeric materials or spring-damper assemblies to absorb and dissipate vibrational energy. Similarly, acoustic enclosures can be used to contain and attenuate noise emissions, particularly in environments where stringent noise limits are imposed. In recent years, advancements in control technology have opened new avenues for improving noise and vibration performance. Adaptive control methods, for instance, enable real-time adjustments to the motor's operating parameters based on feedback from vibration and noise sensors. By dynamically tuning the control algorithm to mitigate specific disturbances, adaptive systems can maintain optimal performance even under varying load conditions and environmental factors.

The integration of AI and machine learning techniques into PMA control systems represents a significant leap forward in noise and vibration mitigation. AI-driven algorithms can analyze vast amounts of operational data to identify patterns and correlations that contribute to noise and vibration. By learning from this data, these systems can develop highly effective control strategies tailored to the specific characteristics of the motor and its operating environment. For example, neural networks can be trained to predict and compensate for vibration-inducing disturbances, while reinforcement learning algorithms can optimize the motor's operation to minimize noise.

Another important consideration in achieving low noise and vibration performance is the role of thermal management. Excessive heat can exacerbate mechanical deformations and material fatigue, leading to increased vibration levels over time. Effective thermal management strategies, such as optimized cooling systems and the use of thermally stable materials, are therefore essential to maintaining consistent performance. Additionally, regular maintenance and condition monitoring are crucial for identifying and addressing emerging issues before they escalate. Predictive maintenance systems, which use sensor data and analytics to forecast potential failures, can play a vital role in preserving low noise and vibration performance throughout the motor's lifecycle.

The development of low noise and vibration PMA systems also benefits from advancements in simulation and modeling tools. Finite element analysis and multi-body dynamics simulations allow engineers to analyze and optimize motor designs in a virtual environment, identifying potential sources of noise and vibration before physical prototypes are built. These tools enable detailed investigations into the effects of material properties, geometric configurations, and operational conditions, providing valuable insights that inform the design process.

While traditional methods have been effective in addressing noise and vibration challenges, the increasing demands of modern robotic applications require innovative solutions that go beyond conventional approaches. The integration of advanced materials, cutting-edge control algorithms, and intelligent monitoring systems represent the future of low-noise and vibration PMA systems. By leveraging these technologies, engineers can achieve unprecedented levels of performance, ensuring that PMAs meet the stringent requirements of next-generation robotics while delivering a comfortable and reliable user experience.

The following sections introduce the implementation of three high-performance control algorithms: fuzzy adaptive PID control, infinite set MPC, and finite set MPC.

3.2   Fuzzy adaptive PID

As illustrated in Section 3.1, fuzzy adaptive PID controllers are highly advantageous for controlling PMAs due to their ability to dynamically adjust control parameters in response to varying operating conditions and system uncertainties. Unlike traditional PID controllers, which require fixed tuning parameters, fuzzy adaptive PID systems incorporate fuzzy logic to adaptively optimize gains based on real-time feedback. This adaptability significantly enhances steady-state performance by minimizing steady-state errors and improving precision. Additionally, fuzzy adaptive PID controllers excel in dynamic performance, ensuring rapid response to transient disturbances and reducing overshoot and settling time. Their robustness to parameter variations, external disturbances, and non-linearities inherent in PMAs further contributes to their reliability and efficiency. These characteristics make fuzzy adaptive PID an effective solution for achieving the high performance, robustness, and stability demanded by PMA applications in robotics and other advanced systems [5].

3.2.1 The principle of fuzzy control

Fuzzy Logic Control (FLC) is a control method that emerged in modern times, reflecting industrial development and technological advancements. Today, industrial systems have higher demands for control systems, and various complex systems are presented to engineers. Classic control principles are no longer sufficient to meet the requirements of "fast, precise, and stable" control. Motor systems contain many variables, and when one of these variables changes, the system state can change significantly. Additionally, motors are typically strong coupling and nonlinear models. When engineers applied fuzzy control methods to motor control, they achieved excellent experimental results.

The foundation of fuzzy control lies in fuzzy set theory, fuzzy language, and fuzzy reasoning. The concept of fuzzy control was first proposed by Professor L.A. Zadeh from the University of California. After more than 20 years of research and development, fuzzy control theory and its applications have made tremendous progress in the field of control. The basic principle of fuzzy control is shown in Figure 3.1, where the core part is the fuzzy controller, as indicated by the dashed part. The fuzzy controller needs to carry out most of the tasks required for fuzzy control, including fuzzification of the error signal, fuzzy inference based on control rules, and defuzzification.

As seen in Figure 3.1, the fuzzy controller is the main difference between a fuzzy control system and a traditional computer-based digital control system. The performance of a fuzzy control system primarily depends on the physical structure of the fuzzy controller, fuzzy control rules, and the inference algorithm. Generally speaking, the block diagram of a fuzzy controller is shown in Figure 3.2.

1. The input to the fuzzy controller must be a definite value, but the input to the inference mechanism should be a fuzzy quantity. The fuzzification process is the conversion of the definite input value into a fuzzy vector. This process requires determining the input and output ranges, as well as a suitable membership function. Common membership functions include triangular functions, trapezoidal functions, and Gaussian distribution functions.
2. The fuzzy inference process is essentially a form of logical operation where one or more fuzzy input values determine the output fuzzy value. When designing, all possible situations should be considered, and logical overlap should be avoided to ensure that each input corresponds to one output.

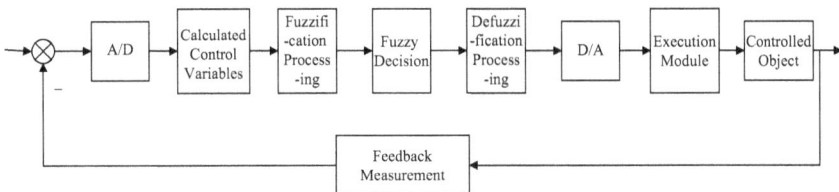

Figure 3.1   Conceptual diagram of fuzzy control

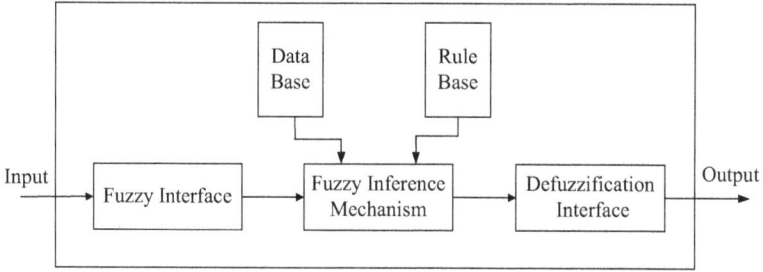

Figure 3.2   Fuzzy controller

3. Fuzzy control rules are based on expert knowledge or the accumulated experience of skilled operators. These rules are expressed in a language format based on human reasoning. Fuzzy control rules typically consist of relational connectors, such as "if-then," "and," and "end." A specific form would be: "if A is NB and B is EC, then U is NB." In practice, the finer the rule division and the more rules there are, it does not necessarily mean that the rule base has higher accuracy. Its accuracy also depends on the precision of the expert knowledge.

4. The output of the fuzzy inference system is also a fuzzy vector, which cannot be directly used as a control signal for the controlled object. The fuzzy vector needs to be converted into specific data in the domain in order to execute the control task. Therefore, a defuzzification process is required. Common defuzzification methods include the median method, the weighted average method, and the maximum membership degree method, with the weighted average method being the most commonly used.

3.2.2   Fuzzy adaptive PID controller

As discussed in the previous section, PID parameters play a significant role in system performance. The parameters of the fuzzy adaptive PID controller can change in real time with the system's state, significantly improving the system's performance. One of the important advantages of fuzzy control is its strong adaptability, as it can change the control variables in real time based on changes in the system's state. PID control, on the other hand, has the advantage of zero steady-state error and high control accuracy. However, traditional PID control processes cannot change parameters in real time. For systems with non-linearities or large variations in system parameters, the control effectiveness may be significantly reduced. Through continuous exploration, researchers have combined PID control with fuzzy control, leveraging their complementary advantages. This control method is called fuzzy adaptive PID (Fuzzy-PID) control. The principle diagram of the fuzzy adaptive PID controller is shown in Figure 3.3.

From Figure 3.3, it can be seen that the Fuzzy-PID controller consists of two main parts: the PID controller and the fuzzy controller. The inputs to the fuzzy controller are the motor speed error e and the rate of change of the speed error Δe.

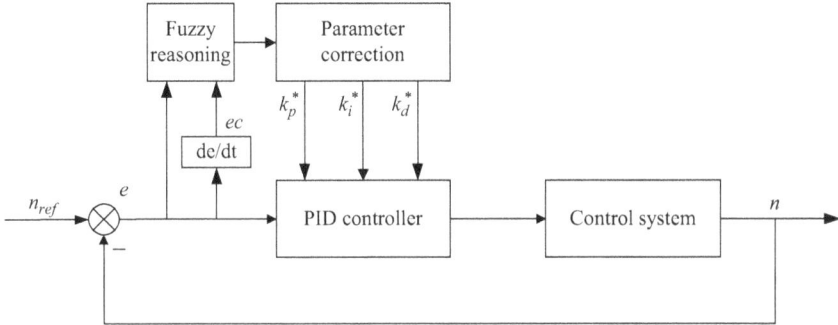

Figure 3.3   The schematic diagram of the fuzzy-PID controller

These two quantities are chosen because they can accurately reflect the dynamic characteristics of the controlled system. After fuzzification and fuzzy inference of e and Δe, the fuzzy control output is obtained. This signal is defuzzified to calculate the correction values for the PID parameters k_p^*, k_i^*, and k_d^*. These correction values are added to the initial PID coefficients k_{p0}, k_{i0}, and k_{d0}, making the current PID parameters for the controller. This entire process achieves adaptive control.

$$
\begin{aligned}
k_p &= k_{p0} + k_p^* \\
k_i &= k_{i0} + k_i^* \\
k_d &= k_{d0} + k_d^*
\end{aligned}
\tag{3.1}
$$

The permanent magnet synchronous motor (PMSM) dual-loop control system includes two PID controllers for the speed loop and current loop. Currently, the most researched method is to replace the speed loop PID with a fuzzy adaptive PID. For the entire system, this is sufficient to achieve the goal of improving system performance. This chapter also focuses on this method, and the design process of the fuzzy adaptive PID controller is detailed below.

3.2.2.1   Fuzzification

Determine the linguistic variables

The motor speed error e and the rate of change of speed error Δe are defined as the input variables, while the parameter adjustment quantities Δk_p, Δk_i, and Δk_d are output variables (Figure 3.4). The linguistic variables are defined as "Speed Error E," "Speed Error Change Rate EC," "Proportional Adjustment k_p^*," "Integral Adjustment k_i^*," and "Differential Adjustment k_d^*."

Determining the fuzzy domain of each linguistic variable

The domain of each linguistic variable can be defined arbitrarily, but it is generally set as integers. This approach not only looks more organized but also facilitates computation. In this case, by using the concept of normalization, the domains of the linguistic variables E, EC, k_p^*, k_i^*, and k_d^* are uniformly set to $[-1,1]$.

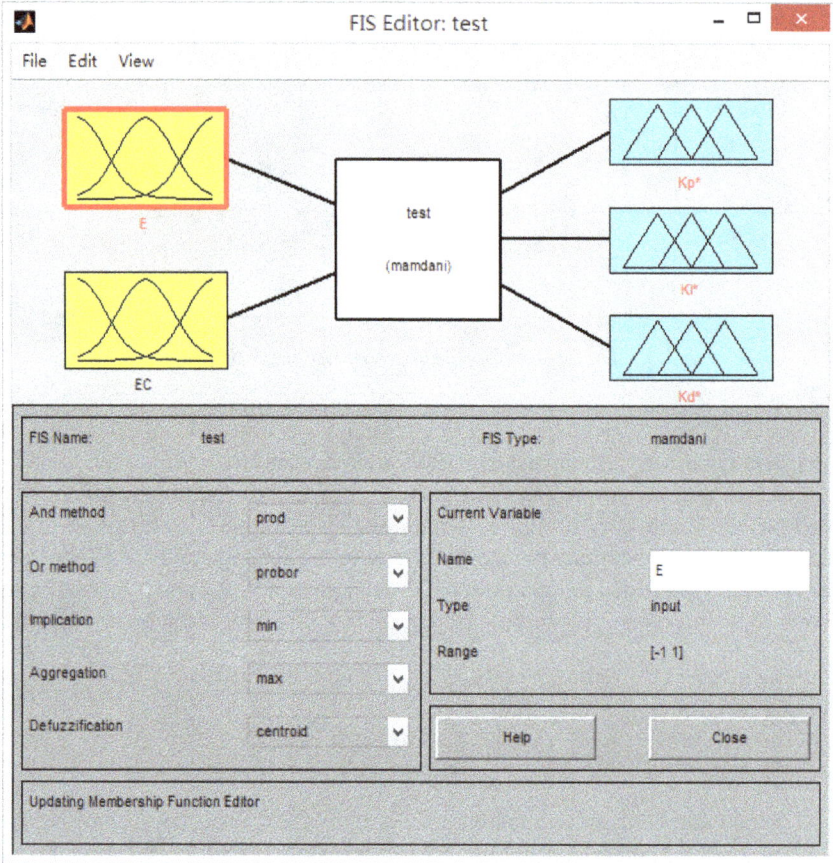

Figure 3.4 Language variable diagram

Determining the input/output scaling factors

The reason why the domains of the linguistic variables can be chosen arbitrarily is due to the role of the scaling factors. For example, if the speed error range is $e = [-e_{max}, e_{max}]$, and its fuzzy domain is $[-1,1]$, then we define:

$$k_e = \frac{1}{e_{max}} \tag{3.2}$$

Thus, K_e is the quantization factor for the speed error input. Similarly, the quantization factor for the speed error rate input, denoted $K_{\triangle e}$, can be defined in a similar manner.

After defining the domains of the output quantities k_p^*, k_i^*, and k_d^*, the fuzzy controller's output range is $[-1,1]$. When the actual range of the proportional adjustment is required to be $\Delta K_p = [-\Delta K_{p\max}, \Delta K_{p\max}]$, we can define:

$$K_p' = \frac{\Delta K_{p\ \max}}{1} \tag{3.3}$$

Thus, K_p' is the quantization factor for the proportional adjustment output. Similarly, the quantization factors for the integral adjustment and differential adjustment can be defined as K_i' and K_d', respectively.

3.2.2.2   Determining the linguistic descriptions of linguistic variables

In this study, the linguistic variables "speed error EEE," "speed error rate ECECEC," "proportional adjustment $kp*k\_p^\wedge*kp*$," "integral adjustment $ki*k\_i^\wedge*ki*$," and "differential adjustment $kd*k\_d^\wedge*kd*$" are consistently described using the seven linguistic terms: "Positive Big (PB)," "Positive Medium (PM)," "Positive Small (PS)," "Zero (ZO)," "Negative Small (NS)," "Negative Medium (NM)," and "Negative Big (NB)."

3.2.2.3   Membership functions

Membership functions map the specific numerical values of the inputs to membership degrees between 0 and 1, transforming the input values into fuzzy linguistic descriptors. In this chapter, triangular membership functions are used, as shown in the equation:

$$u_A(x) = \begin{cases} \dfrac{x-a}{b-a} & a \leq x \leq b \\[2mm] \dfrac{c-x}{c-b} & b \leq x \leq c \\[2mm] 0 & \text{other} \end{cases} \tag{3.4}$$

When selecting membership functions in practice, the following points should be considered:

(a) The distribution of membership functions must cover the entire fuzzy domain; otherwise, gaps may occur, leading to loss of control.
(b) There should be an intersection between adjacent membership functions, with clear boundaries, and no situation where three membership functions intersect simultaneously.
(c) Generally, in areas with large deviations, membership functions with smoother shapes are used, as they provide better stability. In areas with small deviations, sharper membership functions are used, offering higher control sensitivity.
(d) When the intersection between two adjacent membership functions is large, the fuzzy controller exhibits stronger adaptability to parameter changes of the controlled object and better robustness. When the intersection is small, the control sensitivity is higher. To balance sensitivity and robustness, the intersection degree is typically set between 0.4 and 0.8.

Based on these considerations, the membership functions for the five linguistic variables are set as follows:

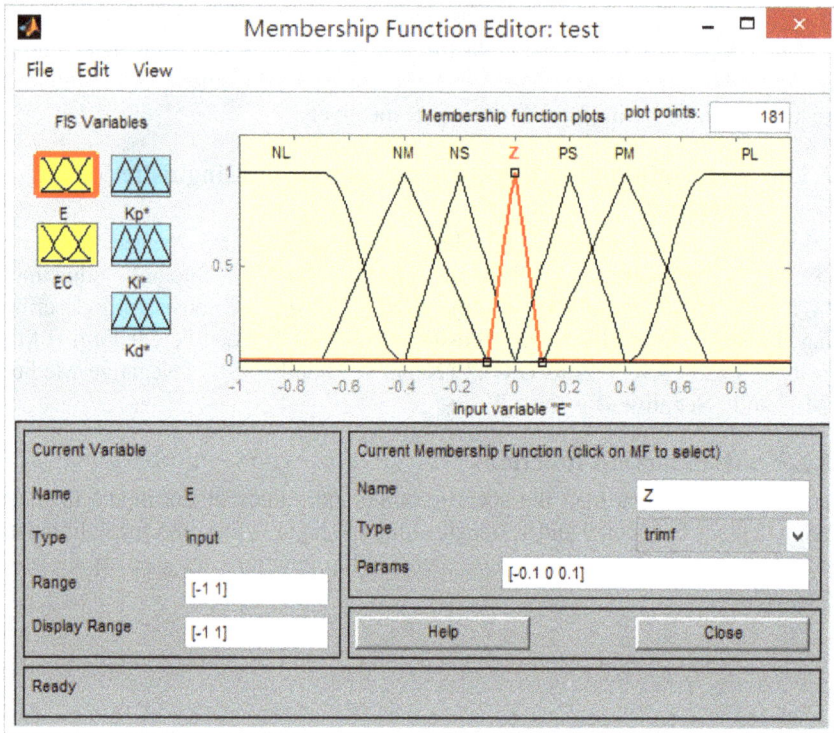

Figure 3.5   Membership function setup diagram

From Figure 3.5, it can be observed that the closer to zero the position, the less the intersection between two membership functions, and the slope of the membership function in the middle position is significantly larger than that of the sides. This way, when the input is very small, the system will respond quickly. If the calculation errors in practice are not considered, the system's control sensitivity is improved. Conversely, when the input value is larger, the system does not need a high response speed. Slower membership functions are very beneficial for maintaining system stability.

3.2.3   Formulating fuzzy rules

Based on experience, the following principles should be followed when formulating fuzzy control rules:

1.   When the error e is large, the actual speed deviates significantly from the set speed. To improve the response speed, k_p should take a larger value. At this point, the integral action cannot be reflected, so to ensure the system's safety, the value of k_i should be kept as small as possible.

2. When *e* is medium-sized, the actual speed is already close to the set speed. At this point, the focus should not be on response speed but rather on ensuring that the system's overshoot is not too large. Therefore, k_p, k_i, and k_d should not be too large and should be kept at moderate values to ensure system performance.
3. When *e* is small, the system is close to a stable state. At this point, the integral part should play its role in eliminating steady-state error. To achieve good steady-state performance, k_i and k_p should be large, but to avoid oscillations and to consider the system's anti-disturbance capability, k_d should be small.
4. When Δe is large, it indicates that the motor is accelerating or decelerating quickly or the speed is fluctuating significantly. To ensure the system's response stability, k_p should not be too large, and k_i should be kept small, while k_d should be relatively large.
5. When Δe is of medium size, it indicates that the motor speed change rate is moderate. k_p and k_i can be increased slightly to accelerate the response speed, and k_d can be set to moderate values to fine-tune the system parameters. Of course, the PID coefficients can also remain unchanged, with k_p, k_i, and k_d all set to 0.
6. When Δe is small, the normal speed change is slow. In this case, the system's response speed should be accelerated, and larger values for k_p and k_i should be chosen while considering the system's anti-disturbance performance. The value of k_d should be chosen moderately.

When formulating fuzzy control rules, the above situations should be comprehensively considered to ensure and improve system performance. Based on the above analysis, the fuzzy control rules used in this chapter are shown in Tables 3.1–3.3.

Fuzzy rule weighting represents the importance of different rules. If the weight of a certain rule is set to 0, then in the aggregation process, the corresponding input will have no corresponding output. As long as the weight is greater than 0, the corresponding input will influence the fuzzy output set. In this chapter, the weight of each rule is set to 1, as shown in Figure 3.6.

Table 3.1 k_p^* control rule table

E EC	NB	NM	NS	ZO	PS	PM	PB
NB	NB	NB	NM	NM	NS	NS	ZO
NM	NB	NB	NM	NM	NS	ZO	PS
NS	NM	NM	NS	NS	ZO	PS	PM
ZO	NM	NM	NS	ZO	PS	PM	PB
PS	NS	NS	ZO	PS	PM	PM	PB
PM	NS	ZO	PS	PS	PM	PM	PB
PB	ZO	ZO	PM	PM	PB	PB	PB

Table 3.2 k_i^* control rule table

E EC	NB	NM	NS	ZO	PS	PM	PB
NB	NB	NB	NM	NM	NS	NS	ZO
NM	NB	NB	NM	NM	NS	ZO	PS
NS	NM	NM	NS	NS	ZO	PS	PM
ZO	NM	NM	NS	ZO	PS	PM	PB
PS	NS	NS	ZO	PS	PM	PM	PB
PM	NS	ZO	PS	PS	PM	PM	PB
PB	ZO	PS	PM	PM	PB	PB	PB

Table 3.3 k_d^* control rule table

E EC	NB	NM	NS	ZO	PS	PM	PB
NB	PB	PB	PM	PM	PS	ZO	ZO
NM	PB	PB	PM	PS	PS	ZO	NS
NS	PB	PB	PM	PS	ZO	ZO	NS
ZO	PM	PS	ZO	ZO	NS	NM	NB
PS	PS	ZO	ZO	NS	NM	NM	NB
PM	ZO	ZO	NS	NM	NM	NB	NB
PB	ZO	NS	NM	NM	NB	NB	NB

3.2.3.1   Defuzzification

The output of the Mamdani-type fuzzy controller is not a single value or an analytical expression, but rather another fuzzy set. Therefore, the output needs to be converted into a single value using some algorithm. The most commonly used defuzzification methods are:

Weighted average method
This method is frequently used in industrial control, where the output value is determined by the following equation:

$$v_o = \frac{\sum_{i=1}^{m} v_i k_i}{\sum_{i=1}^{m} k_i} \tag{3.5}$$

where, v_o represents the output value, and the coefficients k_i are chosen based on the actual situation. Different values of k_i give the system different response characteristics.

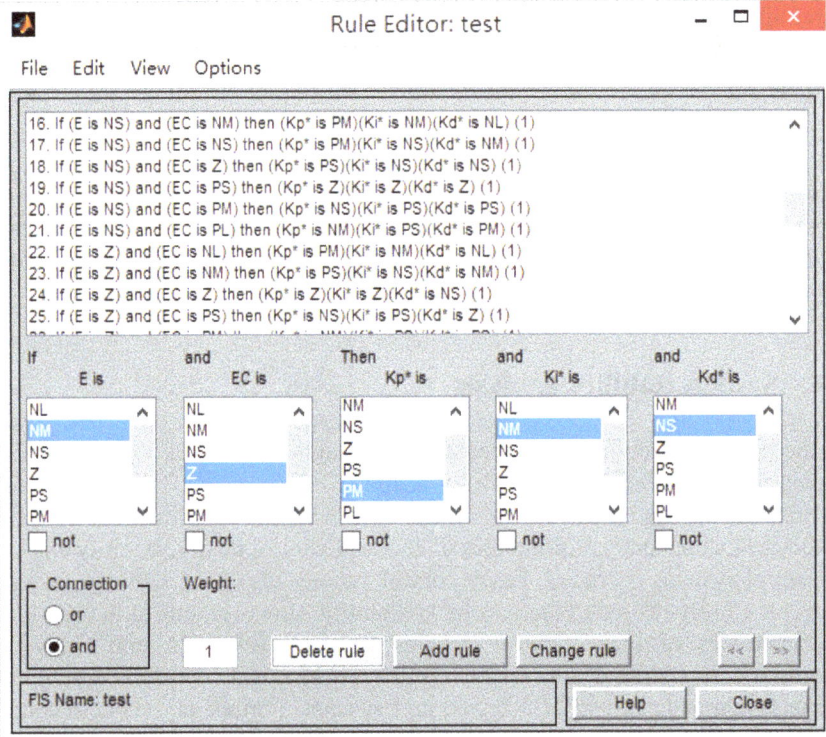

Figure 3.6 Setting fuzzy control rules

3.2.3.2 Maximum membership degree method

This method takes the element with the maximum membership degree in the fuzzy set of the reasoning results as the output value:

$$v_o = \max u_v(v), v \in V \tag{3.6}$$

where, V represents the fuzzy set of the reasoning result. The maximum membership degree method does not consider the shape of the output membership function, only considering the output value at the point of maximum membership degree. Therefore, it may lose a lot of information, but it is very simple to compute, making it suitable for situations where control precision is not critical.

3.2.3.3 Centroid method

To obtain an accurate control value, the defuzzification method should ideally express the result of the output membership function as accurately as possible. The centroid method takes the center of the area formed by the membership function curve and the x-axis as the output value of the fuzzy inference result:

$$v_o = \frac{\int_V v \cdot u_v(v)dv}{\int_V uv(v)dv} \tag{3.7}$$

Compared to the maximum membership degree method, the centroid method provides a smoother output inference result. This means that small changes in the input signal will result in corresponding changes in the output. The centroid method considers the shape of the inference result, containing more information. Therefore, this chapter also uses this method for defuzzification.

3.3   System stability analysis

In the previous section, we demonstrated through simulations that the PMSM vector control system based on fuzzy adaptive PID is stable. However, further theoretical analysis of the system's stability is even more significant. According to the design and analysis principles of classical control theory, stability analysis of control systems is crucial. Fuzzy control systems also need stability analysis, as this is closely related to the system's reliability. However, due to the nonlinear characteristics of fuzzy control systems and the lack of a unified system description, it is difficult for researchers to apply classical control theory directly to the design and analysis of fuzzy control systems. Fortunately, modern control theory has proposed a series of ideas and methods to solve this issue, including: Lyapunov stability theory, small gain theory, phase plane analysis, SMC system analysis, describing function analysis, and circle criterion methods. However, these methods also encounter a series of issues in fuzzy control analysis and applications. To this day, many experts in traditional control theory still cannot fully accept fuzzy control. To address this problem, this chapter seeks an analysis method that aligns with classical control theory and, combined with the characteristics of fuzzy adaptive PID, completes the stability analysis of the PMSM vector control system.

Next, we briefly analyze the PMSM system based on fuzzy adaptive PID. As mentioned earlier, the difference between this system and the traditional PMSM vector control system is that the speed loop PID is replaced by a fuzzy adaptive PID. Although the fuzzy controller is a nonlinear model without a determined expression, its output is clearly bounded, i.e., the output fuzzy domain is $[-1, 1]$, and the output scaling factors are $K'_p = 51$, $K'_I = 0.3$, and $K'_d = 0.15$. Given $k_{p0} = 85$, $k_{i0} = 0.5$, and $k_{d0} = 0.25$, the actual range of values for the fuzzy adaptive PID coefficients are: k_p—[34,136], k_i—[0.2,0.8], and k_d—[0.1,0.4]. At this point, the fuzzy controller can be ignored, and the fuzzy adaptive PID can be treated as a controller where each coefficient varies with the system's operating state. As long as we can prove that the system remains stable within the coefficient variation range, the system's stability is guaranteed. It should be noted that this analysis process is entirely based on classical control theory, making it simple and convenient. Unlike PID controller design, stability analysis first requires knowledge of the entire system's mathematical model. The open-loop transfer function of the motor vector control system is obtained, and then the system stability is determined using Routh-Hurwitz criterion or methods such as MATLAB® to solve the characteristic equation and find the eigenvalues. Below is a detailed introduction to the model of the PMSM dual-loop vector control system in the s-domain.

3.3.1 Motor model

The voltage equation, electromagnetic torque equation, and motion equation of the PMSM are known. When no field weakening control is used, i.e., $i_d = 0$, the motor's output voltage equation and electromagnetic torque equation are as follows:

$$\begin{cases} u_d = -\omega_r L_q i_q \\ u_q = R_s i_q + L_q \dfrac{di_q}{dt} + \omega_r \psi_f \end{cases} \tag{3.8}$$

$$T_e = p_n \psi_f i_q \tag{3.9}$$

To perform stability analysis, the motor model needs to be simplified. First, the motor's voltage-current equations form a strongly coupled model, which typically requires state feedback compensation to achieve decoupling. After decoupling, the actual voltage of the motor is:

$$\begin{cases} u_d^* = u_d - \omega_r L_q i_q \\ u_q^* = u_q + \omega_r \psi_f \end{cases} \tag{3.10}$$

The voltage equation of the motor becomes:

$$\begin{cases} u_d^* = -\omega_r L_q i_q \\ u_q^* = R_s i_q + L_q \dfrac{di_q}{dt} + \omega_r \psi_f \end{cases} \tag{3.11}$$

Equations (3.8), (3.10), and (3.11) are Laplace transformed, and the decoupling model of the permanent magnet motor is obtained in Figure 3.7, where, $K_t = p_n \psi_f$, $L_q = 0.0109$ H, $R_s = 0.29$ Ω, $\psi_f = 0.2339$ Wb, the rotational inertia is $J = 0.008$ kg/m². The viscous friction coefficient is $F = 0.001$ N·m·s.

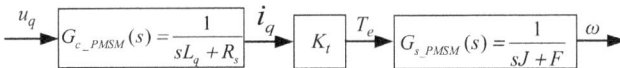

Figure 3.7 Decoupling model of PMSM

3.3.1.1   Inverter model

The voltage inverter model can be equivalent to a first-order inertial link:

$$G_{inv}(s) = \frac{k_{inv}}{sT_s + 1} \tag{3.12}$$

where, T_s is the inverter switching period, $T_s = 0.0005$ s, and k_{inv} is the inverter voltage amplification factor. When controlling with space vector modulation (SVPWM), $k_{inv} = 1$.

3.3.1.2   Current loop feedback model

The current loop feedback can be expressed as an inertial link, and its transfer function can be expressed as:

$$G_{c\_f}(s) = \frac{k_{cf}}{sT_{c\_f} + 1} \tag{3.13}$$

where, k_{cf} represents the amplification factor for the current feedback filter. Since the feedback value is taken from the digital sampling, which represents the actual current value, $k_{cf} = 1$ is the filter time constant of the current feedback channel, with $T_{c\_f} = 10^{-7}$. Since the time constant of this module is very small, at low frequencies, this element can be effectively approximated as a proportional amplifier with an amplification factor of 1.

The current loop and speed loop PID controllers are implemented in the digital controller, so essentially, they are discrete controllers. According to classical control theory, when the sampling period is very short, the additional phase lag caused by the sample-and-hold device is not significant. In this case, the digital part can be approximated by continuous elements, i.e., the discrete system is modeled as a continuous system. Although this is an approximation, the analysis methods for continuous systems are well understood by engineers, and extensive experience has been accumulated in the analysis process, making these methods widely applied. To simplify the analysis process, this chapter models the discrete controller as continuous. Since the three PID coefficients for the current loop are $k_{c\_p} = 4.8$, $k_{c\_i} = 3,200$, and $k_{c\_d} = 0.25$, the transfer functions of the current loop PID are as follows:

$$G_{c\_PID}(s) = 4.8 + \frac{3,200}{s} + 0.25s \tag{3.14}$$

The three PID coefficients of the speed ring are respectively k_{s_p}, k_{s_i}, and k_{s_d}, which vary with the state of the system, and the variation range is shown here as a variable. The transfer function of the PID of the speed ring is:

$$G_{s\_PID}(s) = k_{s\_p} + \frac{k_{s\_i}}{s} + k_{s\_d}s \tag{3.15}$$

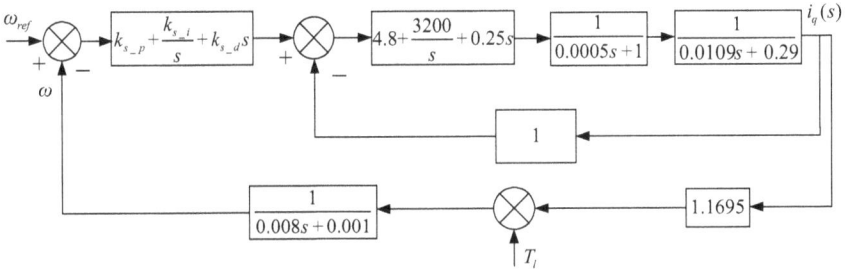

Figure 3.8 The transfer function diagram of the double closed-loop control system of PMSM

According to the above analysis, we can obtain the transfer function of the double closed-loop control system of the PMSM, as shown in Figure 3.8.

As can be seen from Figure 3.8, the open-loop transfer function of the system can be summarized as:

$$G(s) = \frac{0.292k_{s\_d}s^4 + (5.614k_{s\_d} + 0.292k_{s\_p})s^3 + (3742.4k_{s\_d} + 5.614k_{s\_p} + 0.292k_{s\_i})s^2 + (3742.4k_{s\_p} + 5.614k_{s\_i})s + 3742.4k_{s\_i}}{4.36 \times 10^{-8}s^5 + 0.002s^4 + 0.041s^3 + 25.605s^2 + 3.2s}$$

(3.16)

Then, the characteristic equation of the system is:

$$
\begin{aligned}
D(s) = {} & 4.36 \times 10^{-8}s^5 + (0.292k_{s\_d} + 0.002)s^4 \\
& + (5.614k_{s\_d} + 0.292k_{s\_p} + 0.041)s^3 \\
& + (3742.4k_{s\_d} + 5.614k_{s\_p} + 0.292k_{s\_i} + 25.605)s^2 \\
& + (3742.4k_{s\_p} + 5.614k_{s\_i} + 3.2)s + 3742.4k_{s\_i} = 0
\end{aligned}
$$

(3.17)

The system characteristic equation is a 5th-order polynomial containing three unknown variables. If the Routh-Hurwitz criterion is used to determine the system's stability, it becomes very complex due to the large computational workload. The Routh array is shown in Table 3.4.

In contrast, using MATLAB to solve for the characteristic roots of the characteristic equation is much simpler. The characteristic roots of the characteristic equation represent the poles of the closed-loop system. When all system poles lie to the left of the imaginary axis in the complex plane, the system is stable. However, when using this method, we face another issue: currently, MATLAB solves higher-order equations using iterative methods. Mathematical equations of degree four or higher do not have analytical solutions. Therefore, we cannot directly analyze the system's characteristic roots by solving the equation. In fact, as long as it can be proven that the real parts of all the roots of the characteristic equation are less than 0 within the variation range of the PID coefficients, the system can be considered stable. We propose an indirect method to prove this issue using optimization

Table 3.4   Routh table

STEP			
s^5	4.36×10^{-8}	$5.614k_{s\_d} + 0.292k_{s\_p} + 0.041$	$3742.4k_{s\_p}$ $+ 5.614k_{s\_i}$ $+ 3.2$
s^4	$0.292k_{s\_d} + 0.002$	$3742.4k_{s\_d} + 5.614k_{s\_p} + 0.292k_{s\_i}$ $+ 25.605$	$3742.4k_{s\_i}$
s^3	$(0.000584k_{s\_p} + 0.0232k_{s\_d}$ $+ 0.0853k_{s\_p} \cdot k_{s\_d} + 16.393k_{s\_d}$ $+ 0.000082)/(0.292k_{s\_d} + 0.02)$	$(7.485k_{s\_p} + 0.011k_{s\_i} + 0.934k_{s\_d}$ $+ 1092.78k_{s\_p} \cdot k_{s\_d} + 1.639$ $k_{s\_i} \cdot k_{s\_d} + 0.0064)/(0.292k_{s\_d}$ $+ 0.02)$	0
s^2	$0.00044\ k_{s\_p} + 1.488 \times 10^{-6}\ k_{s\_i}$ $+ 0.897k_{s\_d} + \ldots \ldots$	$\ldots \ldots$	0
s^1	$\ldots \ldots$	0	0
s^0	$\ldots \ldots$	0	0

theory: first, we convert the original problem into a constrained optimization problem:

$$\begin{cases} f(s) = \max \mathrm{Re}(s) \\ \text{s.t.} \quad D(s) = 0 \\ \qquad 34 \le k_{s\_p} \le 136 \\ \qquad 0.2 \le k_{s\_i} \le 0.8 \\ \qquad 0.1 \le k_{s\_d} \le 0.4 \end{cases} \tag{3.18}$$

Problem Description: s is the solution to the characteristic equation $D(s) = 0$. The coefficients of the characteristic equation, namely the fourth-order, third-order, second-order, first-order, and constant terms, are determined by k_{s_p}, k_{s_i}, and k_{s_d}, and the variation range of these three variables is known. For each set of k_{s_p}, k_{s_i}, and k_{s_d}, the corresponding characteristic roots can be computed. By comparing the real parts of all characteristic roots, if we can find a set of k_{s_p}, k_{s_i}, and k_{s_d}, such that $\mathrm{Re}(s) > 0$, the system is not fully stable, meaning there are unstable factors. If for any k_{s_p}, k_{s_i}, and k_{s_d} values, the computed $f(s)$ is always less than 0, it can be considered that the system is stable.

In practice, the optimization process is a sampling verification method. Common multi-dimensional optimization methods include the coordinate rotation method, step-size acceleration method, Powell method, and random search method. This chapter uses the random search method, which is introduced as follows.

The random search method does not require starting from a specified point to gradually find the optimal point. Instead, it uses the concept of randomly selecting points in probability and statistics to find the optimal point.

In this case, the search space is a three-dimensional space, represented in vector form as $X = (k_{s\_p}, k_{s\_i}, k_{s\_d})$. Suppose that the proportion of points satisfying

Table 3.5   Relationship between the probability of finding satisfying
points and the number of sampled points

f_n $P(f)$	0.2	0.1	0.05	0.025	0.01	0.005
0.8	7	16	32	64	161	322
0.9	10	22	45	91	230	460
0.95	14	29	59	119	299	598
0.99	20	44	90	192	459	919

$Re(s) > 0$ in the entire search space is $p = 20\%$, so the points that do not satisfy this condition account for 80% of the space. By randomly selecting n points, the probability of encountering points satisfying $Re(s) > 0$ is given by $P(f) = 1 - 0.8n$. As n becomes very large, the probability of finding the optimal point increases. Table 3.5 shows the relationship between the probability of finding satisfying points and the number of points sampled, where f represents the proportion of satisfying points, n is the number of samples, and $P(f)$ is the probability of obtaining satisfying points.

From the analysis of the table data, when the points satisfying $Re(s) > 0$ account for more than 0.5% of the entire region, if the number of experiments exceeds 1,000, the probability of finding satisfying results exceeds 99%. To ensure that the process can find the satisfying points, it is essential that the point selection is random; otherwise, the probability cannot be guaranteed.

The objective function can be arbitrary and may have multiple peaks. This method has a broad applicability. When the objective function has several local peaks (multi-modal), this method can easily find the general location of the global optimum. Once located, other optimization methods can be used to determine the exact global optimum. Based on the above explanation of the random search method and in combination with the problem we need to solve, the method is processed as follows:

(a) Define the objective. We aim to use the random search method to solve the system's stability problem. If a point X exists such that $f(s)$, it can be concluded that the system has an instability risk. However, if after many optimization iterations, no point X satisfying $f(s) > 0$ can be found, due to the discrete nature of the optimization, we cannot confirm the system's stability with 100% certainty. Based on statistical knowledge, we define the system's stability confidence: The entire ks_p, k_{s_i}, k_{s_d} region defines a three-dimensional interval. After taking samples, the sampling points $X_1 = (k_{s\_p1}, k_{s\_i1}, k_{s\_d1})$, $X_2, \ldots, X_n$, are obtained. When the proportion p of the area where $Re(s) > 0$ is known, the system's stability confidence is given by $u = P(f) = 1 - pn$.

(b) Assumption of Existence of X. Here, we assume that there exists an X such that the real part of the characteristic equation's root is greater than 0. During the verification of this assumption, we need to find $\max Re(s)$. However, using the

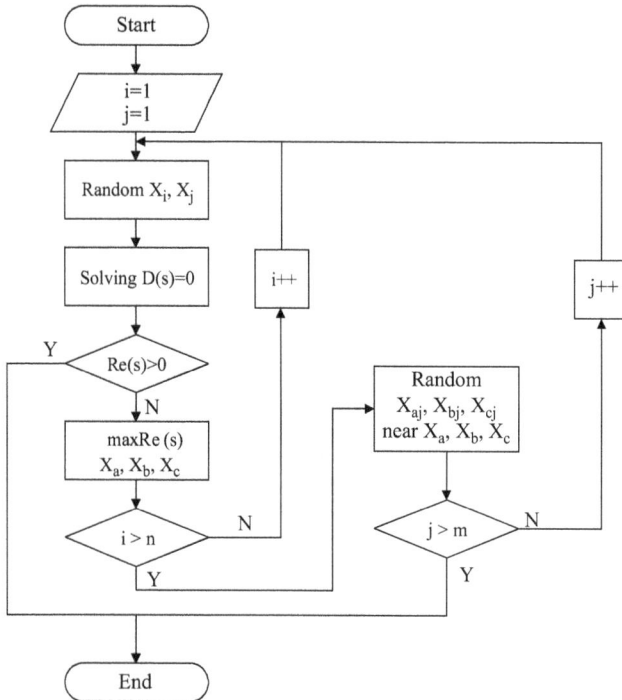

Figure 3.9   Block diagram of the random search algorithm

random search method means that, based on the sample data and small probability theory, we make a decision to accept or reject the hypothesis. Due to the
randomness of the sample, when judging based on a single sample, we might
make a "type I error" (rejecting a true hypothesis). To avoid this, we use
multiple sampling: If the first sampling test shows that max$Re(s) < 0$, we
record the last three points where max$Re(s)$ is updated: Xa, Xb, and Xc. The
probability of finding a point where max$Re(s) > 0$ near these three points is
greatly increased. In the three local regions (with each coordinate varying
by $\pm 10\%$), we use the random search method again (taking m points) for
optimization analysis, and finally, we make a decision on the hypothesis
(Figure 3.9 and Table 3.6).

   Although this chapter assumes the existence of X such that the real part of the
characteristic equation's root is greater than 0, the proportion p of the region where
$Re(s) > 0$ is completely unknown. Even so, we can assume that $p < 0.005$, meaning
that only a very small number of X values result in $Re(s) > 0$. A higher system
stability confidence requires more sampling points, which will make it more likely
to make the correct judgment. Here, let $u \geq 0.99$, we define the first number of
sample points $n \geq 919$ (as 950). If none of these points satisfy $Re(s) > 0$, we take
$m = 200$ additional points. This way, the system stability confidence is guaranteed

Table 3.6   Maximum real part values of the five characteristic roots for different X values

Items	k_{s_p}	k_{s_i}	k_{s_d}	$Re(s)$
1	37.6426	0.7094	0.3802	−0.0188
2	103.2309	0.6546	0.3229	−0.0063
3	74.0072	0.5933	0.1514	−0.0080
4	106.02	0.2191	0.1831	−0.0021
5	38.7095	0.2583	0.3470	−0.0066
6	104.8725	0.3903	0.3851	−0.0037
7	37.5135	0.4632	0.2145	−0.0123
......	......	......	......	......

Table 3.7   Maximum real part values of the characteristic roots for different X values

Items	k_{s_p}	k_{s_i}	k_{s_d}	$Re(s)$
1	134.8679	0.2435	0.2247	−0.0018
2	138.1762	0.2794	0.2443	−0.0020
3	105.3570	0.1911	0.2223	−0.0018
4	103.0774	0.1901	0.1771	−0.0018
5	124.1747	0.3317	0.2997	−0.0027
6	126.3021	0.3385	0.3605	−0.0026
......	......	......	......	......

to be greater than 0.99. The points are generated using the MATLAB function rand (), which ensures they are random.

After the first optimization process, no X satisfying $Re(s) > 0$ was found. The recorded points are: $X_a = (135.6, 0.2469, 0.2328)$; $X_b = (106.02, 0.2191, 0.1831)$; $X_c = (127.55, 0.3715, 0.3272)$. Then, $m = 200$ points were taken near each of the three points, and max$Re(s)$ was recalculated. The results are shown in Table 3.7.

Based on the calculation results, it can be observed that there is no X in the entire region that satisfies $Re(s) > 0$, meaning the system is stable with a confidence level greater than 99%. Furthermore, the variation range of k_{s_p} is the largest, and it has the most significant effect on the characteristic root values. However, it is not enough to cause the characteristic roots to shift to the right half of the imaginary axis. On the other hand, the variation ranges of k_{s_i} and k_{s_d} are very small, so their effect on the characteristic root values is minimal.

In conclusion, we can make the following judgment: the variation in the PID coefficients of the speed loop does not affect the stability of the system, only adjusting the system's dynamic performance. Therefore, the PMSM vector control system based on fuzzy adaptive PID is stable.

3.4   Finite set model predictive control

MPC is applied to control problems with optimization needs. In the past, predictive control was mainly used for slow dynamic systems, but in recent years, increasingly faster systems have started to adopt predictive control methods to improve constraint handling and control performance. In the field of power electronics, due to the rapid development of microelectronics technology, CPU speeds have greatly increased. A series of high-performance processors have been successfully manufactured, making it possible for MPC to be applied in real-time control systems.

MPC uses a system model to predict the future states of all possible control actions and then selects the best control behavior by minimizing a cost function. This method features real-time, online rolling optimization and simple algorithm implementation, making it widely applied.

For PMSM drive systems, an MPC-based controller can be used to implement torque control, flux and power control, speed regulation, and other functions. Another application is Model Predictive Current Control (MPCC), which benefits from rapid dynamic control performance. Generally, MPCC has two implementation methods: One method is an extension of traditional vector control, where an MPC-based controller replaces the current control loop but retains the modulator. The other method completely eliminates the modulator, and the output of the MPC-based controller directly determines the optimal switching states of the inverter, called Finite Set MPCC (FCS-MPCC).

In comparison, the latter method has a simpler topology and is a promising alternative for PMSM control systems. Using a given voltage vector, FCS-MPCC uses the motor model to calculate the predicted future current. The driving value is chosen and applied to the motor, minimizing the cost function to achieve the integrated control objective. The optimization process in each sampling cycle requires repeated use of new measurement data.

While this control algorithm removes the modulation module, it still requires significant computation compared to traditional PI control schemes, resulting in longer computation times. If delays between the actual measurements and control drives are not considered, the performance of the PMSM system can decrease, reducing the effectiveness of optimal control.

3.4.1   Basic principles of FCS-MPC

MPC uses a system's mathematical model to predict the future state under each possible control action and selects the best control behavior by minimizing a cost function. This method features real-time online rolling optimization and simple algorithm implementation, making it widely applied.

All MPC implementations can be summarized into four steps: prediction model, reference trajectory, rolling optimization, and error correction. The structure diagram of predictive control is shown in Figure 3.10.

This chapter mainly discusses FCS-MPC, which effectively combines the finite discrete switching states of the converter with the model's predictive control

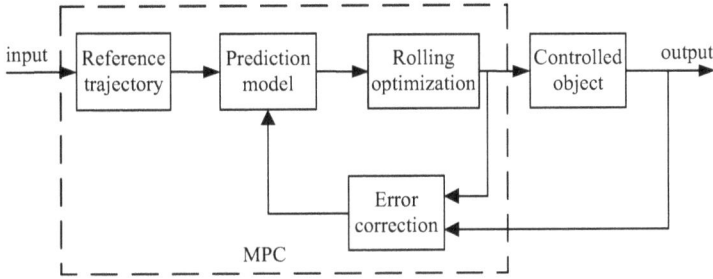

Figure 3.10 Flowchart of model predictive control

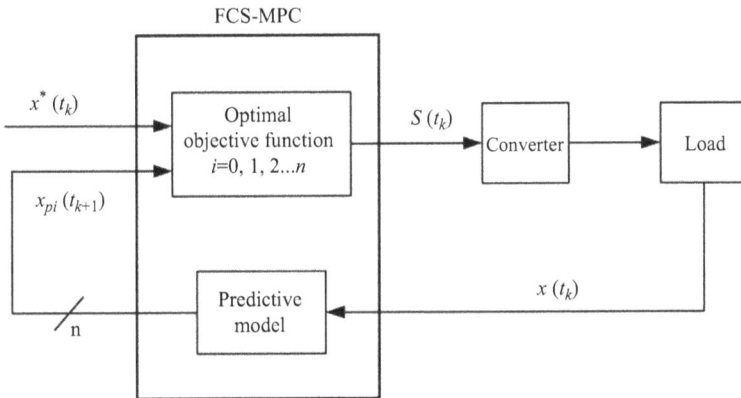

Figure 3.11 FCS-MPC control block diagram

objectives. Figure 3.11 shows the simplified control block diagram of real-time FCS-MPC implementation. $X(t)$ represents the general system variables. At the same time, this control method is not limited to a single variable; it can handle multiple variables, system constraints, disturbances, and saturation. Essentially, all features that can be mathematically modeled and measured can be included in the prediction model and the objective function. This is the foundation of FCS-MPC's immense flexibility and control potential. FCS-MPC requires only the discrete model of the control system to enable a simpler and more direct design of the controller.

The control problem of this algorithm, regardless of whether the actuator or controller changes is essentially about determining the appropriate control signal S (t). It attempts to give a reference value $x^*(t)$ for the state variable $x(t)$ of the system, as close as possible to the state $x(t)$ considered at a sampling time t_k. In the ideal case, it is assumed that the system's measurement and control actions are executed immediately. However, in practice, a sample period T_s exists, and the system has n control actions. Thus, $x(t_k)$ represents the state at the sampling time, and the calculation and control actions are carried out immediately.

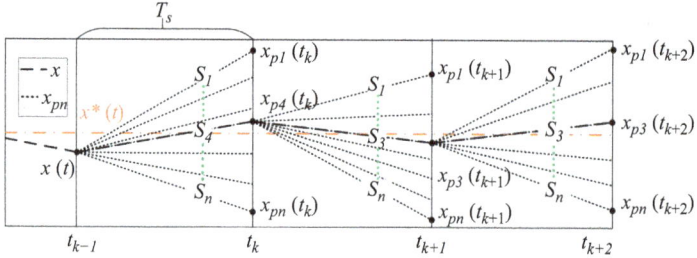

Figure 3.12   Implementing the FCS-MPC principle

Given that the control actions $S_i(I = 0,1,2,\ldots,n)$ and the control set have a finite total number, it is possible to predict the state for the next sampling time using the control actions and the state value $x(t_k)$ at the current time. The state prediction function can be expressed as: $x_{pi}(t_{k+1}) = f_p(x(t_k))$, $(i = 0,1,\ldots,n)$. The prediction model comes from the control system's discrete model and its relevant parameters. Simultaneously, because of the small dynamic response time of the system compared to T_s, the predicted state $x^*(t)$ is very close to the actual state. Therefore, the parameters in the prediction model are assumed to be very small within T_s. To determine the control actions S_i, based on the control objective, we need to define a cost function f_g, since the cost function directly describes the control objective and the weight relationships between control targets and objectives. Hence, it is essential to select the optimal control function to perform control based on the control objectives.

Assume the cost function is given by: $f_g = |x^*(t_{k+1}) - x_{pi}(t_{k+1})|$. As shown in Figure 3.12, the state prediction value $x_{p4}(t_{k+1})$ is the closest approximation to $x^*(t)$. This is because the cost function value is minimized at this point, so the control action S_4 is chosen to operate at sampling point t_k.

During the relatively simple control steps, FCS-MPC also has the advantage of controlling multiple objectives, and the control variables and constraints can include the same target function, such as current, voltage, speed, and other classic control variables. However, when the variable group in the cost function has different characteristics (i.e., different units or scaling), it is necessary to add specific weighting factors to each cost function term to adjust the balance with other control objectives or cost functions.

3.4.2   PMSM model predictive control

PMSM has a slower electrical response compared to electrical variables, with longer periods. Therefore, it is possible to adjust the system and control the electrical variables by choosing appropriate controllers for each. For faster dynamic responses, limited predictive models can be used to improve PI controllers. Additionally, by using FOC with a constant $i_d = 0$ a complete PMSM control system is built for prediction-based control [6].

3.4.2.1 Prediction model

The voltage equation for PMSM in the *d-q* reference frame is as follows:

$$\begin{cases} u_d = Rt_d + L_dpt_d - L_q\omega t_q \\ u_q = Rt_q + L_qpt_q + L_d\omega t_d + \omega\psi_f \end{cases} \tag{3.19}$$

The predictive model of FCS-MPC is usually a discretized model of the controlled system, which calculates the desired state or the future value of the output through the input-output relationship of the system. The simplest and most commonly used discretization method is the first-order Euler method, which approximates the current differential operation as a forward difference operation:

$$pi = \frac{i(k+1) - i(k)}{T_s} \tag{3.20}$$

Then (3.19) is obtained after discretization by the Euler method:

$$\begin{cases} u_d(k) = Ri_d(k) + L_d\dfrac{i_d(k+1) - i_d(k)}{T_s} - L_q\omega i_q(k) \\ u_q(k) = Ri_q(k) + L_q\dfrac{i_q(k+1) - i_q(k)}{T_s} + L_d\omega i_d(k) + \omega\psi_f \end{cases} \tag{3.21}$$

Move the current at time *k*+1 to the left of the (3.21) and get:

$$\begin{cases} i_d(k+1) = i_d(k) + \dfrac{T_s}{L_d}\left[u_d(k) + L_q\omega i_q(k) - Ri_d(k)\right] \\ i_q(k+1) = i_q(k) + \dfrac{T_s}{L_q}\left[u_q(k) - L_d\omega i_d(k) - Ri_q(k) - \omega\psi_f\right] \end{cases} \tag{3.22}$$

According to the above equations, calculate the predicted stator current components corresponding to the eight voltage vectors generated by the inverter. Without considering optimization measures, the computational load increases exponentially with the prediction steps. If single-step prediction requires calculating eight voltage vectors, then two-step prediction would require calculating 64 steps, and for N-step prediction, it would require 8^N steps. This presents a huge burden for the processor, and in typical cases, the marginal benefit brought by the additional computation is relatively limited. Besides the computational load, factors such as system error accumulation and changes in the system model must also be considered when determining the prediction steps. Therefore, this chapter only discusses single-step prediction FCS-MPC.

3.4.2.2 Cost function

Compared to some classical control methods, the cost function more intuitively describes the relationship between the control objectives, constraints, and targets. Therefore, the selection of the cost function is one of the most important steps in designing FCS-MPC. The control objectives can be chosen based on demand and can include all constraints of the control variables. The common objective functions for PMSM FCS-MPC control are expressed in the following two forms:

$$f_g = \left| x^*(u) - x_p(u) \right| \tag{3.23}$$

$$f_g = \left[x^*(u) - x_p(u) \right]^2 \tag{3.24}$$

For the current loop, the control target should not only track the current instructions generated by the outer loop but also limit the current amplitude due to hardware limitations such as inverter capacity. The form of the cost function of the limiting target is:

$$f_{glim} = \begin{cases} 0 & x \in [x_{\min}, x_{\max}] \\ \infty & x \notin [x_{\min}, x_{\max}] \end{cases} \tag{3.25}$$

Once a control action results in the controlled variable exceeding the threshold, the cost function value becomes infinitely large, indicating that the control action is invalid. This approach is suitable for scenarios where the saturation condition must be strictly met. When the controlled object remains within the saturation range, the cost function value is zero.

For the PMSM FCS-MPC current control, the essence is to track the stator current q-axis component. In this chapter, considering the $i_d = 0$ control method for the PMSM, the d-axis current component directly affects the reactive power of the motor. To reduce reactive power, improve current utilization, and minimize energy loss, the predicted d-axis current component should be kept small. Additionally, since excessive current could lead to issues such as inverter overmodulation and voltage harmonics, current limitation control should be applied.

Thus, the cost function is designed as follows:

$$f_g = \left[i_{dp}(k+1) \right]^2 + \lambda \left[i_q^* - i_{qp}(k+1) \right]^2 + f_{glim} \left[i_{dp}(k+1), i_{qp}(k+1) \right] \tag{3.26}$$

$$f_{glim} = \begin{cases} 0 & i_{dp}(k+1) \in [i_{d\min}, i_{d\max}] \cap i_{qp}(k+1) \in [i_{q\min}, i_{q\max}] \\ \infty & i_{dp}(k+1) \notin [i_{d\min}, i_{d\max}] \cup i_{qp}(k+1) \notin [i_{q\min}, i_{q\max}] \end{cases} \tag{3.27}$$

where, λ is the weight factor of i_q tracking performance.

The specific flow of FCS-MPC current control of PMSM is shown in Figure 3.13.

3.4.3 Comparison of FCS-MPC and classical control methods

This section will use MATLAB/Simulink® to build simulation models of the vector control system and the MPC system for the PMSM.

Simulink is a scientific environment with high-performance numerical computation and visualization capabilities. Due to its ease of use, high computational precision, and ability to handle most scientific computations, it is widely recognized in both engineering applications and academic research. Simulink is a graphical simulation tool based on the MATLAB block diagram design environment,

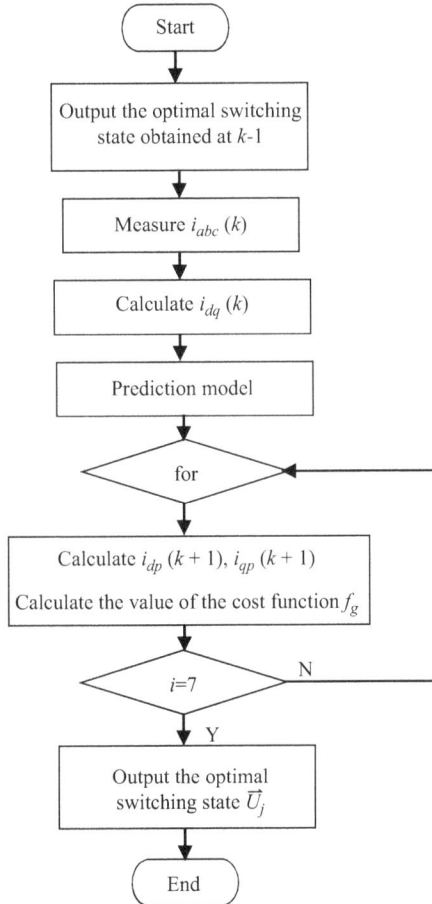

Figure 3.13   FCS-MPC current control flow chart of PMSM

with excellent simulation effects for communications, control, signal processing, video processing, and image processing. Due to its user-friendly, visual nature, it is widely used in the modeling and simulation of linear systems, nonlinear systems, digital control, and digital signal processing. For large-scale and complex systems, Simulink allows simplification by grouping some modules into subsystems. This way, a bottom-up approach can be used during the design process, creating a model that is logically organized and easier to identify errors, as well as enabling easy modification of system functionalities. For the experiments in this chapter, Simulink provides an interactive graphical environment and customizable module libraries for design, simulation, execution, and testing.

From the above description, we can see that using Simulink to simulate a rotor position estimation method for sensorless control based on mathematical models not only verifies the feasibility of the system but also helps identify potential issues

Table 3.8   PMSM parameters

Parameter	Value	Parameter	Value
Rated voltage U_N	100 V	Magnetic flux ϕ	0.0876 Wb
Rated speed N	2,000 rpm	Mechanical moment of inertia J	$0.402*10^{-4}$ kg*m^2
Number of pole-pairs P_n	3	Phase resistance R_s	0.055 Ω
Friction coefficient μ_0	0.003	Inductance L_d	0.8 mH

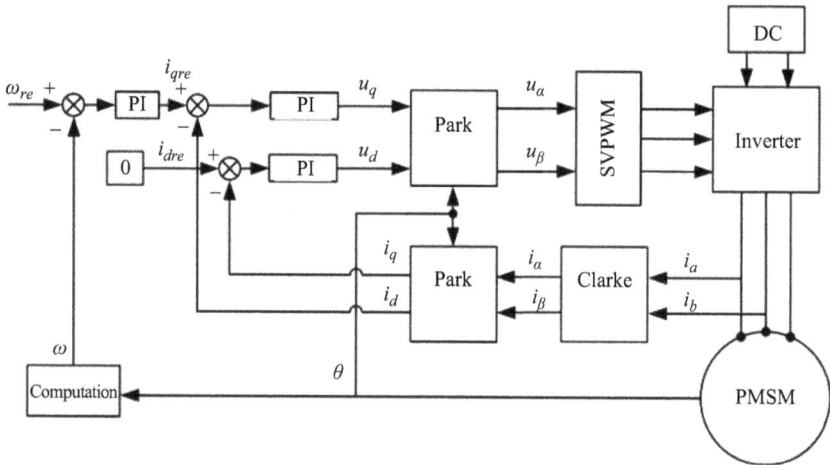

Figure 3.14   PMSM vector control system simulation model

in the system design. Additionally, it can determine optimal system parameters, and the modular simulation analysis can provide a clear programming framework for future object-oriented programming control system designs. The MATLAB version used for simulation in this chapter is R2018b. During the simulation, the motor model is parameterized, setting the motor to be a three-phase motor with a sinusoidal back electromotive force (EMF). The simulation motor parameters are shown in Table 3.8.

The overall simulation model of the PMSM vector control system is shown in Figure 3.14. In the model, basic modules from the Simulink library are used to construct the control circuit, including the Park transformation module and its inverse transformation module, the Clark transformation module, the SVPWM module, and so on. The PMSM model is directly called from the Sim Power Systems library as the controlled object.

The overall simulation model of the PMSM MPC system is shown in Figure 3.15. Unlike the vector control system, it removes the SVPWM module and instead implements the FCS-MPC module program.

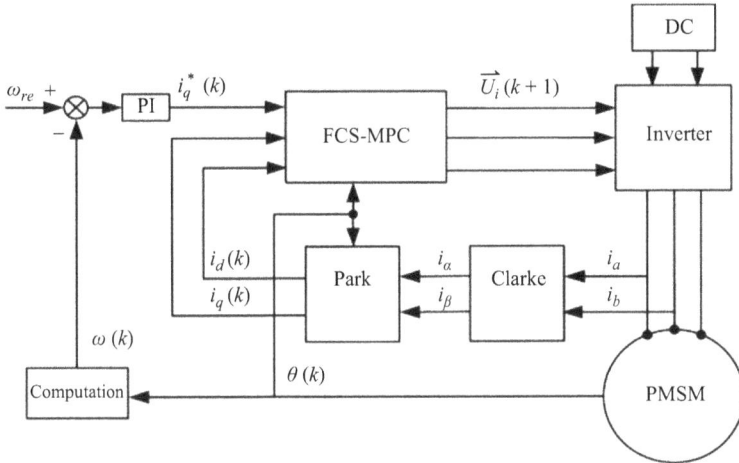

Figure 3.15   PMSM MPC system simulation model

The control frequency is related to the control period and represents the frequency at which updated voltage vectors are used to adjust the motor's state. For FCS-MPCC, during each PWM carrier cycle, a new voltage is selected to adjust the motor's state. Therefore, the control frequency of FCS-MPCC is equal to the PWM carrier frequency.

In the traditional SVPWM method introduced earlier, a combination voltage vector is also selected to adjust the machine's state during each PWM carrier cycle, so the control frequency is also equal to the PWM carrier frequency. Additionally, the switching frequency refers to the frequency at which the three-phase switches change states. During each switching cycle, a switch in the power module experiences both the "on" and "off" states (one cycle). For SVPWM, each control cycle (PWM carrier cycle) uses three basic voltage vectors, so each switch in the power module first turns on and then off (equivalent to one cycle). Therefore, the switching frequency equals the control frequency and the PWM carrier frequency. However, for the FCS-MPCC method, since only one voltage vector is selected and applied during each control cycle, each switch in the power module experiences one of the following states: "on," "off," or "no change" (less than one cycle). Therefore, the switching frequency of the FCS-MPCC method will be lower than the control frequency (PWM carrier frequency).

Compared to the classic PI vector control, FCS-MPC does not require a modulator and typically requires a higher control frequency to implement stator current trajectory control. The comparison of the two control methods is more intuitively presented in the simulation figures. As shown in Figures 3.16 and 3.17, when the control frequency is 5 kHz, the torque, stator current, and speed ripple in the MPC system are significantly larger, while the simulation curves for vector control are smoother and exhibit better performance. As the system control

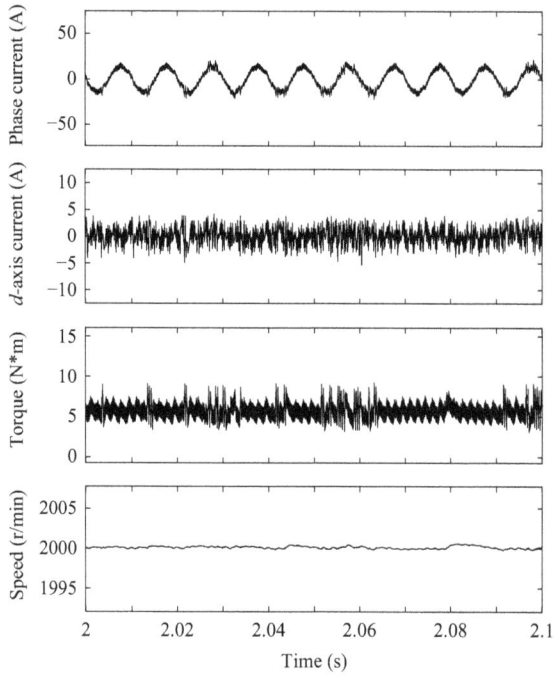

(a)

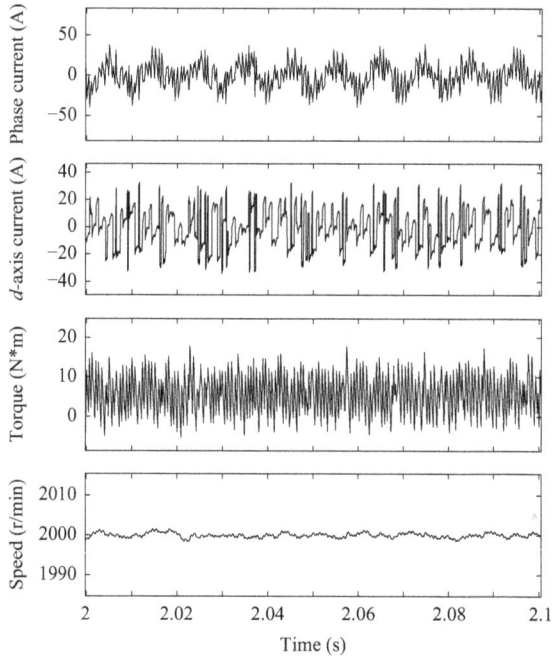

(b)

Figure 3.16   Simulation results with a control frequency of 5 kHz: (a) classic vector control and (b) model predictive control

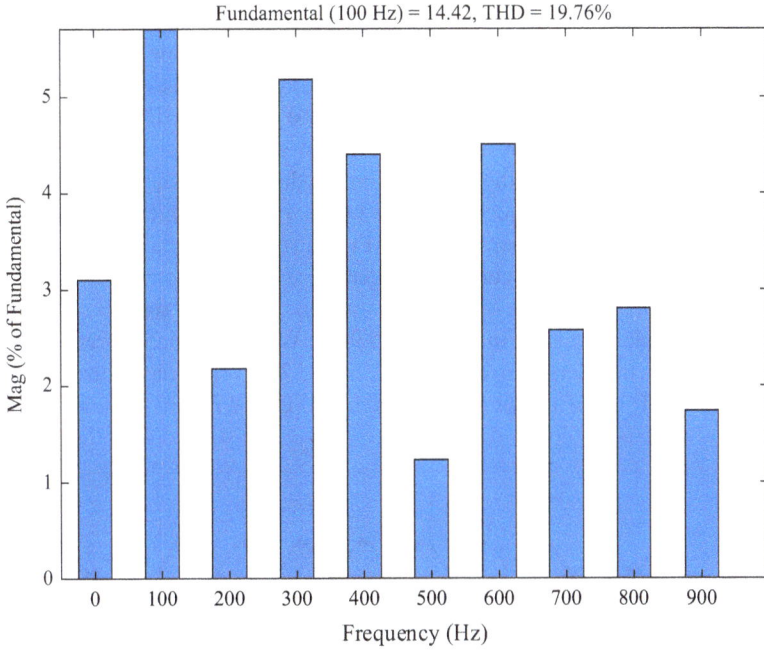

(a)

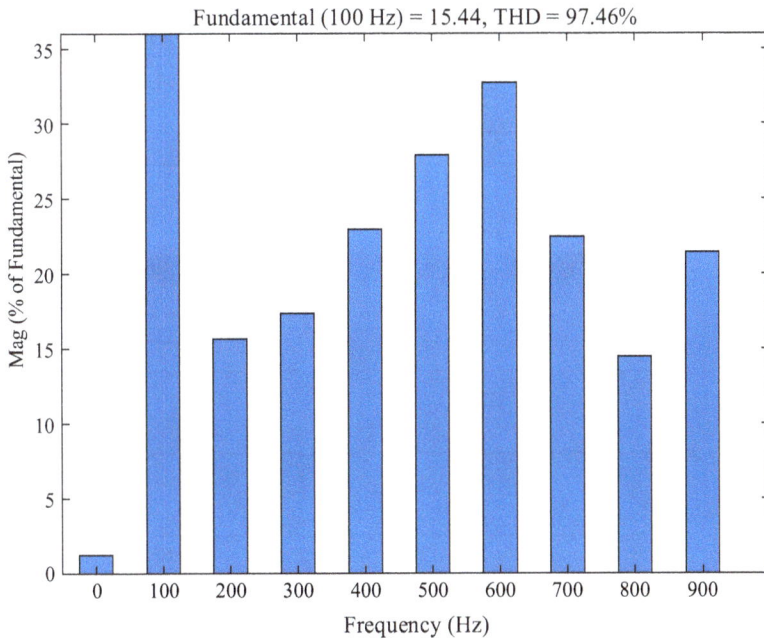

(b)

Figure 3.17 *Phase current analysis at the control frequency of 5 kHz: (a) classic vector control and (b) model predictive control*

frequency increases to 10 kHz, as seen in Figures 3.18 and 3.19, the performance of the MPC system improves as well, with a speed steady-state error of only 0.5 revolutions and a total harmonic distortion (THD) of the phase current of 55.20%. Meanwhile, under traditional PI control, the phase current THD is 18.31%.

The control performance of the motor is to some extent determined by the switching frequency. From the comparison of the relationship between control frequency and switching frequency, it can be seen that to achieve performance in the MPC system that is comparable to the PI control system, the control frequency should be higher than the control frequency of the PI control system. From Figures 3.20 to 3.23, it can be observed that when the control frequency is 20 kHz, the THD of the phase current in MPC is 25.93%. When the control frequency is 25 kHz, the THD of the phase current in MPC is 21.32%. When the control frequency is 33 kHz, the THD of the phase current in MPC is 17.16%. When the control

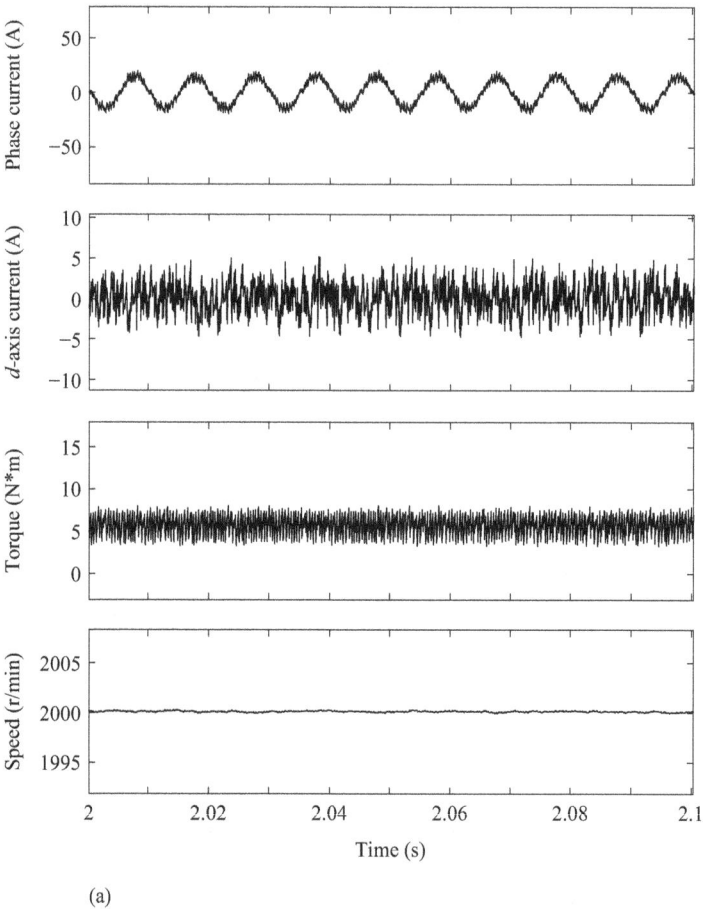

(a)

Figure 3.18   Simulation results at the control frequency of 10 kHz: (a) classic vector control and (b) model predictive control

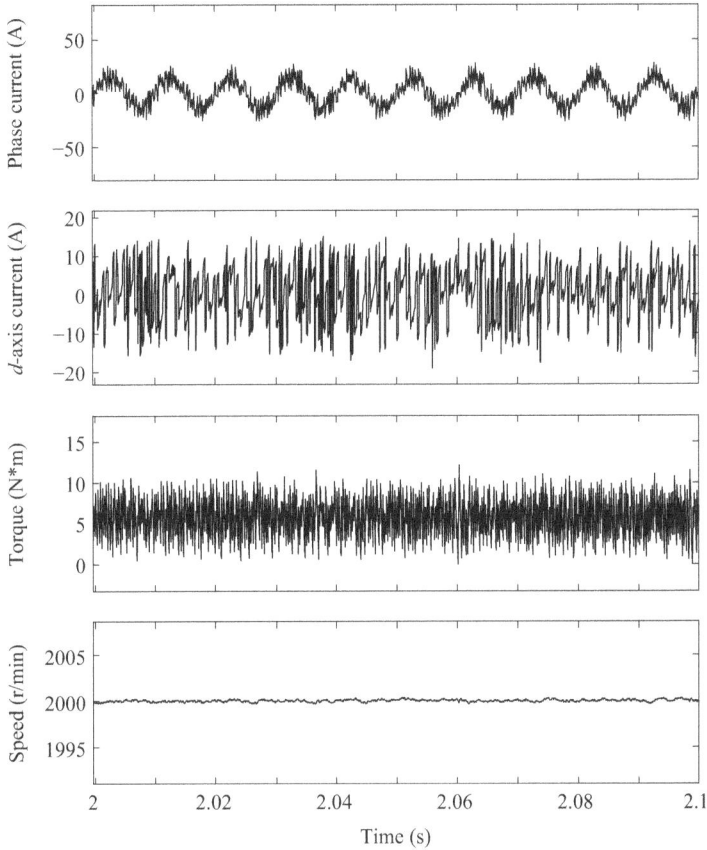

(b)

Figure 3.18 (Continued)

frequency is 50 kHz, the THD of the phase current in MPC is 14.98%. The control accuracy of the motor improves as the control frequency increases. When the control frequency is between 2.5 times and 3.3 times the PI control frequency, the control accuracy is similar to that of the classic PI control, indicating that the switching frequency of the MPC system at this point is close to 10 kHz (Figure 3.24).

From the above simulation results, it can be seen that due to the much lower switching frequency of the FCS-MPC compared to the control frequency, the control frequency can be set very high in experiments. When the control frequency is 25 kHz, under FCS-MPC control, the motor can quickly reach the reference speed with no overshoot. When the load torque changes, the electromagnetic torque responds rapidly without affecting the motor speed. Therefore, in the subsequent research on the motor control system, the control frequency of FCS-MPC is set to 25 kHz.

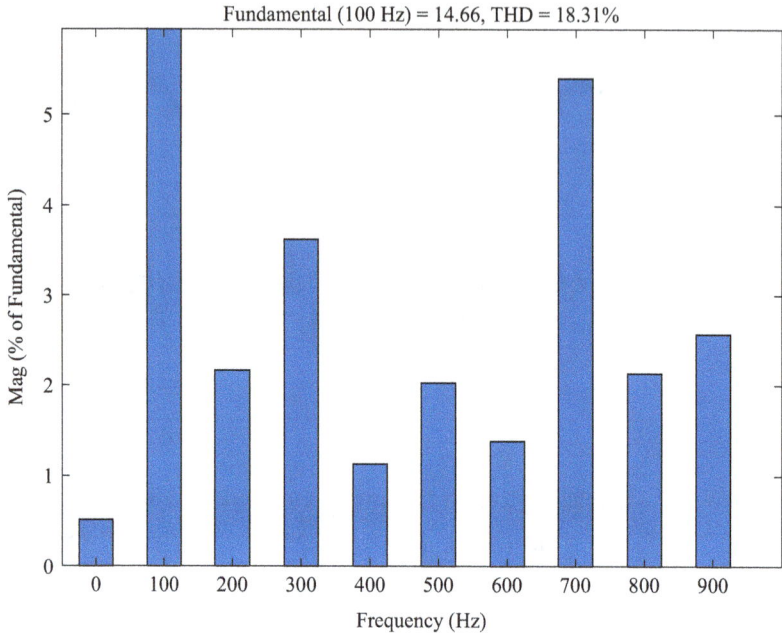

(a)

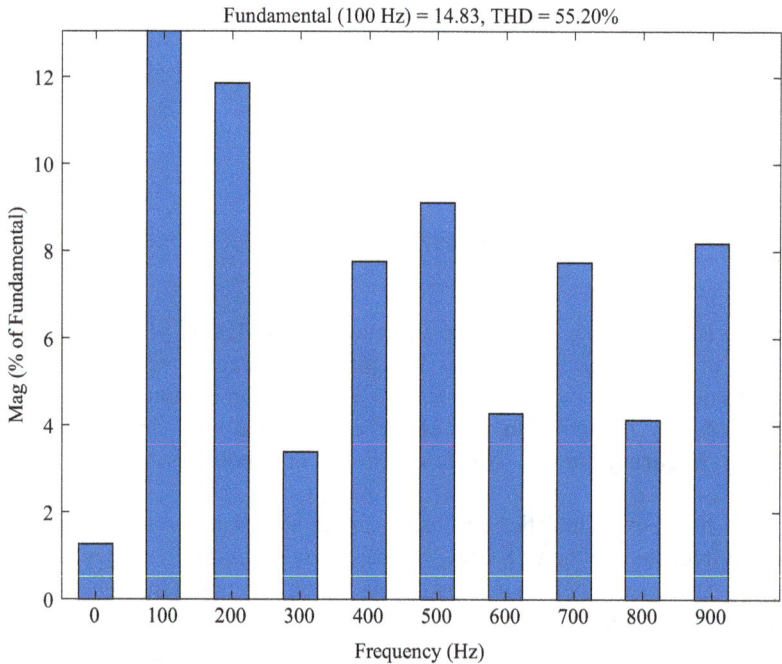

(b)

Figure 3.19   Phase current analysis at the control frequency of 10 kHz: (a) classic vector control and (b) model predictive control

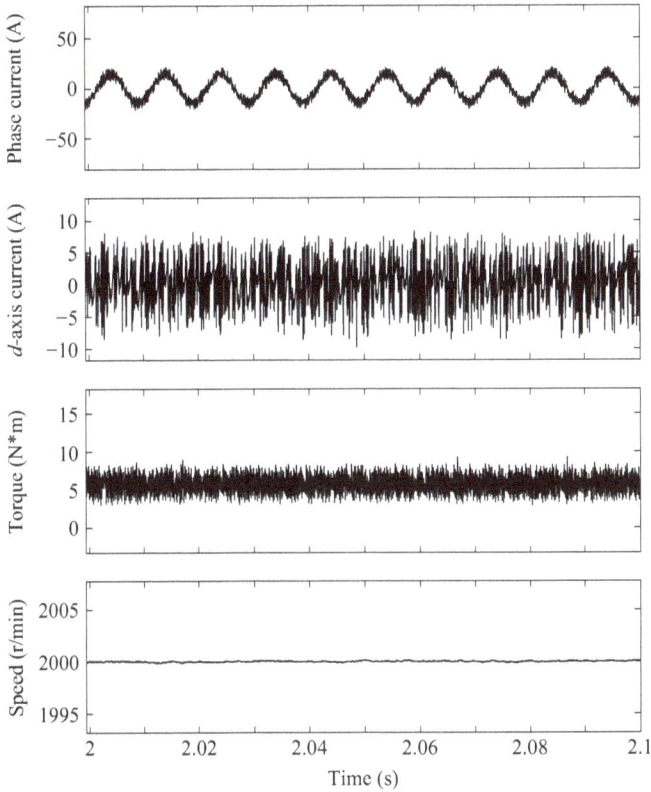

(a)

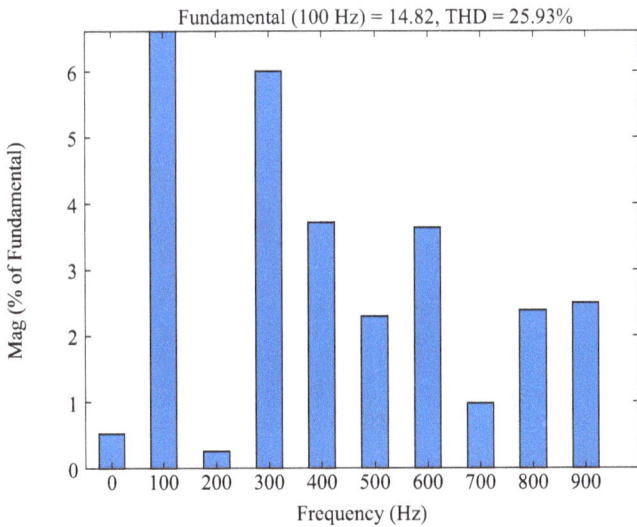

(b)

Figure 3.20   Simulation results of the control frequency 20 kHz: (a) classic vector
control and (b) model predictive control

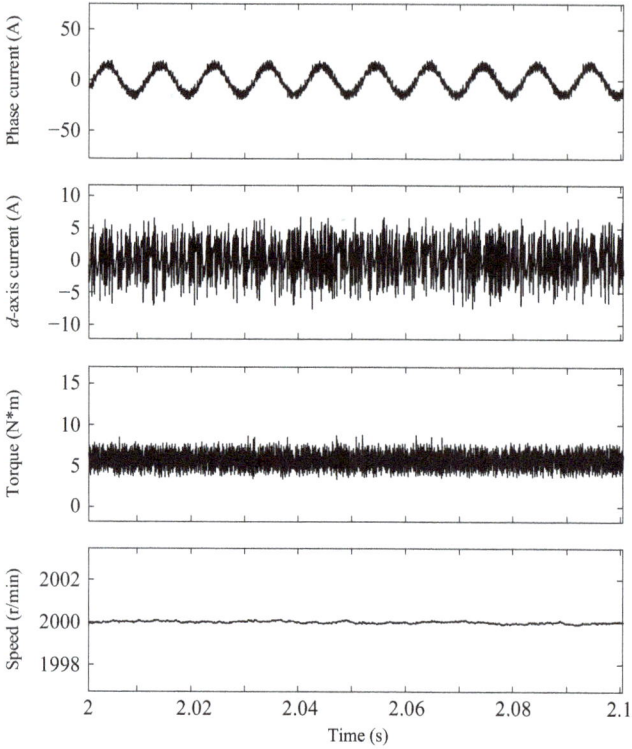

(a)

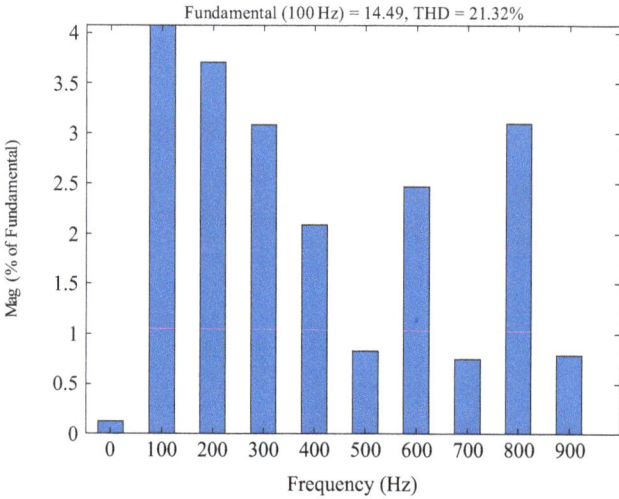

(b)

*Figure 3.21   Simulation results of the control frequency 25 kHz: (a) classic vector
control and (b) model predictive control*

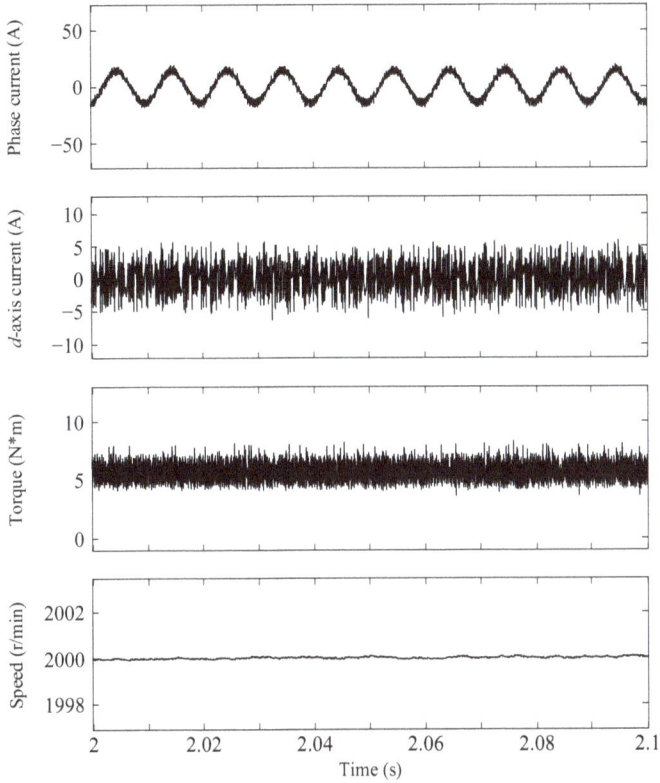

(a)

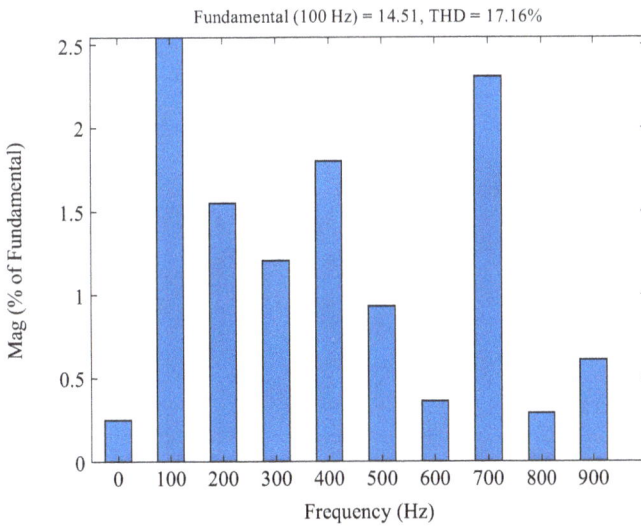

(b)

Figure 3.22 *Simulation results of the control frequency 33 kHz: (a) classic vector control and (b) model predictive control*

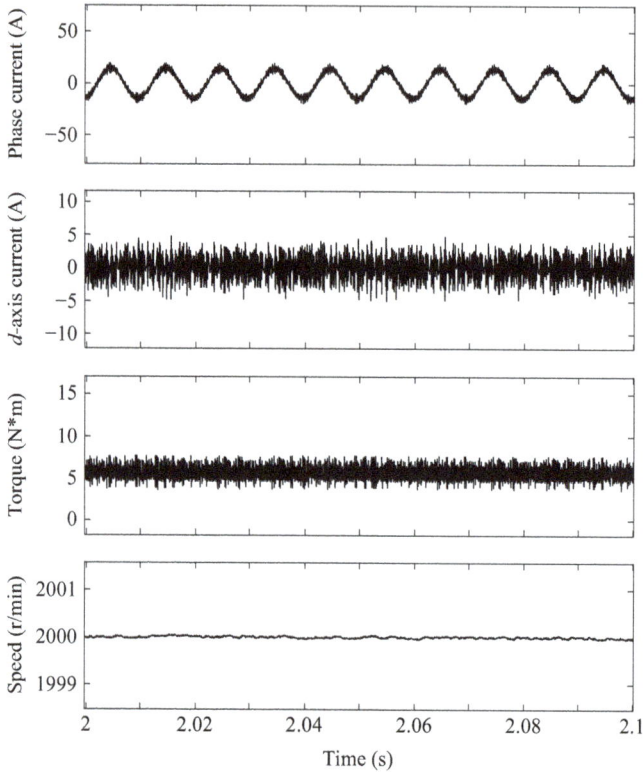

FCS-MPC simulation results

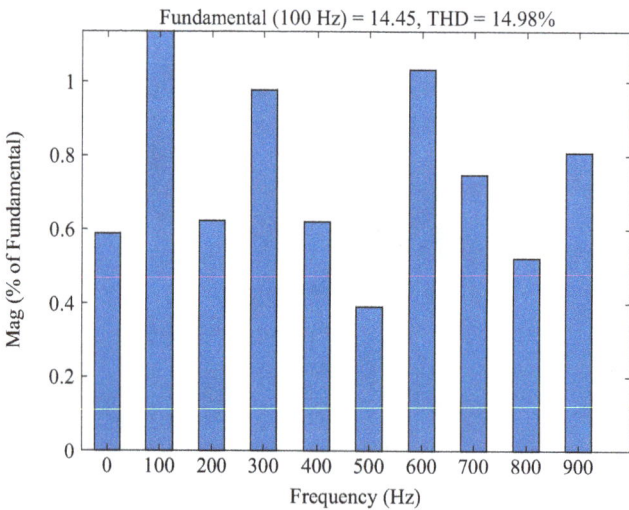

FCS-MPC phase current analysis

Figure 3.23   *Simulation results of the control frequency 33 kHz: (a) classic vector control and (b) model predictive control*

Moreover, whether PI vector control or MPC is used, rotor position information is crucial. Rotor position detection errors will prevent accurate rotation field positioning for vector control and may lead to mistakes in voltage vector selection for MPC. Therefore, the rotor position estimation algorithm plays a key role in the motor's control performance.

Additionally, although the MPC algorithm eliminates the modulation module, compared with the classic PI control scheme, it still requires extensive computation, resulting in longer computation times. If the delay between the actual detection and control drive is not considered, combined with the sensorless algorithm, the computation time will be further extended. This will reduce the performance of the PMSM system, thereby decreasing the optimal control effect.

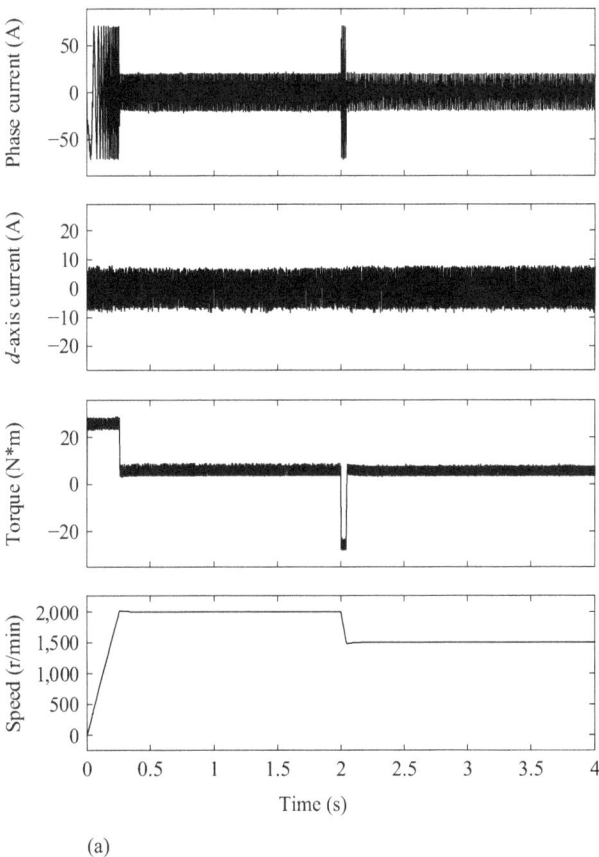

(a)

Figure 3.24   FCS-MPC control frequency 25 kHz: (a) classic vector control and (b) model predictive control

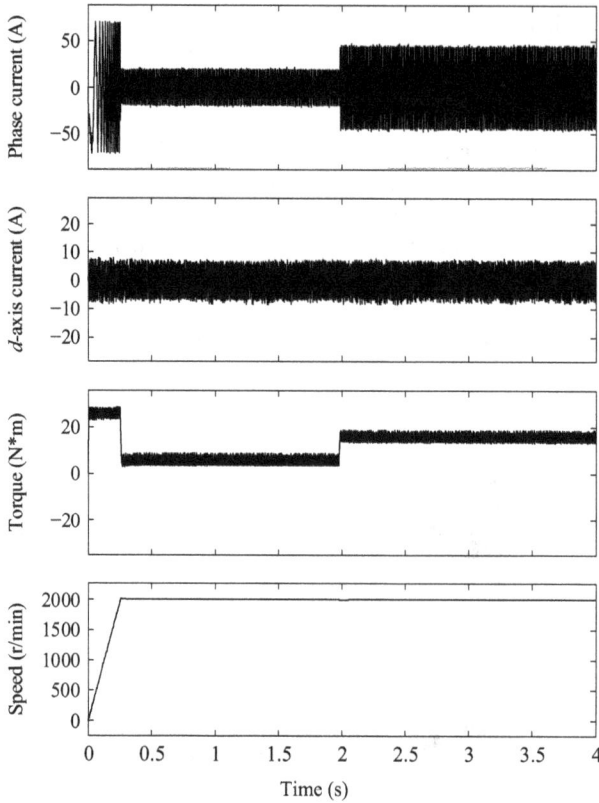

(b)

Figure 3.24   (Continued)

3.4.4   FCS-MPC delay issue solution method

Based on the PMSM FCS-MPC, the sensorless control method includes the following steps:

1. State Detection: Use the traditional sensor-based detection to measure the motor current $i_a(k)$, $i_b(k)$, $i_c(k)$, etc.
2. Position Estimation: Based on the equation, the rotor position is estimated through the calculation of the position $\theta(k)$ and the motor speed $\omega_m(k)$.
3. Coordinate Transformation: Based on $\theta(k)$ the motor currents $i_a(k)$, $i_b(k)$, $i_c(k)$ are transformed into the dq-axis currents $i_d(k)$, $i_q(k)$.
4. Current Prediction: The current $i_{dpn}(t_{k+1})$, $i_{qpn}(t_{k+1})$ at the next sampling time t_{k+1} are estimated for all possible voltage vectors using the discrete model of the motor.
5. Evaluation: Evaluate the total cost function for all predicted currents.

6. Voltage Error: Choose the voltage vector that minimizes the error in the cost function.
7. Switching State: Based on the estimated state of the motor, update the control according to the selected voltage vector.

For simplicity, we implemented the current prediction using a simplified method. The calculation time is optimized (using the formula t_d for time estimation). However, as the FCS-MPC is a real-time control method, it does not involve the delay process in the traditional control approach. Steps (4) and (5) of the calculation will introduce delay t_d. Compared to Figure 3.25 (a), Figure 3.25 (b) shows that between t_{k-1} and t_k, taking into account the delay, the selected voltage is applied to the motor at $t_{k-1} + t_{dk-1}$, resulting in the current not being controlled as intended, and a large deviation after the end of the switching cycle T_s. Unless compensatory measures are taken, this phenomenon cannot repair itself afterward.

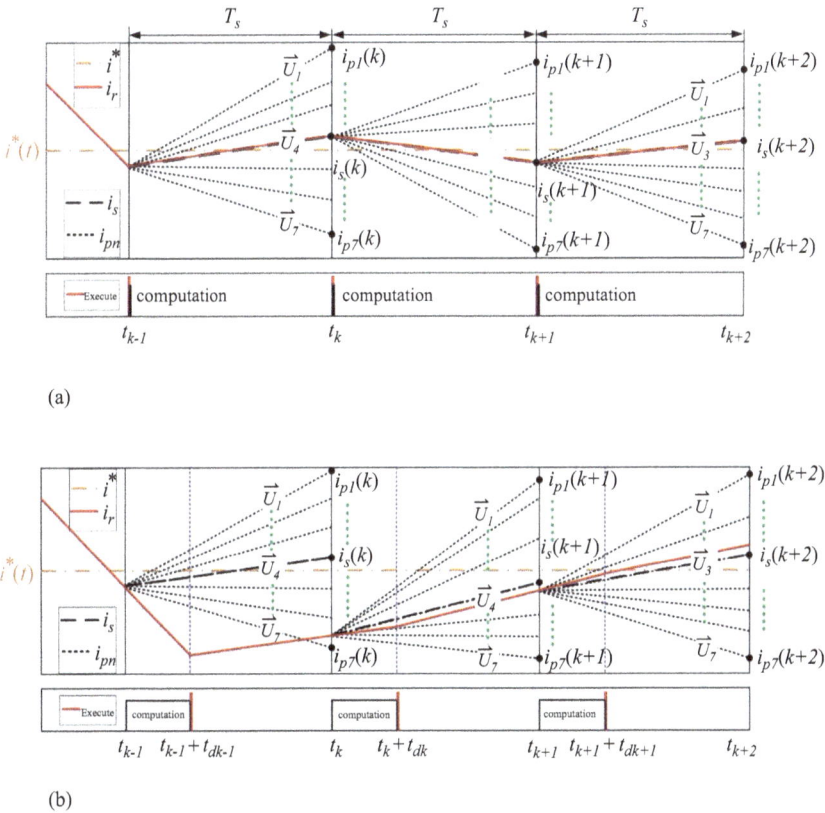

(a)

(b)

Figure 3.25 Implementation of the finite-set model predictive current control algorithm: (a) the ideal case of ignoring delay and (b) considering the actual situation of delay

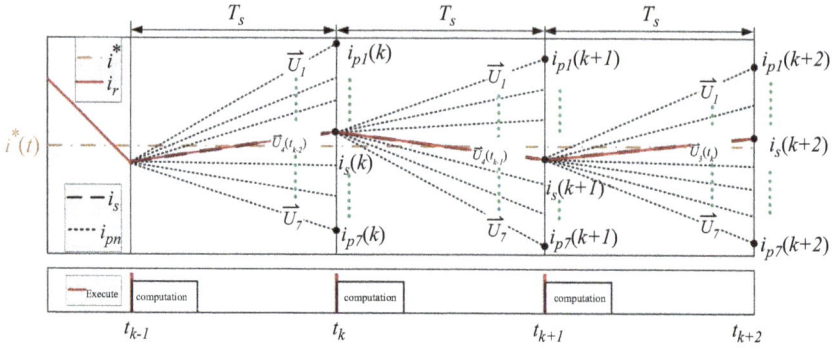

Figure 3.26   Traditional TSP compensation method

Traditional two-step prediction (TSP) compensation strategy of operation as shown in Figure 3.26, t_{k-1} cycle with t_k, for example, illustrates this process. A simple way to compensate for this delay is to consider the computation time and apply the selected switching state after the next sampling instant. In this way, the control algorithm is modified as follows [7]:

1. State detection: Use sensors to detect phase current $i_a(k)$, $i_b(k)$, $i_c(k)$, and other physical quantities in real time.
2. Execute switch: Apply the optimal switch state calculated at t_{k-1} time to the system at t_k.
3. Position estimation: Rotor position $\theta(k)$ and motor speed $\omega_m(k)$ are calculated according to the positionless control algorithm.
4. Coordinate system transformation: According to $\theta(k)$, the measured values $i_a(k)$, $i_b(k)$, $i_c(k)$ are converted into dq-axis current $i_d(k)$, $i_q(k)$.
5. State calculation: Under the control of the switch state, calculate the state value of t_{k+1} time with the state value of t_k time.
6. Current prediction: For all possible voltage vectors, the current $i_{dpn}(t_{k+2})$, $i_{qpn}(t_{k+2})$ at the sampling time t_{k+2} are estimated by using the discrete model of the motor.
7. Evaluation: All predicted currents are substituted into the cost function f_g successively.
8. Voltage vector: Select the voltage vector that minimizes the cost function.

As shown in Figure 3.26, with t_k, continue to apply the previously calculated switching state and measure the system state. Then, the voltage vector is used to estimate the current $i_s(k+1)$ of t_{k+1} at the next sampling instant. In addition, taking $i_s(k+1)$ as the actual current at t_{k+1}, we estimate the current at t_{k+1} for all voltage vectors $(i_{pn}(k+2))$, and use the cost function to select the optimal voltage vector that produces the smallest current deviation (the current is $i'_s(k+1)$). Finally, the switching state that generates the selected voltage is applied.

The new delay compensation method as shown in Figure 3.27, can be divided into two parts: time delay calculation and compensation control.

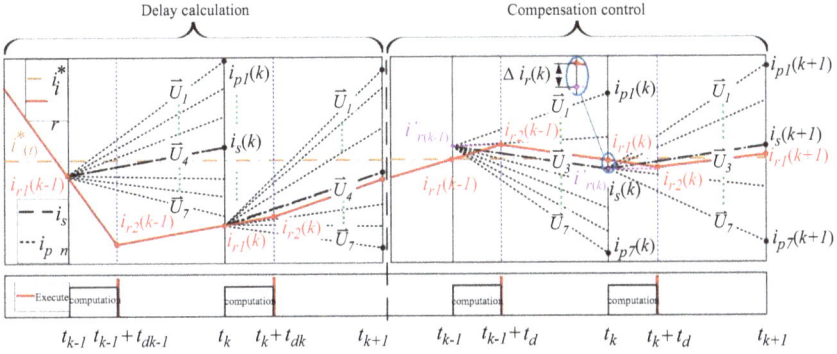

Figure 3.27 New prediction compensation method

3.4.4.1 Delay time calculation

Since the total computation amount does not vary significantly within each cycle, it is reasonable to assume that the calculation delay in the control process is the same, i.e., $t_d = t_{dk-1} = t_{dk}$. To estimate the delay time, the above FCS-MPC algorithm is implemented without any delay compensation. However, in each switching cycle, there are two sampling points: one is similar to stage (1), which occurs at the beginning of the cycle, and the other occurs after the voltage selection ends but before the switching state is applied. In Figure 3.27, the sampled currents during the periods $t_{k-1}-t_k$ and t_k-t_{k+1} are $i_{r1}(k-1)$, $i_{r2}(k-1)$, $i_{r1}(k)$, $i_{r2}(k)$ respectively. Therefore, the calculated delay time is:

$$t_d = \frac{|i_{r2}(k) - i_{r1}(k)|}{|i_{r2}(k) - i_{r1}(k)|} \cdot T_s \qquad (3.28)$$

3.4.4.2 Compensation control

After estimating t_d, the compensation method shown in Figure 3.27 is executed at t_k by performing the following steps:

1. State Measurement: Use sensors to detect the phase currents, rotor position, and speed of the motor in real time.
2. Position Estimation: Calculate the rotor position $\theta(k)$ and motor speed $\omega_m(k)$ based on the sensorless control algorithm.
3. Coordinate Transformation: Use $\theta(k)$ to transform the measured values $i_a(k)$, $i_b(k)$, $i_c(k)$ into dq-axis currents $i_d(k)$, $i_q(k)$.
4. Compensation: Predict the current change $\Delta i_r(k)$ at t_d using the following equation:

$$\Delta i_r(k) = \frac{i_{r1}(k) - i_{r2}(k-1)}{T_s - t_d} \cdot t_d \qquad (3.29)$$

where, $i_{r2}(k\text{-}1)$ is the sampled current when the voltage selection ends in the subsequent time interval, and by adding $\Delta i_r(k)$, we compensate for the sampled current $i_{r1}(k)$:

$$i'_r(k) = i_{r_1}(k) + \Delta i_r(k) \tag{3.30}$$

5. Current Prediction: Use the compensated current to estimate the current at the next sampling instant t_{k+1} for all possible voltage vectors.
6. Evaluation: Substitute all predicted currents into the cost function one by one.
7. Voltage Selection: Determine the optimal voltage vector.
8. Resampling: Repeat steps (1) and (3) to detect the actual current $i_{r2}(k)$.
9. Switching State Application: Select the corresponding switching state and apply it to the drive control.

Theoretically, the actual current at $t_{k+1}+t_d$ is $i_s(k+1)$, generated by the optimal switching state between t_k and t_{k+1}. This means that the method adheres to the optimal control rule.

This section also compares the proposed compensation algorithm based on delay time prediction through simulations. The simulation setup is as follows: the motor speed is stabilized at 2,000 rpm, and the load is set to 5 Nm. The motor position is estimated by a second-order sliding mode observer. Before 1.5 s, the motor is controlled using the FCS-MPCC algorithm without delay compensation. During this period, the delay time t_d is calculated. Between 1.5 s and 3.0 s, the system is controlled by the FCS-MPCC algorithm with traditional TSP compensation. After 3.0 s, a control strategy based on the new compensation algorithm is applied to the motor. Once the system reaches a steady state, the performance of the three control strategies is compared. Figures 4.4, 4.5, and 4.6 show the results of phase current, q-axis current, and position estimation error from 0–1.5 s, 1.5–3.0 s, and after 3.0 s, respectively.

When no delay compensation is applied, the motor current fluctuates significantly, with the phase current fluctuation amplitude reaching up to 40 A. The q-axis current fluctuates in the range of -8 A to 47 A, with a fluctuation size of 55 A. Although the simulation shows that the motor can still operate normally, the current fluctuation leads to large electromagnetic torque fluctuations, reducing the steady-state control performance of the motor. Furthermore, due to the current fluctuation, the position estimation accuracy is also degraded. In the case of no delay compensation, the position estimation error ranges from -0.11 rad to 0.1 rad.

Compared to Figure 3.28, when the traditional TSP compensation algorithm is used, as shown in Figure 3.29, the motor's steady-state current fluctuations significantly decrease. The maximum amplitude of the phase current is reduced to only 30 A (a 25% decrease), and the q-axis current fluctuation is also reduced to 35 A, a decrease of 25.6%. Compared to the traditional FCS-MPC control strategy, applying TSP delay compensation greatly improves the motor's steady-state performance and reduces torque ripple. As the current ripple decreases, the position estimation accuracy improves as well. The position estimation error range is now reduced from -0.08 rad to 0.09 rad.

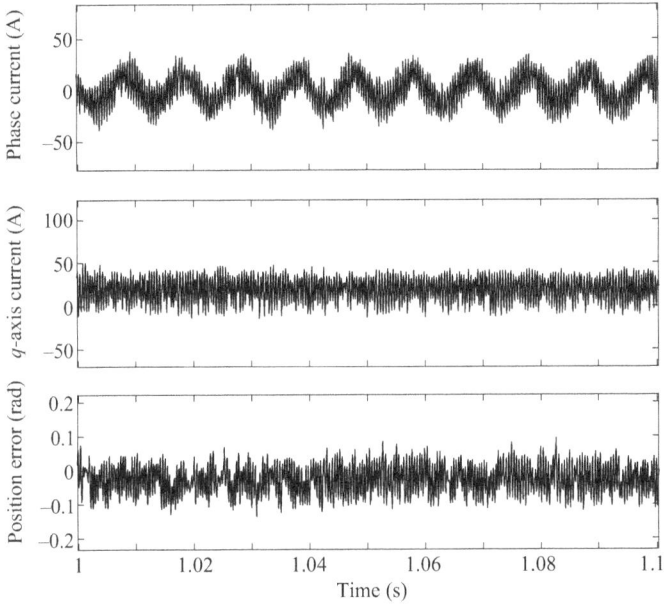

Figure 3.28   *The phase current, q-axis current, and position estimation error for*
FCS-MPC without delay compensation

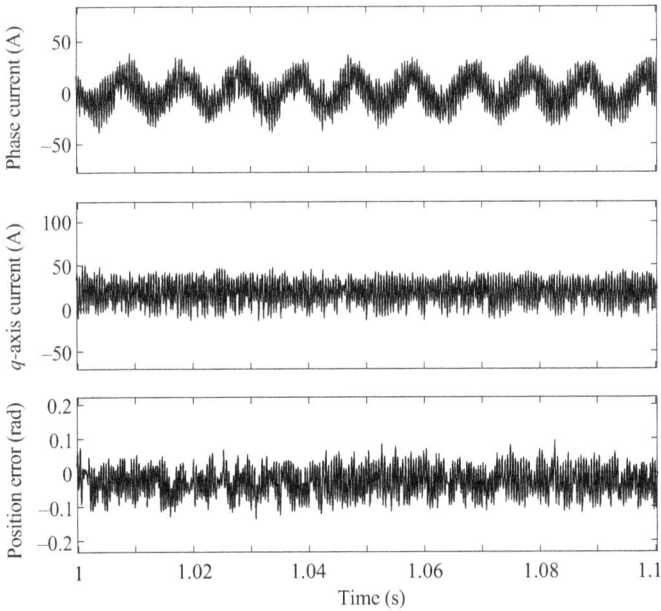

Figure 3.29   *The phase current, q-axis current, and position estimation error for*
FCS-MPC with TSP delay compensation

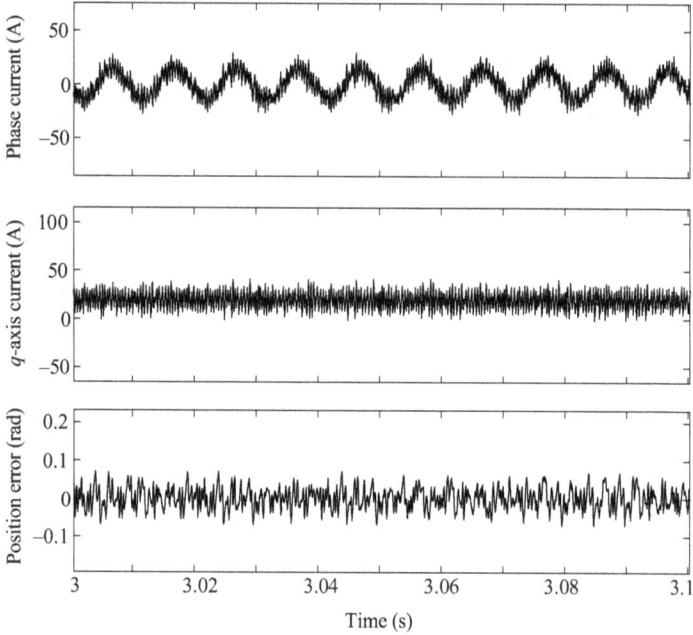

Figure 3.30   The phase current, q-axis current, and position estimation error for FCS-MPC with new delay compensation

Compared to the performance of FCS-MPCC control without delay compensation, the new compensation algorithm based on delay time estimation also significantly improves the system's control performance. From Figure 3.30, we can see that the amplitude of phase current fluctuation is about 28 A, the range of q-axis current fluctuation is 0–36 A (with a fluctuation size of 36 A), and the position estimation error fluctuation range is from −0.07 to 0.08 rad. These results are very similar to those based on TSP delay compensation, indicating that the new compensation strategy has the same compensating ability as the traditional method.

It is worth noting that the traditional compensation strategy uses the best voltage vector for the current cycle and applies it in the next cycle to the motor, while the new compensation algorithm selects and applies the optimal voltage within the current cycle. The two compensation algorithms are fundamentally different. Therefore, the compensation algorithm based on delay time estimation enriches the theoretical framework of MPC and has significant theoretical value.

Chapter 4

Non-position control approaches for permanent magnet actuators

Non-position control approaches have emerged as a transformative methodology in the field of permanent magnet actuators (PMAs), offering an innovative alternative to traditional systems that rely on physical position sensors. Often grouped under the broader category of sensorless control techniques, these approaches have gained significant traction due to their ability to simplify system design, reduce costs, and improve overall system robustness. As illustrated in Figure 2.5, conventional PMA systems typically require at least two position sensors: one to monitor the output position of the reducer and another to track the rotor position of the permanent magnet motor. While the output position sensor at the reducer remains essential for ensuring precise motion and control at the system level, the rotor position sensor can, in many scenarios, be substituted with advanced position estimation algorithms. This substitution marks the core of PMA non-position control, enabling the development of actuator systems that are both economically viable and operationally efficient.

The advancement of non-position control methodologies has been driven by a variety of factors, all of which underscore their potential to revolutionize PMA systems. Traditional rotor position sensors, such as encoders and resolvers, contribute significantly to the overall cost, size, and complexity of PMAs. These sensors also introduce several limitations, including susceptibility to environmental disturbances (e.g., temperature and humidity fluctuations), vulnerability to electromagnetic interference, and maintenance challenges arising from mechanical wear. Non-position control addresses these challenges by eliminating the dependence on physical rotor sensors, thereby reducing the system's physical footprint, cost, and maintenance requirements. Another critical advantage lies in the enhanced robustness and reliability of actuator systems. Non-position control mitigates the risk of performance degradation caused by sensor failure or inaccuracies, which are common in harsh operating environments. By leveraging sensorless techniques, PMA systems achieve greater fault tolerance, making them suitable for applications where reliability is paramount, such as industrial robotics, automotive systems, and aerospace applications.

The core functionality of non-position control revolves around the accurate estimation of rotor position and speed in real time using advanced computational algorithms. These algorithms rely on motor models that describe the physical and electrical behavior of PMAs, along with measurements of readily accessible electrical parameters, such as phase currents and terminal voltages. By processing this data, the rotor's position and velocity can be reconstructed with high precision, effectively replacing the need for physical sensors. Non-position control techniques can be broadly categorized into two main approaches, each tailored to specific operating conditions:

1. Low-Speed Range Methods: At low speeds, where back electromotive force-based methods struggle, alternative techniques such as model reference adaptive systems (MRAS), observer-based approaches, and high-frequency signal injection are employed. These methods leverage different aspects of the motor's electrical characteristics to estimate rotor position and speed with high accuracy.
2. High-Speed Range Methods: These methods are particularly effective for high-speed operation. They exploit the relationship between rotor motion and the induced back electromotive force (EMF) in the motor windings. By analyzing the back-EMF signal, the rotor's position and speed can be inferred. However, these methods are less effective at low speeds, where the back-EMF signal becomes too weak to be reliably detected.

4.1   Introduction to traditional non-position control methods

4.1.1   *Rotor initial position detection*

The detection of the initial rotor position in PMAs is crucial for the successful and smooth startup of the motor. Incorrect detection of the rotor's initial position can reduce the motor's startup torque and even cause motor reversal during startup. Rotor initial position detection falls under the low-speed range non-position control methods, which are typically based on the salient pole characteristics of the permanent magnet synchronous motor (PMSM). Methods for acquiring the initial rotor position include the rotor initial pre-positioning method, inductance parameter matrix calculation method, voltage pulse vector method, high-frequency pulsating voltage injection method, and rotor micro-motion-based method.

4.1.1.1   Rotor initial pre-positioning method

Before the motor startup, a constant voltage vector can be applied to the motor for a certain duration to rotate the rotor in a predetermined direction under the action of electromagnetic torque. This pre-positioning method does not require complex analysis or calculations, making it very simple and easy to implement in practical applications. After applying the constant voltage vector to the motor, a synthesized current vector is formed in the stator windings. The synthesized current vector generates electromagnetic torque, causing the rotor to begin rotating. The schematic diagram of the rotor's initial pre-positioning is shown in Figure 4.1. In the

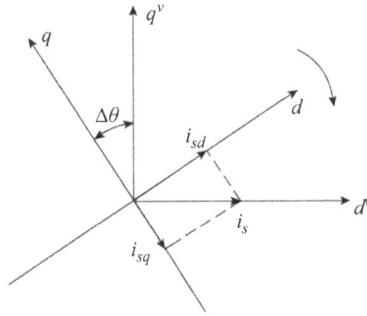

Figure 4.1   The schematic diagram of rotor initial pre-positioning

figure, the dq coordinate system represents the actual rotor position, while the $d^v q^v$ coordinate system represents the pre-positioning target position. The vector i_s is the stator-synthesized current vector. i_s is decomposed into the dq coordinate system representing the actual rotor position, resulting in current components i_{sd} and i_{sq}. In most cases, this pre-positioning method can successfully position the rotor to the target location; however, some dead zones still exist. For instance, if the rotor's actual position is either π or close to π, the component of the synthesized current vector along the rotor's q-axis is either 0 or close to 0, causing the rotor to fail to rotate to the target position. Although the pre-positioning method is simple and reliable, the time required to achieve rotor position alignment is relatively long. During the pre-positioning process, the rotor's position may change, potentially leading to rotor reversal.

4.1.1.2   Voltage pulse vector method

Interior permanent magnet synchronous motors (IPMSMs) have structural salient poles, while surface-mounted PMSMs (SPMSMs) have saturated salient poles. By applying a series of voltage pulse vectors with equal amplitude but opposite directions to the stator windings, rotor position information can be extracted from the current response values. Extracting position information is relatively more challenging for surface-mounted PMSMs, so here we will analyze it using the SPMSM as an example.

Figure 4.2 shows the current values obtained by applying forward and reverse voltage vectors at different rotor positions. In Figure 4.2, the actual rotor position is at $0°$ (electrical angle), where the magnetic pole position coincides with the α-phase winding. Here, I_{ap} represents the current response value obtained by applying a forward voltage vector, and I_{an} represents the current response value obtained by applying a reverse voltage vector. Figure 4.2 represent the current response waveforms when the rotor's actual position is $120°$ and $240°$, respectively. Figure 4.3 shows the difference between the current responses obtained with the forward voltage vector and the reverse voltage vector at different rotor positions.

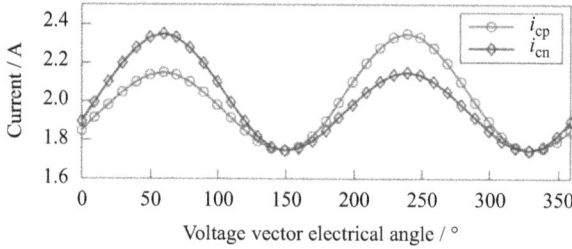

Figure 4.2   Current values corresponding to forward and reverse voltage vectors at different rotor positions

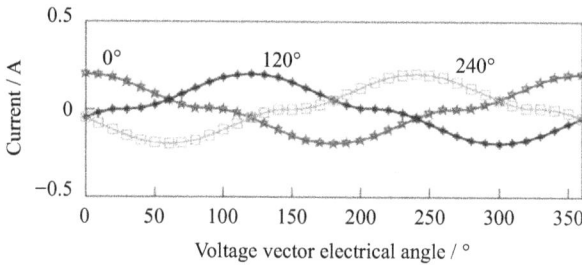

Figure 4.3   Current difference corresponding to forward and reverse voltage vectors at different rotor positions

From Figures 4.2 and 4.3, it can be seen that the current difference between the forward and reverse voltage vectors is the largest at the magnetic pole position. Assuming the rotor is at any position, by applying a series of forward and reverse voltage pulse vectors to the stator windings, the direction with the largest current difference indicates the location of the rotor magnetic pole. This method can achieve high position detection accuracy, but it also places high demands on the accuracy of the sampling circuit. Additionally, during the application of pulse voltage vectors, the inverter may perform multiple switching actions, which can interfere with the sampling circuit and affect sampling accuracy.

4.1.1.3   High-frequency pulsating voltage injection method

Both SPMSMs and IPMSMs can use the high-frequency pulsating voltage injection method to detect the rotor's initial position. A high-frequency pulsating voltage $U_h\cos(\omega_h t)$ is injected along the virtual synchronous coordinate system's d^v-axis. The rotor position is estimated by detecting the stator winding current response. Here, U_h is the injected voltage amplitude, and ω_h is the angular frequency of the injected voltage. The pulsating voltage injection is illustrated in Figure 4.4, where $\Delta\theta$ represents the error between the virtual synchronous coordinate system and the rotor's actual electrical angle. Its expression is written as:

$$\Delta\theta = \theta_{re} - \theta_{pe}^v \tag{4.1}$$

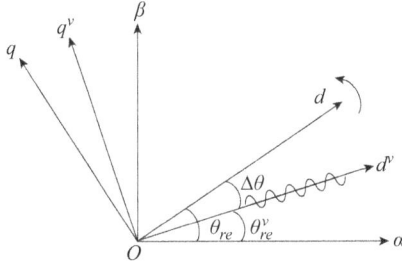

Figure 4.4   Virtual synchronous coordinate system and rotor position

Under high-frequency pulsating signal injection, the stationary motor stator windings can be equivalently represented as a pure inductance, and its mathematical model is:

$$\frac{d}{dt}\begin{bmatrix} i_{dh} \\ i_{qh} \end{bmatrix} = \begin{bmatrix} \dfrac{1}{L_{dh}} & 0 \\ 0 & \dfrac{1}{L_{qh}} \end{bmatrix} \begin{bmatrix} u_{dh} \\ u_{qh} \end{bmatrix} \tag{4.2}$$

where, L_{dh} and L_{qh} represent the stator winding inductance along the d-axis and q-axis under high-frequency signal excitation, respectively. Through a rotational transformation, the variables in the actual synchronous coordinate system of the motor are transformed into the virtual synchronous coordinate system. Further substitution of the high-frequency pulsating voltage excitation along the d^v-axis yields the high-frequency current response in the virtual synchronous coordinate system as:

$$\begin{bmatrix} i_{dh}^v \\ i_{qh}^v \end{bmatrix} = \frac{U_h \sin(\omega_h t)}{\omega_h(L_{sa}^2 - L_{sd}^2)} \begin{bmatrix} I_{sa} + I_{sd}\cos(2\Delta\theta) \\ L_{sd}\sin(2\Delta\theta) \end{bmatrix} \tag{4.3}$$

Here,

$$I_{sa} = \frac{I_{qh} + L_{dh}}{2}, I_{sd} = \frac{I_{qh} - L_{dh}}{2}.$$

The current response on the q^v-axis can be multiplied by the coefficient $2\sin(\omega_h t)$ to modulate the current response, thereby obtaining the magnitude of the current response on the q^v-axis in the form of a DC component. From this, it can be concluded that when the current response magnitude along the q^v-axis is zero, it indicates that the electrical angle error in rotor position tracking is either 0 or π. The rotor position can be determined, but the polarity of the rotor magnetic poles cannot yet be determined. When the rotor position is accurately estimated, the high-frequency current response magnitude along the d^v-axis is:

$$\left| i_{dh}^v \right| = \frac{U_h}{\omega_h(L_{sa}^2 - L_{sd}^2)}[L_{sa} + L_{sd}\cos(2\Delta\theta)] \approx \frac{U_h}{\omega_h L_{dh}} \tag{4.4}$$

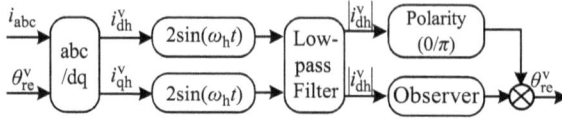

Figure 4.5   High-frequency pulsating voltage injection method for rotor initial position detection

The DC current present along the d^v-axis can affect the magnetic saturation level of the stator iron core, which in turn influences the d-axis inductance L_{dh}, and consequently impacts the high-frequency current response magnitude along the d^v-axis. Therefore, by observing the effect of the DC current along the d^v-axis on the high-frequency response current magnitude, the polarity of the rotor can be determined, i.e., whether the rotor position electrical angle error is 0 or π. The algorithm structure for calculating the rotor position and determining the magnetic pole direction based on the current response obtained from high-frequency pulsating voltage injection is shown in Figure 4.5. The high-frequency pulsating voltage injection method can accurately detect the rotor's initial position, but the algorithm is relatively complex and requires high sampling accuracy and computational capability.

4.1.1.4   Inductance parameter matrix calculation method

Based on the principle of dq-axis inductance differences, the inductance parameter matrix can be calculated by applying two linearly independent voltage vectors to the stator windings and measuring the corresponding transient current responses. This inductance parameter matrix contains rotor position information. When the motor is in a stationary state, its back-EMF is zero. In the two-phase stationary coordinate system, the mathematical model of the PMSM is:

$$\begin{bmatrix} u_\alpha \\ u_\beta \end{bmatrix} = R_s \begin{bmatrix} i_\alpha \\ i_\beta \end{bmatrix} + \begin{bmatrix} L_{11} & L_{12} \\ L_{21} & L_{22} \end{bmatrix} \frac{d}{dt} \begin{bmatrix} i_\alpha \\ i_\beta \end{bmatrix} \tag{4.5}$$

Here,

$$\begin{bmatrix} L_{11} & L_{12} \\ L_{21} & L_{22} \end{bmatrix} = \begin{bmatrix} L_1 + L_2 \cos(2\theta_{re}) & L_2 \sin(2\theta_{re}) \\ L_2 \sin(2\theta_{re}) & L_1 - L_2 \cos(2\theta_{re}) \end{bmatrix} \tag{4.6}$$

$$\begin{cases} L_1 = \dfrac{L_d + L_q}{2} \\ L_2 = \dfrac{L_d - L_q}{2} \end{cases} \tag{4.7}$$

where, L_1 represents the average inductance of L_d and L_q, while L_2 represents the half-difference inductance of L_d and L_q. Equation (4.6) is the inductance parameter matrix that contains rotor position information. By applying the voltage vectors $u_{\alpha 1}$, $u_{\beta 1}$ to the stator windings, the transient current responses $i_{\alpha 1}$, $i_{\beta 1}$ are obtained. Similarly, by applying the voltage vectors $u_{\alpha 2}$, $u_{\beta 2}$, the transient current responses $i_{\alpha 2}$, $i_{\beta 2}$ are obtained. The two applied voltage vectors are linearly independent, allowing for the calculation of the inductance parameter matrix. Based on this matrix, the rotor position expression can be further derived as:

$$\theta_{\text{re}} = \frac{1}{2} \arctan \frac{L_{12} + L_{21}}{L_{11} - L_{22}} \tag{4.8}$$

4.1.1.5 Rotor micro-motion-based method

A low-frequency rotating voltage vector is applied to the stator windings, generating a pulsating torque that induces micro-motion in the rotor. This micro-motion affects the motor's current response, allowing rotor position information to be extracted from the current response. When the frequency of the injected rotating voltage is relatively low, the phase difference between the q-axis current response and the motor's speed pulsation is minimal, meaning that the phase of the q-axis current response can reflect the rotor's position information. This method presents two challenges:

- During detection, the rotor undergoes micro-motion. After detecting the rotor position and stopping the application of the rotating voltage, the rotor may stop at any nearby position due to inertia, leading to significant detection errors. The only way to minimize stopping errors is by gradually reducing the injected voltage.
- The motor oscillates continuously around its equilibrium position during detection, which not only affects the motor's lifespan but also limits the applicability of this method in certain scenarios.

4.1.2 Low-speed range control methods

At low speeds, the rotor salient pole position is typically tracked using the high-frequency signal injection method. However, the high-frequency injection method has drawbacks, including computational complexity and increased motor losses. Additionally, open-loop speed operation control is an important operating strategy for PMSMs, offering simplicity, reliability, and low cost [8].

4.1.2.1 Open-loop voltage-to-frequency (V/F) control

Although open-loop control does not achieve closed-loop speed control, it is still a type of sensorless control technique and serves as a cost-effective control method for PMSMs. Open-loop voltage-to-frequency (V/F) control does not rely on rotor position detection and features a relatively simple algorithm. The principle of open-loop V/F control is to maintain a constant V/F ratio, ensuring constant flux for motor speed regulation. However, PMSMs operating under open-loop V/F control are prone to instability and oscillations, as this method does not directly control the electromagnetic torque, affecting the motor's load-handling capability and disturbance rejection ability. Small variations in voltage amplitude within V/F control can lead to significant changes in stator current amplitude, potentially causing motor overcurrent. The V/F control curve of a specific PMSM is shown in Figure 4.6, demonstrating a narrow adjustment range and requiring repeated parameter tuning.

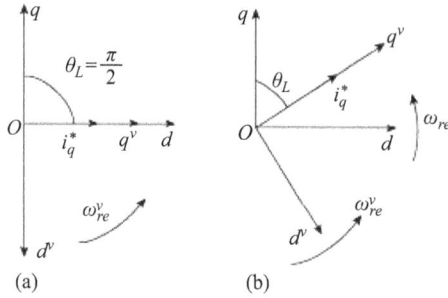

Figure 4.6   I/F control scheme of PMSM: (a) startup and (b) running

4.1.2.2   Current-to-frequency (I/F) operation control

Current-to-frequency (I/F) control stabilizes the current through a closed-loop current regulation process during variable-frequency speed control, preventing excessive or insufficient current. I/F control allows direct control of torque current, enhancing the matching capability between motor output torque and load torque while avoiding low-frequency oscillations during operation. In V/F control, both speed and current are open-loop, whereas in I/F control, speed remains open-loop while current is closed-loop. The closed-loop current regulation not only provides better control over electromagnetic torque but also prevents overcurrent protection activation in the stator current.

Figure 4.6 illustrates the startup and operation principles of PMSMs under I/F control. At the initial startup, the virtual synchronous coordinate system lags behind the synchronous coordinate system by an electrical angle of $\pi/2$. As the virtual synchronous coordinate system rotates, the motor rotor begins to follow this rotation. The motor's output electromagnetic torque is determined by the phase difference between the two coordinate systems, and the motor's electromagnetic torque is given by:

$$T_e \frac{3}{2} n_p \psi_f i_q^* \cos \theta_L \tag{4.9}$$

The torque balance equation of the PMSMs during acceleration is:

$$T_e - T_L = \frac{J}{n_p} \cdot \frac{d\omega_{re}}{dt} = \frac{J}{n_p} \cdot \frac{d^2\theta_{re}}{dt} \tag{4.10}$$

where, i_q^* is the current given value; J is the moment of inertia; T_L is the load torque.

4.1.2.3   High-frequency rotating voltage signal injection method

A high-frequency rotating voltage signal is injected into the two-phase stationary coordinate system, as illustrated in Figure 4.7. The rotor position signal can be resolved from the high-frequency current response signal.

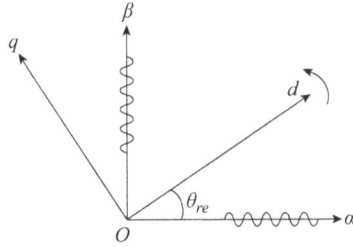

Figure 4.7 The schematic diagram of high-frequency rotating voltage signal injection into a stationary coordinate system

The selected high-frequency rotating signal is:

$$\begin{bmatrix} v_{\alpha h} \\ v_{\beta h} \end{bmatrix} = U_h \begin{bmatrix} \cos(\omega_h t) \\ \sin(\omega_h t) \end{bmatrix} \tag{4.11}$$

The high-frequency current response signal of the stator windings can be expressed in terms of the rotor position and the d-axis and q-axis inductances of the stator windings under high-frequency signal excitation as:

$$\begin{bmatrix} i_{\alpha h} \\ i_{\beta h} \end{bmatrix} = \begin{bmatrix} I_p \cos(\omega_h t - \pi/2) \\ I_p \sin(\omega_h t - \pi/2) \end{bmatrix} + \begin{bmatrix} I_n \cos(-\omega_h t + 2\theta_{re} + \theta_m + \pi/2) \\ I_n \sin(-\omega_h t + \theta_m + \pi/2) \end{bmatrix} \tag{4.12}$$

where, I_p and I_n represent the positive-sequence and negative-sequence components of the high-frequency current response in the stator windings, respectively. The expressions can be written as:

$$\begin{cases} I_p = \dfrac{U_h}{\omega_h} \cdot \dfrac{L_{sa}}{L_{dh}L_{qh} - L_{dqh}^2} \\[3mm] I_n = \dfrac{U_h}{\omega_h} \cdot \dfrac{\sqrt{L_{sd}^2 + L_{dqh}^2}}{L_{dh}L_{qh} - L_{dqh}^2} \end{cases} \tag{4.13}$$

$$\theta_m = \arctan \frac{-L_{dqh}}{L_{sd}} \tag{4.14}$$

During motor operation, when pulsating voltage injection is performed, the influence of cross-saturation effects must be considered. The cross-saturation effect is related to the motor load current and is represented in the mathematical model by the mutual inductance L_{dqh}, which is generally small. θ_m is the compensation angle caused by the cross-saturation effect. It can be seen that the negative-sequence component of the high-frequency current response in the stator windings contains rotor position information. To extract this negative-sequence current component, the fundamental frequency component and the positive-sequence high-frequency current component in the stator winding current must be filtered out.

4.1.2.4   Carrier frequency component method

This is a carrier frequency component method sensorless control strategy for IPMSMs. The method uses the high-frequency carrier signal of the inverter as the high-frequency excitation and calculates the rotor position by analyzing the carrier component in the high-frequency current. This method takes advantage of the motor's structural salient pole effect, making it unsuitable for SPMSMs.

Under carrier frequency excitation, the stator resistance voltage drop can be neglected, and the stator's carrier frequency component current response is:

$$
\begin{bmatrix} i_{\alpha h} \\ i_{\beta h} \end{bmatrix} = \begin{bmatrix} L_1 - L_2 \cos(2\theta_{re}) & -L_2 \sin(2\theta_{re}) \\ -L_2 \sin(2\theta_{re}) & L_1 + L_2 \cos(2\theta_{re}) \end{bmatrix} \cdot \frac{1}{L_1^2 - L_2^2} \cdot \begin{bmatrix} \int u_{\alpha h} dt \\ \int u_{\beta h} dt \end{bmatrix} \tag{4.15}
$$

Additionally, a new two-phase stationary coordinate system $\gamma\delta$-axis is established, where the two axes are perpendicular to each other and lead the $\alpha\beta$-axis system by an electrical angle of $\pi/4$. The carrier frequency component's current response is:

$$
\begin{bmatrix} i_{\gamma h} \\ i_{\delta h} \end{bmatrix} = \begin{bmatrix} L_1 - L_2 \sin(2\theta_{re}) & L_2 \cos(2\theta_{re}) \\ L_2 \cos(2\theta_{re}) & L_1 + L_2 \sin(2\theta_{re}) \end{bmatrix} \cdot \frac{1}{L_1^2 - L_2^2} \cdot \begin{bmatrix} \int u_{\gamma h} dt \\ \int u_{\delta h} dt \end{bmatrix} \tag{4.16}
$$

Based on (4.15) and (4.16), the rotor position can be calculated as shown in (4.17) where the four variables represent the peak values of the carrier frequency component currents on the $\alpha\beta$- and $\gamma\delta$- axes.

$$
\theta_{re} = \frac{1}{2} \arctan \frac{|i_{\gamma h}|_{peak}^2 - |i_{\delta h}|_{peak}^2}{|i_{\alpha h}|_{peak}^2 - |i_{\beta h}|_{peak}^2} \tag{4.17}
$$

4.1.2.5   High-frequency pulsating voltage signal injection method

The high-frequency pulsating voltage signal injection method was introduced in Section 4.1. It involves injecting a high-frequency pulsating voltage signal $U_h \cos(\omega_h t)$ into the virtual synchronous coordinate system's d^v-axis. The high-frequency pulsating signal injection method during the motor's low-speed operation is similar to the injection method when the rotor is stationary, but with two key differences:

- During low-speed operation, the high-frequency response current needs to be extracted using a high-pass filter.
- During operation, the influence of cross-saturation effects must be considered.

The mathematical model of high-frequency injection for the motor's low-speed operation, considering cross-saturation effects, can be expressed as:

$$\frac{d}{dt}\begin{bmatrix} i_{dh} \\ i_{qh} \end{bmatrix} = \begin{bmatrix} \dfrac{L_{qh}}{L_{dh}L_{qh} - L_{dqh}^2} & \dfrac{-L_{dqh}}{L_{dh}L_{qh} - L_{dqh}^2} \\ \dfrac{-L_{dqh}}{L_{dh}L_{qh} - L_{dqh}^2} & \dfrac{L_{dh}}{L_{dh}L_{qh} - L_{dqh}^2} \end{bmatrix}\begin{bmatrix} u_{dh} \\ u_{qh} \end{bmatrix} \tag{4.18}$$

The mutual inductance L_{dqh} is generally small. If the motor's operating speed is very low, close to zero speed, the mutual inductance L_{dqh} can be approximated as zero. When this mutual inductance is approximated as zero, and considering the cross-saturation effect, the high-frequency current response of the stator windings is:

$$\begin{bmatrix} i_{dh}^v \\ i_{qh}^v \end{bmatrix} = \sin{(\omega_h t)}\begin{bmatrix} I_p + I_n\cos{(2\Delta\theta + \theta_m)} \\ I_n\sin{(2\Delta\theta + \theta_m)} \end{bmatrix} \tag{4.19}$$

where I_p and I_n are shown in (4.19). Similar to the high-frequency rotating voltage injection method, to accurately estimate the rotor position, compensation for the phase difference θ_m caused by the cross-saturation effect is required. After determining the rotor position, the polarity of the magnetic poles must also be assessed.

4.1.3 High-speed range control methods

In PMSMs operating at high speeds, the stator winding current and back-EMF are relatively easy to detect, and the rotor position can be estimated based on the motor's fundamental waveform model. These types of algorithms include the third-harmonic back-EMF method, observer-based algorithms, and flux-linkage estimation methods, among others.

4.1.3.1 Third-harmonic back-EMF method

Due to the magnetic saturation effect in PMSMs or the design of third harmonic components in the PM flux-linkage, there is a significant third harmonic back-EMF component in the stator windings. Figure 4.8 illustrates the measurement circuit for the third-harmonic back-EMF. By analyzing the third-harmonic back-EMF shown in Figure 4.8, the motor's rotor position can be estimated. The rotor position can be estimated by detecting the zero-crossing points of the third-harmonic back-EMF or by analyzing the continuous signal of the third-harmonic

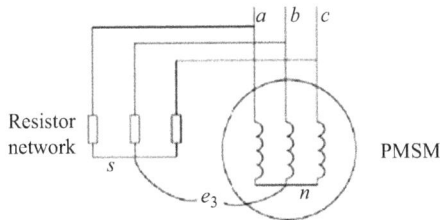

Figure 4.8 The measurement circuit for the third-harmonic back-EMF

back-EMF and introducing a Phase Locked Loop (PLL) for rotor position estimation. Both methods can also be applied simultaneously to improve the accuracy of rotor position detection.

4.1.3.2   Model reference adaptive observer

A model reference adaptive observer can be used to estimate rotor speed and rotor
position. The motor itself can serve as the reference model, while the stator voltage
equation, with the estimated rotor speed as a parameter, acts as the adjustable model.
The adaptive observer constructs an error term based on the difference between the
actual current from the reference model and the estimated current from the
adjustable model. This error term is then input into a Proportional-Integral (PI) controller to estimate the speed. When the adaptive observer accurately estimates the
speed, the error between the reference model and the adjustable model tends to be zero.

The adjustable model selects the voltage equation of the PMSM to observe the
stator flux-linkage and stator current, as follows:

$$\begin{cases} \dfrac{d\hat{\psi}_{sd}}{dt} = u_{sd} - R_s\hat{i}_{sd} + \hat{\omega}_{re}\hat{\psi}_{sq} + \lambda\left(i_{sd} - \hat{i}_{sd}\right) \\[2mm] \dfrac{d\hat{\psi}_{sq}}{dt} = u_{sq} - R_s\hat{i}_{sq} - \hat{\omega}_{re}\hat{\psi}_{sd} + \lambda\left(i_{sq} - \hat{i}_{sq}\right) \end{cases} \tag{4.20}$$

$$\begin{cases} \hat{i}_{sd} = \dfrac{\hat{\psi}_{sd} - \psi_f}{L_d} \\[2mm] \hat{i}_{sq} = \dfrac{\hat{\psi}_{sq}}{L_q} \end{cases} \tag{4.21}$$

where $\hat{i}$ represents the observed current, and λ denotes the observer gain. An error
term $i - \hat{i}$ is constructed by comparing the observed current with the sampled current
to reflect the rotor speed estimation error. The rotor speed is then obtained through a
PI adaptive. By integrating the rotor speed, the rotor position can be determined.

4.1.3.3   Extended Kalman filter

Kalman filtering is an efficient recursive filter used to estimate the state of a
dynamic system. The original Kalman filter was only applicable to linear systems. The Extended Kalman Filter (EKF) can linearize nonlinear systems and
then apply Kalman filtering. EKF has optimization and adaptive capabilities,
allowing it to suppress noise from measurements and disturbances, providing
good system state estimation capabilities. The EKF algorithm consists of three
stages: prediction, calculation of Kalman gain, and correction. When considering
system noise and measurement noise, the mathematical model of the nonlinear
system is:

$$\begin{cases} \dfrac{dx(t)}{dt} = f[x(t)] + Bu(t) + V(t) \\[2mm] y(t) = h[x(t)] + W(t) \end{cases} \tag{4.22}$$

where $x(t)$ is the state vector; $u(t)$ is the input vector; $y(t)$ is the output vector; $V(t)$ and $W(t)$ represent the system noise and measurement noise, respectively. Both $V(t)$ and $W(t)$ are zero-mean white noise, with covariance matrices Q and R, respectively. The two vectors used in the EKF are defined as:

$$F_k = \left. \frac{\partial f(x)}{\partial x} \right|_{x=x_k^g} \tag{4.23}$$

$$H_{k+1} = \left. \frac{\partial h(x)}{\partial x} \right|_{x=x_{k+1}^c} \tag{4.24}$$

First, the definitions of the three types of data are explained: actual values, estimated values, and predicted values. The actual value is the value obtained through system sampling; the predicted value is the uncorrected computed value, denoted by a superscript c; the estimated value is the corrected computed value, denoted by a superscript g. The EKF algorithm can be expressed as:

$$x_{k+1}^c = x_k^g + \left| f\left(x_k^g\right) + Bu_k \right| T_s \tag{4.25}$$

$$P_{k+1}^c = P_k^g + \left[F_k P_k^g + P_k^g F_k^T \right] T_s + Q \tag{4.26}$$

$$K_{k+1} = P_{k+1}^c H_{k+1}^T \left(H_{k+1} P_{k+1}^c H_{k+1}^T + R \right)^{-1} \tag{4.27}$$

$$x_{k+1}^g = x_{k+1}^c + K_{k+1} \left[y_{k+1} - h\left(x_{k+1}^c\right) \right] \tag{4.28}$$

$$P_{k+1}^g = P_{k+1}^c - K_{k+1} H_{k+1} P_{k+1}^c \tag{4.29}$$

where T_s represents the control cycle, and P_k denotes the covariance matrix for the calculation error.

4.1.4 Full-speed range control methods

Full-Speed Range Sensorless control for PMSM typically combines low-speed and medium-high-speed control algorithms, leveraging the advantages of different methods at distinct speed stages to achieve complementary performance. Current research refers to such strategies as hybrid or composite control algorithms.

4.1.4.1 I/F control startup with flux-linkage estimation method

After determining the rotor's initial position, the I/F control drives the motor startup. Once the motor reaches a certain speed and the stator current, back-EMF becomes detectable, the system switches to a sensorless control strategy based on flux-linkage estimation. By carefully managing the transition from open-loop to closed-loop control, oscillations during switching are mitigated. This method has been successfully applied to machines, ensuring smooth switching between strategies.

4.1.4.2   High-frequency rotating voltage injection combined with flux-linkage estimation method

To achieve full-speed-range operation for PMSM, the literature integrates high-frequency rotating voltage injection and flux linkage estimation. Both methods calculate the rotor's PM flux, and the results are weighted to determine the final output. At low speeds, the high-frequency injection dominates, while flux linkage estimation takes precedence at medium-high speeds. In addition to the above two, others include combination methods: (1) High-Frequency Square-Wave Injection with Model Reference Adaptive System; (2) I/F control startup with Model Reference Adaptive System, etc.

4.2   Implementation of low-speed range methods

The high-frequency rotating sinusoidal injection method injects high-frequency voltage into the stationary coordinate system. Compared to the high-frequency voltage signal injection method, which uses high-frequency voltage signals for estimating the d-axis, the sinusoidal injection method has a more stable injection structure. However, system delays can still cause phase shifts in the induced currents, leading to inaccurate rotor position estimation. To address this, scholars have optimized the method of injecting high-frequency signals into the stationary coordinate system, maintaining the advantages of stable injection structures while improving position estimation accuracy.

Tang Qipeng proposed a bidirectional rotating high-frequency carrier signal injection strategy. They used four independent equations to obtain the frequency components of the induced currents, which weakened the impact of stator resistance variation, system delays, and inverter nonlinearity. However, the signal separation process requires a large number of low-pass filters (LPF). Wang Gaolin proposed a high-frequency orthogonal square wave injection method, which discretized and summed the product of the induced current and demodulation function to obtain intermediate variables containing rotor position and error angle, then used phase compensation methods to eliminate delay errors. Xu Peilin improved the demodulation process of the high-frequency rotating injection method by using a regulator to track the error phase and use it as the demodulation function's phase angle, suppressing position errors caused by system delays. However, the additional error tracking regulator increased system complexity. Toso and others proposed an elliptical fitting method based on recursive least squares (RLS) algorithms to identify rotor motion trajectories, making the demodulation algorithm unaffected by signal processing delay effects, but the multiple orthogonal-triangular (QR) decompositions and Givens matrix calculations are relatively cumbersome.

Existing methods use LPF or error compensation algorithms to eliminate the effects of system delay. However, the use of LPF reduces the system's bandwidth, while error compensation algorithms increase the complexity of the estimation algorithm. Therefore, this chapter proposes a position estimation method based on the decoupling of the positive and negative sequence components of the induced current. This method can simultaneously eliminate the effects of the high-pass filter and system delays, with the position estimation system using only high-frequency induced currents as input, without the need for demodulation functions and LPF. First, since both the positive and negative sequence components contain phase errors generated by the high-pass filter and system delay, and their amplitudes are the same but with opposite signs, the positive and negative sequence components are chosen as state variables to demodulate the rotor position without error terms. Next, the induced currents at adjacent moments are chosen as computational variables to decouple the state variables. The decoupling process of state variables is transformed into the solution of a linear non-homogeneous system of equations, and the necessary and sufficient conditions for a unique solution for state variables are derived for both sine and square wave injection waveforms. This proposed method is applicable to both the rotating sine injection method and the orthogonal square wave injection method. Since the rotating injection method requires the motor's inductance to have structural salient pole effects, this method is not suitable for SPMSMs that completely lack structural salient pole effects.

4.2.1   High-frequency rotating voltage signal injection principle

This section presents the voltage models for high-frequency rotating sine injection and high-frequency orthogonal square wave injection. The basic voltage balance equation for PMSM in the two-phase stationary coordinate system is:

$$u_{\alpha\beta} = R_s i_{\alpha\beta} + p(L_{\alpha\beta} i_{\alpha\beta}) + j\omega_r \psi_f e^{j\theta_e} \qquad (4.30)$$

$$L_{\alpha\beta} = \begin{bmatrix} L_0 + L_1 \cos 2\theta_e & L_1 \sin 2\theta_e \\ L_1 \sin 2\theta_e & L_0 - L_1 \cos 2\theta_e \end{bmatrix} \qquad (4.31)$$

where, $u_{\alpha\beta} = [u_\alpha, u_\beta]T$ is the voltage vector in a two-phase stationary coordinate system; $i\alpha\beta = [i\alpha, i\beta]T$ is the motor current vector in a two-phase stationary coordinate system; $L_0 = \frac{L_d + L_q}{2}, L_1 = \frac{L_d - L_q}{2}$; ϑ_e is the motor rotor position electrical angle.

For the motor's low-speed operation, the injection frequency is much higher than the current fundamental frequency, so the stator resistance Rs and back-EMF terms can be neglected. Additionally, the second term in (4.30) is the derivative of the high-frequency current signal, and its value is much larger than the other two terms. Therefore, the voltage balance equation under high-frequency injection can be simplified as:

$$u_{\alpha\beta h} = p(L_{\alpha\beta} i_{\alpha\beta h}) \qquad (4.32)$$

where, $u_{\alpha\beta h} = [u_{\alpha h}, u_{\beta h}]^T$ is the high-frequency voltage vector in a two-phase stationary coordinate system; $i_{\alpha\beta h} = [i_{\alpha h}, i_{\beta h}]^T$ is the high-frequency induced current vector in a two-phase stationary coordinate system.

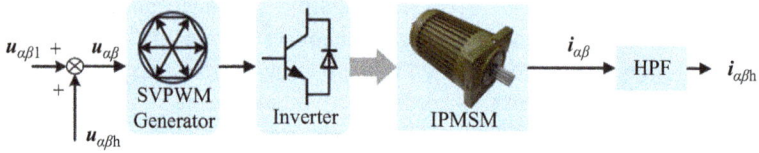

Figure 4.9   The principle of high-frequency voltage signal injection

After the high-frequency voltage and fundamental voltage are superimposed, the input for Space Vector Pulse Width Modulation (SVPWM) is obtained, and then it is fed into the motor's stator windings through the inverter. Finally, high-frequency induced current can be obtained through current sampling and a high-pass filter. The basic principle block diagram for high-frequency voltage injection and induced current extraction is shown in Figure 4.9.

When high-frequency rotating sinusoidal voltage vectors are injected in the $\alpha\beta$ coordinate system, high-frequency induced current can be obtained. The high-frequency rotating sinusoidal voltage injection (HRSI) vector can be expressed as:

$$u_{\alpha\beta h} = V_h e^{j\omega_h t} \tag{4.33}$$

The induced current vector obtained from HRSI is

$$i_{\alpha\beta h} = i^p_{\alpha\beta h} + i^n_{\alpha\beta h} = I_p e^{j(\omega_h t - \pi/2)} + I_n e^{j(-\omega_h t + \pi/2 + 2\theta_e)} \tag{4.34}$$

where $i^p_{\alpha\beta h}$ is the positive-sequence current vector corresponding to the injected sinusoidal wave; $i^n_{\alpha\beta h}$ is the negative-sequence current vector corresponding to the injected sinusoidal; $I_p = \dfrac{V_h L}{\omega_h (L_0^2 - L_1^2)}$, $I_n = -\dfrac{V_h \Delta L}{\omega_h (L_0^2 - L_1^2)}$; I_p is the positive sequence component of induced current; I_n is the negative sequence component of induced current.

When high-frequency orthogonal square-wave voltage vectors are injected in the $\alpha\beta$ coordinate system, high-frequency induced current can also be obtained. The high-frequency orthogonal square-wave voltage injection (HOSI) vector can be expressed as:

$$u_{\alpha\beta h squ} = \frac{4V_h}{\pi}\left[\sum_{m=1}^{+\infty} \frac{(-1)^m}{2} \frac{1}{m-1} e^{j(-1)^m(2m-1)\omega_h t}\right] \tag{4.35}$$

where, $u_{\alpha\beta h\_squ}$ is a high-frequency square wave voltage vector.

$$i_{\alpha\beta h squ} = i^p_{\alpha\beta h_s qu} + i^n_{\alpha\beta h_s qu}$$

$$= \sum_{m=1}^{+\infty} \frac{4I_p}{\pi(2m-1)^2} e^{j[(-1)^m(2m-1)\omega_h t - 0.5\pi]}$$

$$+ \sum_{m=1}^{+\infty} \frac{4I_n}{\pi(2m-1)^2} e^{j[(-1)^{m+1}(2m-1)\omega_h t + 2\theta_e + 0.5\pi]} \tag{4.36}$$

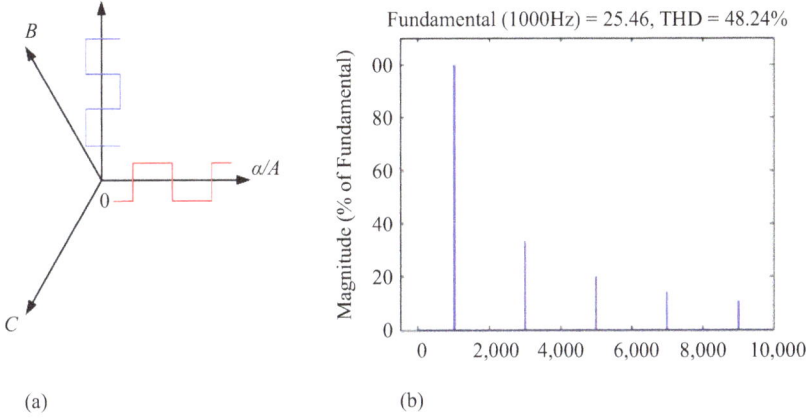

Figure 4.10 *Square wave injection form and FFT analysis: (a) square wave injection form and (b) FFT of α-axis square wave voltage*

where, $i_{\alpha\beta h\_squ}$ is the induced current vector corresponding to the injected square wave; $i^p_{\alpha\beta h\_squ}$ is the positive-sequence current vector corresponding to the injected square wave; $i^n_{\alpha\beta h\_squ}$ is the negative-sequence current vector corresponding to the injected square wave.

From (4.34) and (4.36), it can be seen that both HRSI and HOSI can extract the rotor position angle from the phase of the current negative-sequence component.

Figure 4.10(a) shows the injection form of HOSI in the stationary coordinate system. Figure 4.10(b) presents a Fast Fourier Transform (FFT) analysis of the α-axis square wave voltage with an injection frequency of 1 kHz. It can be observed that the square wave contains harmonic components of different orders.

4.3   Phase error analysis of induced current in rotating injection method

In motor drive systems, it is difficult to obtain high-frequency current with an ideal phase as shown in (4.34) and (4.36). On one hand, due to system delay effects, there is a phase delay between the ideal induced current and the actual induced current. On the other hand, separating high-frequency and low-frequency currents requires a digital high-pass filter, which can also introduce phase errors in the current. Therefore, the actual high-frequency induced current will experience phase shifts, and this section will analyze this in detail.

4.3.1   Impact of system delay on sampled current

System delay causes a phase delay δ_d between the ideal induced current and the actual induced current. Equation (4.37) below gives the induced current under

HRSI with the effect of system delay:

$$i_{\alpha\beta h} = i_{\alpha\beta h_d}^p + i_{\alpha\beta h_d}^n = I_p e^{j(\omega_h t - \delta_d - \pi/2)} + I_n e^{j(-\omega_h t + \delta_d + 2\theta_e + \pi/2)} \quad (4.37)$$

The delay angle δ_d causes position estimation errors in the HRSI when un-compensated.

4.3.2   Impact of high-pass filter on high-frequency current phase

As shown in Figure 4.9, in vector control, there is a basic voltage vector $u_{\alpha\beta l}$ that maintains the motor's rotation, which, together with the injected high-frequency voltage vector $u_{\alpha\beta h}$, is sent to the motor's stator windings. Therefore, the stator current of the motor contains two components: one corresponding to the basic voltage vector, the low-frequency current component $i_{\alpha\beta l}$, and the other being the high-frequency induced current $i_{\alpha\beta h}$, as shown in equations. A discrete digital high-pass filter (DHPF) is a commonly used method for extracting high-frequency signals. DHPF can also introduce phase errors in the induced current, which can affect the position estimation results. This section will analyze the phase errors introduced by DHPF to the high-frequency induced current [9].

DHPF is divided into recursive filters (Infinite Impulse Response, IIR) and non-recursive filters (Finite Impulse Response, FIR), depending on the presence of feedback structures with output signals. The transfer functions of IIR and FIR filters are represented by the following equations:

$$H_{IIR}(z) = \frac{\sum_{k=0}^{M} b_k z^{-k}}{1 - \sum_{k=0}^{N} a_k z^{-k}} \quad (4.38)$$

$$H_{FIR}(z) = \sum_{k=0}^{N-1} h_k z^{-k} \quad (4.39)$$

To analyze the phase characteristics of the DHPF, different parameters and types of DHPFs were designed, and the amplitude-phase characteristic curves were plotted separately. Based on the amplitude-phase characteristic curves of the different DHPFs, the phase shift caused by the DHPF on the high-frequency induced current was analyzed, which in turn allows for the analysis of position estimation errors caused by the current phase shift. The designed FIR and IIR filters both have a filter order of 20, with a cutoff frequency of 400 Hz and sampling frequencies of 8 kHz, 6 kHz, and 4 kHz. Figure 4.11 shows the amplitude-phase characteristic curves of the designed DHPFs under these parameters, assuming the injected high-frequency sinusoidal signal has a frequency of 1 kHz.

From Figure 4.11, when the sampling frequency is 8 kHz, the phase shift for FIR is -1.58 rad, and for IIR is 4.96 rad; when the sampling frequency is 6 kHz, the phase shift for FIR is -4.22 rad, and for IIR is 4.76 rad; when the sampling

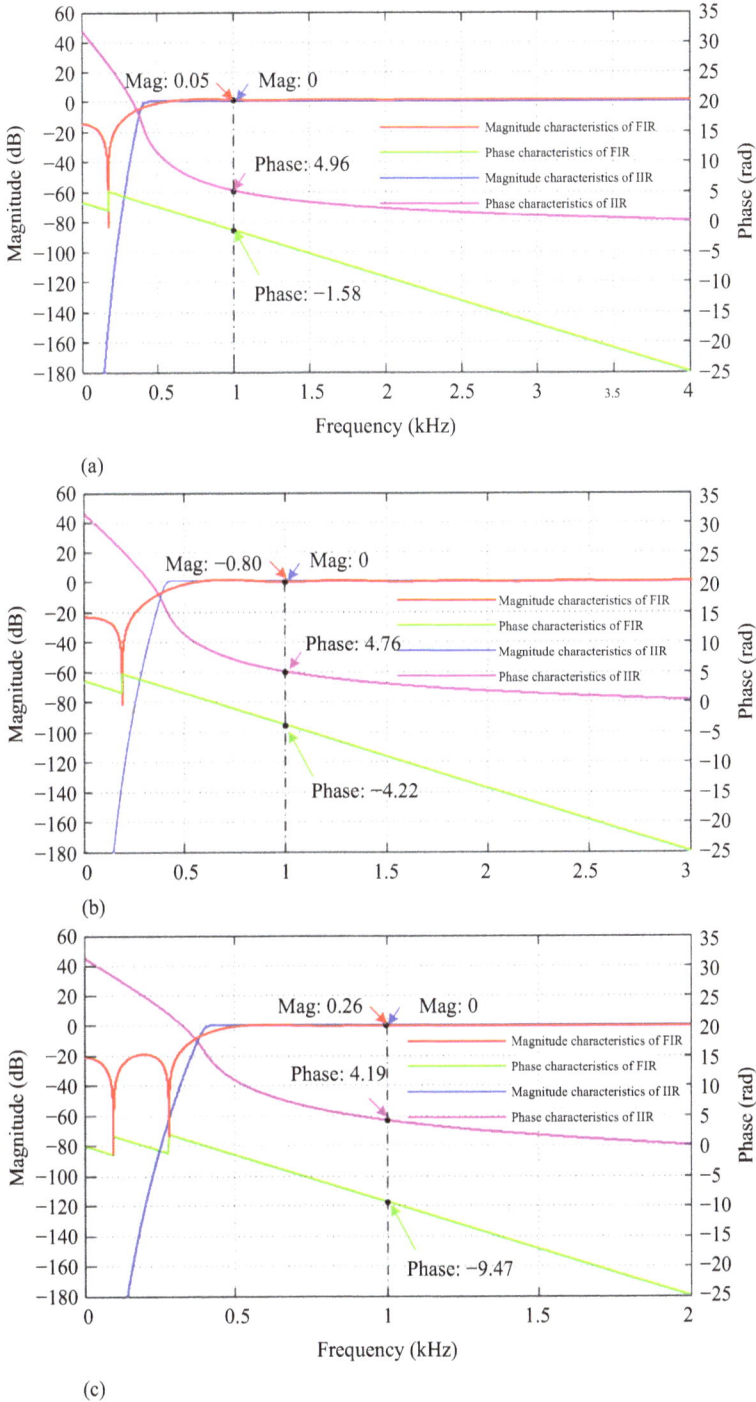

Figure 4.11   *Magnitude-phase characteristic curves of FIR and IIR DHPFs at different sampling frequencies: (a) sampling frequency of 8 kHz, (b) sampling frequency of 6 kHz, and (c) sampling frequency of 4 kHz.*

frequency is 4 kHz, the phase shift for FIR is −9.47 rad, and for IIR is 4.19 rad. It can be observed that regardless of the type of DHPF, the filtered signal will produce a phase lead angle δ_H, and δ_H will vary with changes in the injected signal frequency and sampling frequency. Moreover, for the FIR filter, δ_H shows a linear relationship with the injected frequency, while for the IIR filter, δ_H does not exhibit a linear relationship with the injected frequency.

In conclusion, the injected and sampling frequencies, filter type, and parameters all influence δ_H. Therefore, while offline compensation methods can be applied to offset delays caused by mathematical filters, offline compensation needs to consider the filter and injection waveform parameters, which increases the complexity and difficulty of compensation. From the magnitude curve in Figure 4.11, it can be seen that the magnitude response of these filters results in an output-to-input ratio greater than 0.99 (Mag ≤ 0.8 dB), so the magnitude attenuation caused by the filters can be ignored.

Considering the phase shift caused by system delay and the DHPF, (4.37) can be re-derived, yielding the actual high-frequency induced current as:

$$i_{\alpha\beta h} = i_{\alpha\beta h}^{p} + i_{\alpha\beta h}^{n} = I_p e^{j(\omega_h t - \delta_d + \delta_H - \pi/2)} + I_n e^{j(-\omega_h t + \delta_d - \delta_H + 2\theta_e + \pi/2)} \qquad (4.40)$$

From Eq. (4.40), it can be observed that due to the effects of system delay and digital filtering, the phase of both the positive and negative sequence components of the induced current includes additional phase shifts δ_d and δ_H. Since the rotor position angle is also embedded in the phase of the induced current, it is typically estimated by extracting the phase of the induced current. However, if the phase shifts caused by system delay and digital filtering are not considered and traditional phase extraction methods are used, the estimated rotor position will be inaccurate. This inaccuracy will further affect the performance of vector control.

A phase transformation operation was performed using a demodulation function $e^{j\omega_h t}$ on the induced current $i_{\alpha\beta h}$, as shown in the equation, yielding the state variable $i_{\alpha\beta 1}$. Then, an LPF was applied to remove the positive-sequence components from $i_{\alpha\beta 1}$, resulting in the orthogonal vector $i_{\alpha\beta 0}$. This demodulation method is referred to as the synchronous-axis rotating filtering method.

$$\begin{aligned}
i_{\alpha\beta 0} &= \mathrm{LPF}\left(\begin{bmatrix} \cos(\omega_h t) & -\sin(\omega_h t) \\ \sin(\omega_h t) & \cos(\omega_h t) \end{bmatrix} \begin{bmatrix} i_{\alpha h} \\ i_{\beta h} \end{bmatrix}\right) \\
&= \mathrm{LPF}(i_{\alpha\beta 1}) \\
&= \mathrm{LPF}\left(I_p e^{j\left(2\omega_h t - \delta_d + \delta_H - \frac{\pi}{2}\right)} + I_n e^{j\left(\delta_d - \delta_H + 2\theta_e + \frac{\pi}{2}\right)}\right) = I_n e^{j\left(\delta_d - \delta_H + 2\theta_e + \frac{\pi}{2}\right)}
\end{aligned}$$

$$\hspace{12cm} (4.41)$$

where, $i_{\alpha\beta 1} = I_p e^{j(2\omega_h t - \delta_d + \delta_H - \pi/2)} + I_n e^{j(\delta_d - \delta_H + 2\theta_e + \pi/2)}$ is the state variable obtained by the phase transformation operation.

The actual components involved in the phase transformation are $\sin(\omega_h t)$ and $\cos(\omega_h t)$, which are commonly used as demodulation functions in traditional demodulation algorithms. After obtaining the orthogonal vector $i_{\alpha\beta 0}$, the

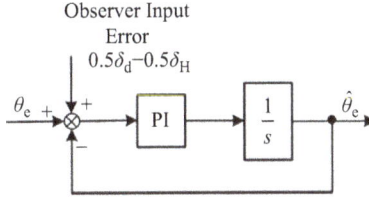

Figure 4.12 Equivalent position tracker block diagram obtained by the synchronous-axis rotating filtering method

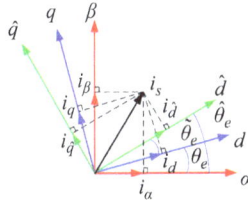

Figure 4.13 Influence of position error on field orientation

heterodyne method and a second-order PLL are used to estimate the rotor position as:

$$\hat{\theta}_e = \theta_e + 0.5\delta_d - 0.5\delta_H \tag{4.42}$$

The equivalent position tracker block diagram obtained by the synchronous-axis rotating filtering method is shown in Figure 4.12, with the rotor position error given as:

$$\tilde{\theta}_e = \theta_e - \hat{\theta}_e = 0.5\delta_H - 0.5\delta_d \tag{4.43}$$

From (4.43), it can be seen that if the phase of the induced current contains delay components δ_d and δ_H, using the above demodulation method will result in an error of $0.5\delta_H-0.5\delta_d$ between the actual rotor position and the estimated rotor position. An inaccurate rotor position estimation will affect the performance of vector control. The specific analysis is as follows.

The *dq*-axis reference coordinate system is constructed based on the estimated rotor position, as shown in Figure 4.13. Both flux and torque are related to the *dq*-axis current. When the estimated rotor position is accurate, the *dq*-axis current approximately equals the $\hat{d}\hat{q}$-axis current and flux and torque can be decoupled and controlled in the actual *dq* coordinate system. However, when there is an error $\tilde{\theta}_e$ between the estimated rotor position and the actual rotor position, the relationship between the $\hat{d}\hat{q}$-axis current and the *dq*-axis current can be expressed as:

$$\begin{bmatrix} i_d \\ i_q \end{bmatrix} = \begin{bmatrix} \cos\tilde{\theta}_e & \sin\tilde{\theta}_e \\ -\sin\tilde{\theta}_e & \cos\tilde{\theta}_e \end{bmatrix} \begin{bmatrix} \hat{i}_d \\ \hat{i}_q \end{bmatrix} \tag{4.44}$$

From the above equation, it can be seen that both $\hat{d}\hat{q}$-axis current $\hat{i}_d$ and $\hat{i}_q$ simultaneously affect the magnetic field current i_d and torque current i_q, causing coupling between flux linkage and torque.

If the estimated rotor position angle $\hat{\theta}_e$ obtained from the equation is used for Park and IPark transformations, it will lead to deviations in the dq-axis feedback current obtained through IPark transformation, as well as deviations in the $\alpha\beta$-axis output voltage generated through Park transformation. Inaccurate field orientation will degrade the system's dynamic performance, leading to current oscillations and other instability phenomena.

4.4 Position estimation method based on self-decoupling of induced current positive and negative sequence components

4.4.1 Basic principle of induced current positive and negative sequence self-decoupling

Given that the injected voltage is known, the input of the position estimation system is the high-frequency induced current, and the output is the rotor position. According to Eq. (4.40), when θ_e is taken as the independent variable and $i_{\alpha\beta h}$ as the dependent variable, $i_{\alpha\beta h}$ is an implicit function of θ_e. Since θ_e is embedded in the phase angle of $i_{\alpha\beta h}$, directly solving for it is challenging. Typically, intermediate variables such as computational variables and state variables are used to obtain it indirectly, as shown in Figure 4.14. In this figure, computational variables are known quantities, consisting of observable variables and demodulation functions. State variables are obtained after demodulation operations on computational variables, and the rotor position angle θ_e is derived from state variables. The correlation between state variables and θ_e affects both the computational efficiency and estimation accuracy of the estimation system.

Therefore, this section analyzes the shortcomings of existing methods and establishes new principles for selecting computational variables and state variables. The selected computational variables are the high-frequency induced current $i_{\alpha\beta h}$ and the demodulation function. The solved state variable is $i_{\alpha\beta 1}$, but $i_{\alpha\beta 1}$ contains not only the low-frequency negative sequence components related to the rotor position but also high-frequency components. To extract the low-frequency

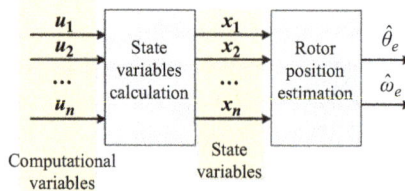

Figure 4.14   Basic calculation block diagram

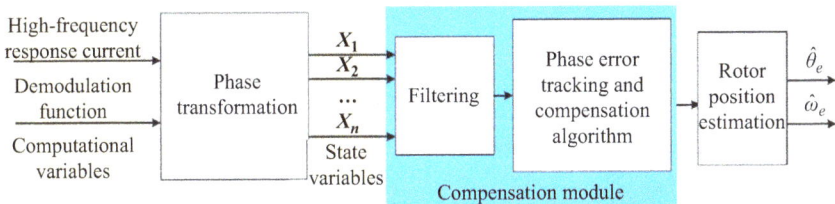

Figure 4.15   Calculation block diagram with compensation module

components, an LPF is needed to remove the high-frequency components, and the phase angle of the low-frequency negative sequence component also contains δ_d and δ_H. Therefore, additional algorithms, such as filtering, error tracking, or compensation, are needed to eliminate the negative effects of δ_d and δ_H. The basic block diagram of these compensation methods is shown in Figure 4.15. These auxiliary algorithms improve the accuracy of the rotating injection method but increase the system's complexity and computational load. The direct cause is that the phase transformation operation of the computational variables has not eliminated the error phase, which results in the phase angle of the state variable still containing error terms. These must be eliminated by auxiliary algorithms. Thus, the selection of computational variables and state variables is crucial for reducing the performance and complexity of the position estimation method. By considering the phase relationships of the computational variables, selecting those that can directly eliminate error terms in the phase transformation can yield state variables without phase errors. This not only improves the estimation accuracy of the rotating injection method but also reduces the complexity of the estimation algorithm.

Based on the above selection principles, this section analyzes the phase characteristics of the induced current in HRSI and selects a new set of computational variables: the high-frequency induced currents at two adjacent sampling intervals, and a new set of state variables (the sum of the positive and negative sequence current vectors). From (4.40), it can be seen that the sum of the phases of the positive and negative sequence vectors ($i_{\alpha\beta h}^p$ and $i_{\alpha\beta h}^n$) is $2\theta_e$, no longer containing δ_d and δ_H. By using the induced currents at adjacent times, the positive and negative sequence vectors of the induced current can be decoupled, and by performing phase transformation on the positive and negative sequence vectors, the rotor position cosine and sine functions without error terms can be directly obtained. This avoids the need for phase error tracking and compensation in subsequent steps, reducing the computational load, eliminating errors, and eliminating the need for demodulation functions and LPF. The basic principle block diagram of the proposed method is shown in Figure 4.16.

Since different injection/update frequency ratios can lead to situations where the linear non-homogeneous system of equations for solving state variables $i_{\alpha\beta h}^p$ and $i_{\alpha\beta h}^n$ from the computational variables $i_{\alpha\beta h}$ may have no solution, infinite solutions, or a unique solution, only the unique solution meets the system requirements. Therefore, this chapter discusses the conditions for having a unique solution for the state variables.

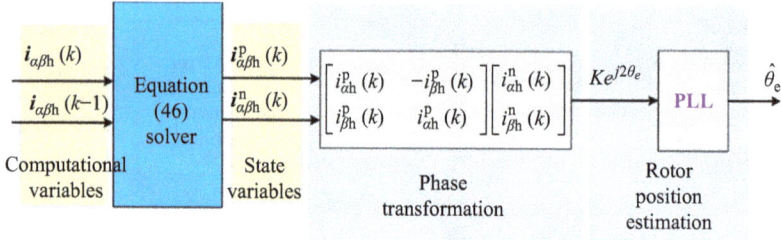

Figure 4.16   Calculation block diagram of the proposed method

4.4.2   System modeling and selection of system variables

First, the position sensorless estimation system is mathematically modeled to determine the system structure and variables. Since current sampling and control are performed in discrete mode, Eq. (4.41) should be discretized. The high-frequency discrete current on the $\alpha\beta$-axis can be expressed as:

$$\begin{cases} i_{\alpha h}(k) = i_{\alpha h}^{\mathrm{p}}(k) + i_{\alpha h}^{\mathrm{n}}(k) = I_{\mathrm{p}}\sin\left(2\pi\gamma k - \delta_{\mathrm{d}} + \delta_H\right) - I_{\mathrm{n}}\sin\left(-2\pi\gamma k + \delta_{\mathrm{d}} - \delta_H + 2\theta_{\mathrm{e}}\right) \\ i_{\beta h}(k) = i_{\beta h}^{\mathrm{p}}(k) + i_{\beta h}^{\mathrm{n}}(k) = -I_{\mathrm{p}}\cos\left(2\pi\gamma k - \delta_{\mathrm{d}} + \delta_H\right) + I_{\mathrm{n}}\cos\left(-2\pi\gamma k + \delta_{\mathrm{d}} - \delta_H + 2\theta_{\mathrm{e}}\right) \end{cases}$$

$$(4.45)$$

where k is the control time instant; $\gamma = f_h/f_s$ is the ratio of the injection frequency to the control frequency. In this section, the control frequency is equal to the sampling frequency.

The phase of a vector can be changed through coordinate transformation. Let the unit vector be denoted as $F_{\alpha\beta} = e^{j\theta}$, which is called the reference vector. The reference vector undergoes coordinate transformation to obtain a new vector $\boldsymbol{F}_{\alpha\beta1}$ with phase $\theta + \theta_1$, which is called the transformed vector. The expression for phase transformation is:

$$\boldsymbol{F}_{\alpha\beta1} = \begin{bmatrix} \cos\theta_1 & -\sin\theta_1 \\ \sin\theta_1 & \cos\theta_1 \end{bmatrix} \begin{bmatrix} F_\alpha \\ F_\beta \end{bmatrix} = e^{j(\theta + \theta_1)} \tag{4.46}$$

where θ_1 is the transformation phase.

Operations in the form of (4.46) are phase summation operations. The angles involved in the phase transformation and the relationships between the vectors before and after the transformation are shown in Figure 4.17.

From (4.45), it can be seen that the sum of the phases of the positive and negative sequence vectors ($i_{\alpha\beta h}^{\mathrm{p}}(k)$ and $i_{\alpha\beta h}^{\mathrm{n}}(k)$) is $2\theta_{\mathrm{e}}$. If the positive and negative sequence vectors can be obtained, the rotor position's sine and cosine functions can be directly derived through phase summation operations. That is, the system's output vector $\boldsymbol{y}$ is:

$$\boldsymbol{y} = \begin{bmatrix} y_1 \\ y_2 \end{bmatrix} = \begin{bmatrix} i_{\alpha r}(k) \\ i_{\beta r}(k) \end{bmatrix} = \begin{bmatrix} i_{\alpha h}^{\mathrm{p}}(k) & -i_{\beta h}^{\mathrm{p}}(k) \\ i_{\beta h}^{\mathrm{p}}(k) & i_{\alpha h}^{\mathrm{p}}(k) \end{bmatrix} \begin{bmatrix} i_{\alpha h}^{\mathrm{n}}(k) \\ i_{\beta h}^{\mathrm{n}}(k) \end{bmatrix} = \begin{bmatrix} I_{\mathrm{p}}I_{\mathrm{n}}\cos 2\theta_{\mathrm{e}} \\ I_{\mathrm{p}}I_{\mathrm{n}}\sin 2\theta_{\mathrm{e}} \end{bmatrix} \tag{4.47}$$

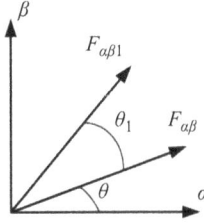

Figure 4.17   Phase transformation diagram

From (4.47), it can be seen that the obtained sine and cosine components of the rotor position are independent of δ_d and δ_H. Therefore, selecting $i_{\alpha h}^p(k)$, $i_{\alpha h}^n(k)$, $i_{\beta h}^p(k)$, and $i_{\beta h}^n(k)$ as state variables, and defining them as a 4-dimensional state vector x, it can be expressed as:

$$x = \begin{bmatrix} x_1 \\ x_2 \\ x_3 \\ x_4 \end{bmatrix} = \begin{bmatrix} i_{\alpha h}^p(k) \\ i_{\beta h}^p(k) \\ i_{\alpha h}^n(k) \\ i_{\beta h}^n(k) \end{bmatrix} \tag{4.48}$$

Due to the special phase relationship of $i_{\alpha \beta h}^p(k)$ and $i_{\alpha \beta h}^n(k)$, the rotor position obtained from the four state variables in Eq. (4.48) will not be affected by sampling and filtering effects. Thus, the problem is transformed into how to obtain these four state variables. Based on the voltage balance equation for high-frequency voltage injection in the stationary coordinate system, the high-frequency current values in the $\alpha\beta$-axis are the most directly known quantities. Defining these as the computational variables u, they can be expressed as:

$$u = \begin{bmatrix} u_1 \\ u_2 \end{bmatrix} = \begin{bmatrix} i_{\alpha h}(k) \\ i_{\beta h}(k) \end{bmatrix} \tag{4.49}$$

According to (4.45), the relationship between the state variables and the computational variables can be transformed into the following non-homogeneous linear equation system:

$$Ax = u \tag{4.50}$$

Here,

$$A = \begin{bmatrix} 1 & 0 & 1 & 0 \\ 0 & 1 & 0 & 1 \end{bmatrix}.$$

In this way, the problem of solving the rotor position is transformed into solving the (4.50). Since rank$(A) = 2 < 4$, the matrix A is not full rank, meaning that x has infinitely many solutions. To ensure a unique solution for x, the coefficient matrix A must be full rank. If the number of rows in the matrix can be increased—that is, by increasing the number of known quantities—so that it reaches 4, then it is possible for A to become full rank. The following section will discuss methods for ensuring A is full rank.

4.4.3 Solution of positive and negative sequence state variables in HRSI method

In motor control systems, measurable physical quantities include voltage, current, torque, temperature, noise, etc. Among these, current is an essential variable in the feedback loop of the control system and is also the most intuitive physical quantity. Measuring other physical quantities requires additional sensors, which increases the hardware cost and system complexity. Therefore, the most direct method to increase the number of known quantities is to use the current values from adjacent control cycles to obtain $\boldsymbol{u}$, which can be expressed as:

$$\boldsymbol{u}^* = \begin{bmatrix} u_1 \\ u_2 \\ u_3 \\ u_4 \end{bmatrix} = \begin{bmatrix} i_{\alpha h}(k) \\ i_{\beta h}(k) \\ i_{\alpha h}(k-1) \\ i_{\beta h}(k-1) \end{bmatrix} \tag{4.51}$$

By replacing k with $k-1$ and substituting it into Eq. (4.45), the high-frequency current response at $k-1$ can be obtained as:

$$\begin{cases} i_{\alpha h}(k-1) = i_{\alpha h}^{\mathrm{p}}(k-1) + i_{\alpha h}^{\mathrm{n}}(k-1) = I_{\mathrm{p}}\sin\left(2\pi\gamma k - 2\pi\gamma - \delta_{\mathrm{d}} + \delta_H\right) \\ - I_{\mathrm{n}}\sin\left(2\pi\gamma - 2\pi\gamma k + \delta_{\mathrm{d}} - \delta_H + 2\theta_{\mathrm{e}}\right) \\ i_{\beta h}(k-1) = i_{\beta h}^{\mathrm{p}}(k-1) + i_{\beta h}^{\mathrm{n}}(k-1) = -I_{\mathrm{p}}\cos\left(2\pi\gamma k - 2\pi\gamma - \delta_{\mathrm{d}} + \delta_H\right) \\ + I_{\mathrm{n}}\cos\left(2\pi\gamma - 2\pi\gamma k + \delta_{\mathrm{d}} - \delta_H + 2\theta_{\mathrm{e}}\right) \end{cases} \tag{4.52}$$

By comparing (4.45) and (4.52), it can be seen that $i_{\alpha h}^{\mathrm{p}}$, $i_{\alpha h}^{\mathrm{n}}$, $i_{\beta h}^{\mathrm{p}}$, and $i_{\beta h}^{\mathrm{n}}$ have the same magnitude but with different phases at time k and $k-1$. At $k-1$, the positive sequence component differs from the positive sequence component at k by $-2\pi\gamma$, while the negative sequence component differs by $2\pi\gamma$. Therefore, when γ is predefined in the estimation system, the phase difference of the positive and negative sequence vectors remains fixed. Using the sum-angle formulas for trigonometric functions, (4.52) can be expanded by replacing $i_{\alpha h}^{\mathrm{p}}(k-1), i_{\alpha h}^{\mathrm{n}}(k-1), i_{\beta h}^{\mathrm{p}}(k-1)$, and $i_{\beta h}^{\mathrm{n}}(k-1)$ with their respective equivalents $i_{\alpha h}^{\mathrm{p}}(k), i_{\alpha h}^{\mathrm{n}}(k), i_{\beta h}^{\mathrm{p}}(k)$, and $i_{\beta h}^{\mathrm{n}}(k)$.

$$\begin{cases} i_{\alpha h}(k-1) = \cos 2\pi\gamma I_{\mathrm{p}}\sin\left(2\pi\gamma k - \delta_{\mathrm{d}} + \delta_H\right) - \sin 2\pi\gamma I_{\mathrm{p}}\cos\left(2\pi\gamma k - \delta_{\mathrm{d}} + \delta_H\right) \\ -\cos 2\pi\gamma I_{\mathrm{n}}\sin\left(-2\pi\gamma k + \delta_{\mathrm{d}} - \delta_H + 2\theta_{\mathrm{e}}\right) - \sin 2\pi\gamma I_{\mathrm{n}}\cos\left(-2\pi\gamma k + \delta_{\mathrm{d}} - \delta_H + 2\theta_{\mathrm{e}}\right) \\ = \cos 2\pi\gamma i_{\alpha h}^{\mathrm{p}}(k) + \sin 2\pi\gamma i_{\beta h}^{\mathrm{p}}(k) + \cos 2\pi\gamma i_{\alpha h}^{\mathrm{n}}(k) - \sin 2\pi\gamma i_{\beta h}^{\mathrm{n}}(k) \\ i_{\beta h}(k-1) = -\cos 2\pi\gamma I_{\mathrm{p}}\cos\left(2\pi\gamma k - \delta_{\mathrm{d}} + \delta_H\right) - \sin 2\pi\gamma I_{\mathrm{p}}\sin\left(2\pi\gamma k - \delta_{\mathrm{d}} + \delta_H\right) \\ + \cos 2\pi\gamma I_{\mathrm{n}}\cos\left(-2\pi\gamma k + \delta_{\mathrm{d}} - \delta_H + 2\theta_{\mathrm{e}}\right) - \sin 2\pi\gamma I_{\mathrm{n}}\sin\left(-2\pi\gamma k + \delta_{\mathrm{d}} - \delta_H + 2\theta_{\mathrm{e}}\right) \\ = \cos 2\pi\gamma i_{\beta h}^{\mathrm{p}}(k) - \sin 2\pi\gamma i_{\alpha h}^{\mathrm{p}}(k) + \cos 2\pi\gamma i_{\beta h}^{\mathrm{n}}(k) + \sin 2\pi\gamma i_{\alpha h}^{\mathrm{n}}(k) \end{cases} \tag{4.53}$$

From (4.54), it can be seen that $i_{\alpha h}(k-1)$ and $i_{\beta h}(k-1)$ are still functions of $i_{\alpha h}^{\mathrm{p}}(k)$, $i_{\alpha h}^{\mathrm{n}}(k)$, $i_{\beta h}^{\mathrm{p}}(k)$ and $i_{\beta h}^{\mathrm{n}}(k)$. Therefore, based on (4.45) and (4.53), the relationship between the state variables and the new computational variables can be derived and expressed as the following system of equations:

$$A^* x = u^* \tag{4.54}$$

where,

$$A^* = \begin{bmatrix} 1 & 0 & 1 & 0 \\ 0 & 1 & 0 & 1 \\ \cos 2\pi\gamma & \sin 2\pi\gamma & \cos 2\pi\gamma & -\sin 2\pi\gamma \\ -\sin 2\pi\gamma & \cos 2\pi\gamma & \sin 2\pi\gamma & \cos 2\pi\gamma \end{bmatrix}$$

is the coefficient matrix in the high-frequency rotating injection method.

From (4.54), it can be seen that A^* is the parameter matrix. Therefore, different values of γ will affect the rank of the A^* matrix, which in turn affects the solution of x. By comparing (4.54) with (4.50), it can be observed that the number of unknowns has not changed and remains 4, but the number of known quantities has increased to 4. After performing elementary matrix transformations on A^* and analyzing the relationship between the range of γ and rank(A^*), the results can be found in Table 4.1.

From Table 4.1, it can be seen that when the inequality $\gamma \neq 0.5(m\text{-}1)$ is satisfied, A^* is full rank. According to the necessary and sufficient conditions for the existence of a solution to the linear non-homogeneous system of equations, when A^* is full rank, (4.54) has a unique solution for x, which can be calculated as:

$$\begin{bmatrix} x_1 \\ x_2 \\ x_3 \\ x_4 \end{bmatrix} = \begin{bmatrix} \dfrac{\sin 2\pi\gamma \cdot u_1 + \cos 2\pi\gamma \cdot u_2 - u_4}{2\sin 2\pi\gamma} \\ \dfrac{-\cos 2\pi\gamma \cdot u_1 + \sin 2\pi\gamma \cdot u_2 + u_3}{2\sin 2\pi\gamma} \\ \dfrac{\sin 2\pi\gamma \cdot u_1 - \cos 2\pi\gamma \cdot u_2 + u_4}{2\sin 2\pi\gamma} \\ \dfrac{\cos 2\pi\gamma \cdot u_1 + \sin 2\pi\gamma \cdot u_2 - u_3}{2\sin 2\pi\gamma} \end{bmatrix} \tag{4.55}$$

After obtaining the state variables x, the sine and cosine components of the rotor position can be determined using (4.47). Throughout the entire calculation process, only two consecutive current values are required, and by setting the parameter γ in advance such that $\gamma \neq 0.5(m\text{-}1)$, the rotor position can be uniquely determined.

Table 4.1   The relationship between γ and rank(A^*)

$|\cos 2\pi\gamma|$	γ	rank(A^*)
0	0.25+0.5(m-1)	4
1	0.5(m-1)	2
$\neq 0, 1$	/	4

where, $m \in N+$.

In practical applications, when γ is too small, it limits the injection signal frequency, especially when the control frequency is low. In such cases, the injection frequency approaches the operating frequency of the motor during low-speed operation, which increases the difficulty of high-frequency signal extraction. Conversely, when γ is too large, it leads to distortion in the waveform of the digitally synthesized sinusoidal signal. Therefore, a compromise must be made between the difficulty of extracting the high-frequency signal and the degree of distortion in the injected sinusoidal wave. In this section, γ is chosen as 1/8 for the HRSI method.

Due to the limitation of the injection voltage frequency by the control frequency, the performance of the sinusoidal injection method is still insufficient in some applications. The high-frequency orthogonal square-wave injection method can achieve higher injection frequencies and broader system bandwidth. As the injection frequency increases, it becomes easier to separate the injection signal from the low-frequency signals in vector control. The proposed induced current positive and negative sequence self-decoupling method can still be applied in the high-frequency orthogonal square-wave injection method to eliminate the negative impacts of system delay and the high-pass filter on position estimation accuracy, thus achieving higher position estimation accuracy. The high-frequency induced current under orthogonal square-wave injection contains higher-order harmonic components. Therefore, it is necessary to analyze the phase errors in the higher-order harmonics of the induced current and, based on this, derive new system equations, computational variables, and state variables.

4.4.4   Phase error analysis in higher-order harmonics of induced current under HOSI method

When the injection waveform is a square wave, the system not only contains the fundamental frequency component of the injection signal but also higher-order harmonic components caused by the square wave signal. Therefore, the injection method using a single-frequency sinusoidal signal cannot be used for analysis. It is necessary to analyze the phase errors caused by system delay and the DHPF for the harmonics of different frequencies. The phase error consists of two parts.

4.4.4.1   System delay

The positive and negative sequence components in different orders of harmonics have equal frequencies but opposite signs. Assuming that the phase error caused by system delay for the positive sequence component in the m-th harmonic is δ_{dm}, the phase error for the negative sequence component is $-\delta_{\mathrm{dm}}$.

4.4.4.2   DHPF

Since the higher-order harmonic frequencies in the induced current exceed the cutoff frequency of the DHPF, the amplitude-frequency response characteristics of the digital filter need to be analyzed under periodic extension. First, the phase curve is an odd function of frequency. Second, the frequency response period is 2π. Once the amplitude-frequency response in the range of $0-\pi$ (corresponding to the analog frequency range from 0 to $0.5f_s$) is determined, the amplitude-frequency response for any frequency can be derived. The absolute value of the phase shift at the same frequency is the same. Therefore, for the m-th harmonic, the phase error caused by the high-pass filter for the positive sequence component is δ_{Hm}, and for the negative sequence component is $-\delta_{Hm}$.

Considering the phase errors caused by system delay and the DHPF, the induced current response under HOSI can be expressed as:

$$i_{\alpha\beta h_q u} = i_{\alpha\beta psqu}^p + i_{\alpha\beta nsqu}^n$$

$$= \sum_{m=1}^{+\infty} \frac{4I_p}{\pi(2m-1)^2} e^{j\left((-1)^m(2m-1)\omega_n + \delta_m - 0.5\pi\right)} + \sum_{m=1}^{+\infty} \frac{4I_n}{\pi(2m-1)^2} e^{j\left((-1)^{m-1}(2m-1)\omega_n t - \delta_m + 2\theta_e + 0.5\pi\right]}$$

$$(4.56)$$

where, $\delta_m = \delta_{dm} + \delta_{Hm}$ is the phase error corresponding to the mth harmonic.

4.4.5 Solution of positive and negative sequence state variables in HOSI method

For the HOSI method, the most direct way to select state variables is to still use $i_{\alpha h_s q u}^p(k)$, $i_{\alpha b_s q u}^n(k)$, $i_{\beta h_s q u}^p(k)$, and $i_{\beta h_s q u}^n(k)$ as the system state variables. Let this be represented as $x_{\_squ}$, which can be expressed as:

$$x_{squ} = \begin{bmatrix} x_{1\_squ} \\ x_{2\_squ} \\ x_{3\_squ} \\ x_{4\_squ} \end{bmatrix} = \begin{bmatrix} i_{\alpha h\_squ}^p(k) \\ i_{\beta h\_squ}^p(k) \\ i_{\alpha h\_squ}^n(k) \\ i_{\beta h\_squ}^n(k) \end{bmatrix} \tag{4.57}$$

Similarly, let the high-frequency induced current at continuous sampling moments be set as the computational variable $u_{\_squ}$, which can be expressed as:

$$u_{\_squ} = \begin{bmatrix} u_{1\_squ} \\ u_{2\_squ} \\ u_{3\_squ} \\ u_{4\_squ} \end{bmatrix} = \begin{bmatrix} i_{\alpha h\_squ}(k) \\ i_{\beta h\_squ}(k) \\ i_{\alpha h\_squ}(k-1) \\ i_{\beta h\_squ}(k-1) \end{bmatrix} \tag{4.58}$$

Then, it is necessary to derive the new matrix equation based on the relationship between the computational variables and the state variables, which can be expressed as:

$$A^{**}x_{\_squ} = u_{\_squ} \tag{4.59}$$

where A^{**} is the coefficient matrix in HOSI.

After discretizing (4.56), the discrete high-frequency current at adjacent moments can be expressed as:

$$i_{\alpha h\_squ}(k) = i^{\mathrm{p}}_{\alpha h\_squ}(k) + i^{\mathrm{n}}_{\alpha ph\_squ}(k)$$

$$= \sum_{m=1}^{+\infty} \frac{4I_p}{\pi(2m-1)^2} \sin\left[(-1)^m(2m-1)2\pi\gamma k + \delta_m\right]$$

$$+ \sum_{m=1}^{+\infty} \frac{4I_n}{\pi(2m-1)^2} \sin\left[2\theta_e + (-1)^{m+1}(2m-1)2\pi\gamma k - \delta_m\right]$$

(4.60)

$$i_{\beta h\_squ}(k) = i^{\mathrm{p}}_{\beta h\_squ}(k) + i^{\mathrm{n}}_{\beta h\_squ}(k)$$

$$= -\sum_{m=1}^{+\infty} \frac{4I_p}{\pi(2m-1)^2} \cos\left[(-1)^m(2m-1)2\pi\gamma k + \delta_m\right]$$

$$+ \sum_{m=1}^{+\infty} \frac{4I_n}{\pi(2m-1)^2} \cos\left[2\theta_e + (-1)^{m+1}(2m-1)2\pi\gamma k - \delta_m\right]$$

(4.61)

$$i_{\alpha h\_squ}(k-1) = i^{\mathrm{p}}_{\alpha h\_squ}(k-1) + i^{\mathrm{n}}_{\alpha h\_squ}(k-1)$$

$$= \sum_{m=1}^{+\infty} \frac{4I_p}{\pi(2m-1)^2} \sin\left[(-1)^m(2m-1)2\pi\gamma k + (-1)^{m+1}(2m-1)2\pi\gamma + \delta_m\right]$$

$$- \sum_{m=1}^{+\infty} \frac{4I_n}{\pi(2m-1)^2} \sin\left[2\theta_e + (-1)^{m+1}(2m-1)2\pi\gamma k + (-1)^m(2m-1) - 2\pi\gamma - \delta_m\right]$$

(4.62)

$$i_{\beta h\_squ}(k-1) = i^{\mathrm{p}}_{\beta h\_squ}(k-1) + i^{\mathrm{n}}_{\beta h\_squ}(k-1)$$

$$= \sum_{m=1}^{+\infty} \frac{4I_p}{\pi(2m-1)^2} \cos\left[(-1)^m(2m-1)2\pi\gamma k + (-1)^{m+1}(2m-1)2\pi\gamma + \delta_m\right]$$

$$- \sum_{m=1}^{+\infty} \frac{4I_n}{\pi(2m-1)^2} \cos\left[2\theta_e + (-1)^{m+1}(2m-1)2\pi\gamma k + (-1)^m(2m-1) - 2\pi\gamma - \delta_m\right]$$

(4.63)

From (4.60) to (4.63), it can be seen that due to the appearance of higher-order harmonics in HOSI, the difference between the harmonic components at $k-1$ and k for the positive sequence components is $(-1)^{m+1}(2m-1)2\pi\gamma$, while for the negative sequence components, the difference is $(-1)^m(2m-1)2\pi\gamma$. Compared to HRSI, this difference not only depends on γ but also on the harmonic order m. If (4.62) and (4.63) are directly expanded in the form of (4.53), the resulting expansion will depend on m, which cannot be directly obtained from the induced current. To eliminate the influence of m, the difference between the harmonic components at time $k-1$ and time k for the positive and negative sequence components is defined as γ^{p}_m and γ^{n}_m, respectively, and through polynomial decomposition of γ^{p}_m, the following can be derived:

$$\gamma_m^p = (-1)^{m+1}(2m-1)2\pi\gamma = 2\pi\gamma + \left[(-1)^{m+1}(2m-1)-1\right]2\pi\gamma \qquad (4.64)$$

$$\gamma_m^n = -\gamma_m^p \qquad (4.65)$$

From (4.64), it can be seen that the coefficient $(-1)^m(2m-1)-1$ of the second term in γ_m^p is a multiple of 4. Therefore, when γ is an integer multiple of 1/4, $\gamma_m^p = 2\pi\gamma + 2\pi q$, where q is an integer. According to the periodicity property of trigonometric functions, $2q\pi$ can be canceled in the calculation. This eliminates the influence of m. At this point, (4.63) and (4.64) can be re-derived as:

$$i_{\alpha h\_squ}(k-1) = i_{\alpha h\_squ}^p(k-1) + i_{\alpha h\_qu}^n(k-1)$$

$$= \sum_{m=1}^{+\infty} \frac{4I_p}{\pi(2m-1)^2} \sin\left[(-1)^m(2m-1)2\pi\gamma k + 2\pi\gamma + \delta_m\right]$$

$$- \sum_{m=1}^{+\infty} \frac{4I_n}{\pi(2m-1)^2} \sin\left[2\theta_e + (-1)^{m+1}(2m-1)2\pi\gamma k - 2\pi\gamma - \delta_m\right]$$

$$\qquad (4.66)$$

$$i_{\beta h\_squ}(k-1) = i_{\beta h\_squ}^p(k-1) + i_{\beta h\_qu}^n(k-1)$$

$$= \sum_{m=1}^{+\infty} \frac{4I_p}{\pi(2m-1)^2} \cos\left[(-1)^m(2m-1)2\pi\gamma k + 2\pi\gamma + \delta_m\right]$$

$$+ \sum_{m=1}^{+\infty} \frac{4I_n}{\pi(2m-1)^2} \cos\left[2\theta_e + (-1)^{m+1}(2m-1)2\pi\gamma k - 2\pi\gamma - \delta_m\right]$$

$$\qquad (4.67)$$

By comparing (4.60) and (4.66), as well as (4.61) and (4.67), it can be observed that for the positive sequence components, the difference between the harmonic components at $k-1$ and k is $2\pi\gamma$, while for the negative sequence components, the difference is $-2\pi\gamma$. This difference is independent of m. At this point, expanding (4.66) and (4.67) and replacing $i_{\alpha h\_squ}^p(k-1), i_{\alpha h\_squ}^n(k-1), i_{\beta h\_squ}^p(k-1)$, and $i_{\beta h\_squ}^n(k-1)$ with $i_{\alpha h\_squ}^p(k), i_{\alpha h\_squ}^n(k), i_{\beta h\_squ}^p(k)$, and $i_{\beta h\_squ}^n(k)$, we get:

$$i_{\alpha h\_squ}(k-1) = \cos 2\pi\gamma i_{\alpha h\_quu}^p(k) - \sin 2\pi\gamma i_{\beta h\_squ}^p(k) + \cos 2\pi\gamma i_{\alpha h\_squ}^n(k)$$

$$+ \sin 2\pi\gamma i_{\beta h\_squ}^n(k)$$

$$\qquad (4.68)$$

$$i_{\beta h\_squ}(k-1) = \sin 2\pi\gamma i_{\alpha h\_squ}^p(k) + \cos 2\pi\gamma i_{\beta h\_squ}^p(k) + \cos 2\pi\gamma i_{\beta h\_squ}^n(k)$$

$$- \sin 2\pi\gamma i_{\alpha h\_squ}^n(k)$$

$$\qquad (4.69)$$

Based on (4.60), (4.61), (4.68), and (4.69), the coefficient matrix A^{**} in (4.59) can be obtained as:

$$A^{**} = \begin{bmatrix} 1 & 0 & 1 & 0 \\ 0 & 1 & 0 & 1 \\ \cos 2\pi\gamma & -\sin 2\pi\gamma & \cos 2\pi\gamma & \sin 2\pi\gamma \\ \sin 2\pi\gamma & \cos 2\pi\gamma & \cos 2\pi\gamma & -\sin 2\pi\gamma \end{bmatrix} \tag{4.70}$$

Based on the analysis of (4.64) and (4.65), to eliminate the influence of m, γ must be an integer multiple of 1/4. The maximum frequency of the injected orthogonal square wave is 1/4 of the control frequency. Therefore, by setting $\gamma=1/4$ and substituting it into (4.70), we obtain:

$$A^{**} = \begin{bmatrix} 1 & 0 & 1 & 0 \\ 0 & 1 & 0 & 1 \\ \cos 0.5\pi & -\sin 0.5\pi & \cos 0.5\pi & \sin 0.5\pi \\ \sin 0.5\pi & \cos 0.5\pi & \cos 0.5\pi & -\sin 0.5\pi \end{bmatrix} = \begin{bmatrix} 1 & 0 & 1 & 0 \\ 0 & 1 & 0 & 1 \\ 0 & -1 & 0 & 1 \\ 1 & 0 & 0 & -1 \end{bmatrix} \tag{4.71}$$

From the above equation, it can be seen that rank(A^{**}) = 4. At this point, x has a unique solution, which can be calculated as:

$$\begin{bmatrix} x_1 \\ x_2 \\ x_3 \\ x_4 \end{bmatrix} = \begin{bmatrix} 0.5(u_4 + u_1) \\ 0.5(u_1 - u_4) \\ 0.5(u_2 - u_3) \\ 0.5(u_3 + u_2) \end{bmatrix} \tag{4.72}$$

After obtaining $i^{p}_{\alpha h\_squ}(k), i^{n}_{\alpha h\_squ}(k), i^{p}_{\beta h\_squ}(k)$, and $i^{n}_{\beta h\_squ}(k)$, based on (4.61) and (4.62), it can be observed that the phase angles of the positive and negative sequence components for each harmonic are equal to $2\theta_e$. Therefore, the rotor position's sine and cosine components can be calculated using a method similar to that in (4.48):

$$y = \begin{bmatrix} y_1 \\ y_2 \end{bmatrix} = \begin{bmatrix} i_{\alpha r\_squ}(k) \\ i_{\beta r\_squ}(k) \end{bmatrix} = \begin{bmatrix} i^{p}_{\alpha h\_squ}(k) & -i^{p}_{\beta h\_squ}(k) \\ i^{p}_{\beta h\_squ}(k) & i^{p}_{\alpha h\_squ}(k) \end{bmatrix} \begin{bmatrix} i^{n}_{\alpha h\_squ}(k) \\ i^{n}_{\beta h\_squ}(k) \end{bmatrix}$$

$$= \begin{bmatrix} G\cos 2\theta_e \\ G\sin 2\theta_e \end{bmatrix} \tag{4.73}$$

where $i_{\alpha\beta r\_squ}$ is an orthogonal vector in HOSI that is related to the position of the rotor.

$$G = \frac{\pi^2}{6} I_p I_n + \frac{32}{\pi^2} I_p I_n \begin{pmatrix} \sum\limits_{n=m+1,m=1}^{+\infty} \frac{1}{(2m-1)^2}\frac{1}{(2n-1)^2}\cos\left[(-1)^m(2m-1)\delta_d - (-1)^n(2n-1)\delta_d\right]+ \\ \sum\limits_{n=m+1,m=2}^{+\infty} \frac{1}{(2m-1)^2}\frac{1}{(2n-1)^2}\cos\left[(-1)^m(2m-1)\delta_d - (-1)^n(2n-1)\delta_d\right]+ \\ \sum\limits_{n=m+1,m=+\infty}^{+\infty} \frac{1}{(2m-1)^2}\frac{1}{(2n-1)^2}\cos\left[(-1)^m(2m-1)\delta_d - (-1)^n(2n-1)\delta_d\right] \end{pmatrix} \tag{4.74}$$

Unlike in (4.48), due to the presence of higher-order harmonics, G is an infinite series. To eliminate the coefficients of the rotor position's sine and cosine components in (4.48) and (4.75), standardization calculations are required, as follows:

$$\begin{bmatrix} \sin 2\theta_e \\ \cos 2\theta_e \end{bmatrix} = \begin{bmatrix} \dfrac{i_{\beta r}(k)}{\sqrt{(i_{\beta r}(k))^2 + (i_{ar}(k))^2}} \\ \dfrac{i_{ar}(k)}{\sqrt{(i_{\beta r}(k))^2 + (i_{ar}(k))^2}} \end{bmatrix}$$

$$= \begin{bmatrix} \dfrac{i_{\beta r_{s\,squ}}(k)}{\sqrt{(i_{\beta r_{scqu}}(k))^2 + (i_{ar\_squ}(k))^2}} \\ \dfrac{i_{ar_{squu}}(k)}{\sqrt{(i_{\beta r\_squ}(k))^2 + (i_{ar\_squ}(k))^2}} \end{bmatrix} \tag{4.75}$$

After the standardization calculation in (4.75), the coefficients of the rotor position's sine and cosine values are 1. After applying a second-order PLL, an open-loop gain of 1 is achieved for the position tracker, thereby providing a stable tracker bandwidth.

The validity of (4.75) requires that both I_p, I_n in (4.47) and G in (4.74) are positive values. From (4.34), it is clear that I_p, I_n are positive. However, G contains an infinite series and requires further determination of its sign. By setting the cosine function values in the equation to the minimum value of -1, we get:

$$G > \frac{\pi^2}{6} I_p I_n - \frac{32}{\pi^2} I_p I_n \sum_{m=1}^{+\infty} \frac{1}{(2m-1)^2} \sum_{m=2}^{+\infty} \frac{1}{(2m-1)^2} \approx 0.71 I_p I_n \tag{4.76}$$

Therefore, the range of values for G satisfies the conditions for the validity of (4.75). Based on the analysis in Sections 4.4.3–4.4.5, the software flowchart for applying the proposed method to HRSI and HOSI is shown in Figure 4.18.

In recent years, scholars have made modifications to rotor position estimation methods to eliminate position estimation errors in high-frequency rotating injection methods. Below, the proposed method is analyzed and compared theoretically with these methods, and a comparison of control structures and algorithm turnaround times is made.

4.4.5.1 Comparison with methods that change the injection form

There are three different methods of injection, including sinusoidal voltage injection at three different frequencies on the *ABC*-axis, bidirectional rotating voltage injection, and orthogonal square-wave injection on the $\alpha\beta$-axis, respectively. The common characteristic of these three methods is that the induced current is first phase-transformed with a demodulation function to obtain the state variables, whose phase contains both the rotor position and error terms. Then, angle extraction algorithms or error compensation algorithms are used to identify and compensate for the errors,

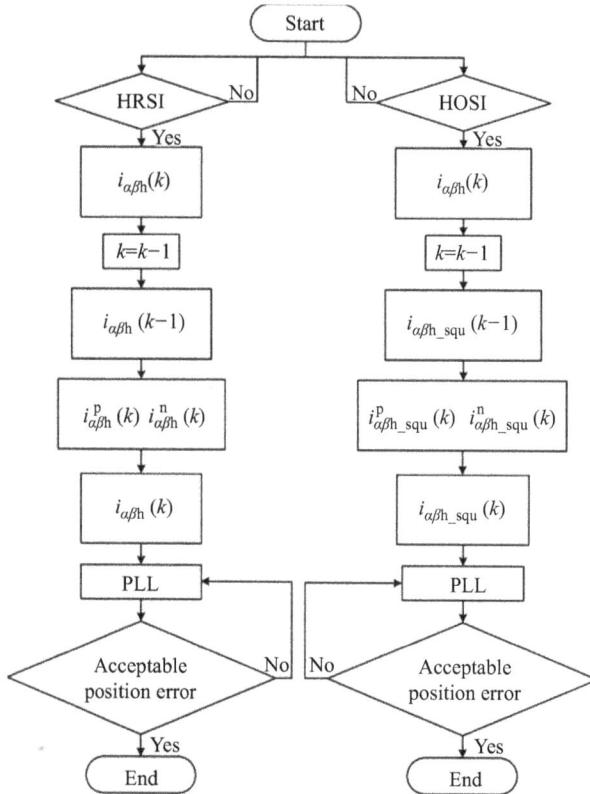

Figure 4.18   Overall flowchart of the proposed method applied to HRSI and HOSI

thereby eliminating their influence. A theoretical comparison of these methods and the proposed method leads to the following conclusions:

1. The proposed method has lower computational complexity than these methods. The latter methods only change the injection form and propose a sensorless solution that is immune to errors such as filtering and system delay. However, they do not specifically analyze the characteristics of the induced current. Although these methods also eliminate errors, they do so by adding computation modules such as LPF, phase extraction, and vector modulus calculation. In contrast, the proposed method only requires a single-phase operation and a single-state variable-solving process, as shown in (4.56) and (4.73), without additional filters and compensation loops.
2. These methods do not change the basic idea of the synchronous axis rotating filtering method. The computational variables chosen are still the induced current and a self-built demodulation function. Since the phase of the induced current contains error terms, while the phase of the demodulation function does not, the phase of the state variables still contains error terms. This error is

eliminated by adding compensation algorithms, making these methods essentially a supplement to the synchronous axis rotating filtering method. In contrast, the proposed method, from the perspective of constructing new computational variables, performs phase transformation operations on the positive and negative sequence components of the current. As a result, the phase of the obtained state variables no longer contains error terms, thus eliminating the need for compensation algorithms.

4.4.5.2 Comparison with methods that modify the initial phase of the demodulation function

In the previous type of method, the initial phase of the demodulation function remains fixed. To eliminate position estimation errors, an error term is introduced at the initial stage of the demodulation function and the PI controller is utilized for error tracking. The tracked error is updated in real time and incorporated into the phase of the demodulation function, effectively adding an online error-tracking system to the system.

When the output of the error phase tracking system converges to φ_1, the position estimation error is eliminated. However, the convergence speed of the error-tracking system directly affects the overall computational efficiency of the system. The structure of this method is illustrated in Figure 4.19.

4.4.5.3 Comparison with the ellipse fitting method

A rotor position estimation method is proposed that replaces the demodulation operation. This method uses the QR recursive least squares algorithm to identify the elliptical trajectory of the $\alpha\beta$-axis current and then calculates the rotor position based on the ellipse parameters. Since the elliptical trajectory does not change due to delay effects, this method is insensitive to position errors caused by delay. The basic block diagram is shown in Figure 4.20. Comparing this method with the proposed method, the following conclusions can be drawn:

1. Sampling Requirement:
   Since this method requires performing QR decomposition on the current matrix A_k ($A_k \in R^{m*n}$), it results in $A_k = Q_k B_k$, where m is the number of consecutive sampling points required, and n is the number of unknowns, with $n = 3$. According to the principles of QR decomposition, $m > n$. Therefore, m must be

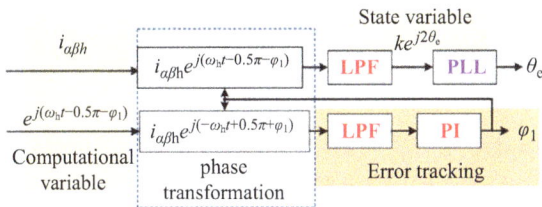

Figure 4.19 *Structure of the position estimation method for changing the initial phase of the demodulation function*

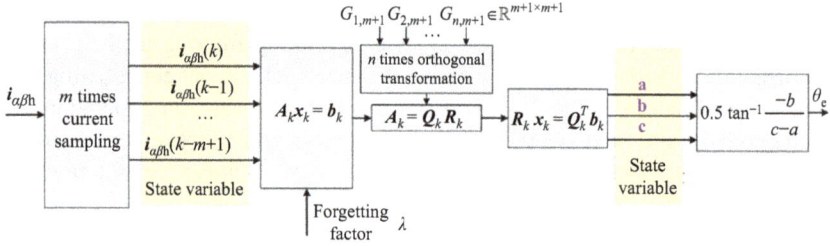

Figure 4.20   The structure diagram of the position estimation method based on ellipse fitting

at least 4, meaning at least four consecutive sampling results are required. In contrast, the proposed method only requires two consecutive sampling results.

2. Computational Complexity:

Since this method uses a recursive approach, the purpose of employing Givens transformations is to make the last row of R_k zero. Therefore, n Givens matrices ($G_{ij} \in R^{m+1*m+1}$) are needed to complete n Givens matrix multiplications. Additionally, before the recursive operation starts, a full QR decomposition of A_k must be performed, which requires $n(n+1)/2$ Givens matrix multiplications. This method involves multiple Givens matrix operations, leading to high computational complexity.

To verify the conclusion that the proposed method reduces computational complexity, the primary structures and algorithm turnaround times of different methods were analyzed and compared. After obtaining the state variables, all methods use arctangent instead of a position tracker to determine the rotor position. In the ellipse fitting method, phase transformation is performed using Givens transformations, where the order of the Givens matrix is at least 3, while in other references, phase transformation operations are all second-order matrix operations. All algorithms were implemented on the DSP28335, and the main structure and actual turnaround times of the algorithms are summarized in Table 4.2. It can be seen that the proposed method has a simpler structure, consisting of only one phase transformation, two normalization operations, four multiplications, and four additions. The entire algorithm does not include LPF, modulus calculations, PI controllers, or demodulation functions. The turnaround time of the proposed method applied to HRSI is 1.6 ms, and for HOSI, it is 1.2 ms, demonstrating an advantage in actual execution time.

4.5   Implementation of high-speed range methods

Taking the Sliding Mode Observer (SMO)-based back-EMF estimation method as an example, this section analyzes high-speed sensorless methods.

Table 4.2   Comparison of the main algorithm structures and turnaround times

Method	Number of demodulation functions	Phase transfor- mation	LPF	PI	Modulus calculation	Normal- ization	Algorithm turnaround time (ms)
Sinusoidal voltage injection on ABC-axis	6	2	6	0	6	0	7.1
Bidirectional rotating voltage injection	4	4	4	0	1	0	8.5
Orthogonal square-wave injection on $\alpha\beta$-axis	2	2	0	0	0	4	2.1
Modified demodulation function initial phase	2	3	2	1	0	0	5
Ellipse fitting method	0	9	0	0	0	0	5.7
Proposed method	0	1	0	0	0	2	1.6/1.2

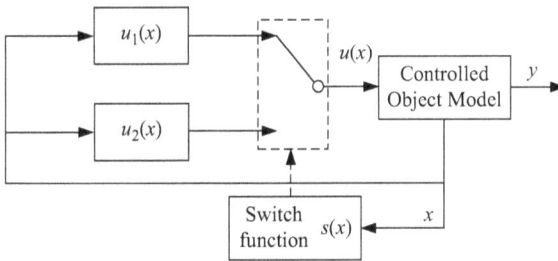

Figure 4.21   Sliding mode control structure diagram

4.5.1   Sliding mode variable structure control theory

As early as the mid-1950s, variable structure control theory began to emerge. Due to its ability to effectively solve many practical engineering problems, it has been continuously developed and improved, becoming a research hotspot. Sliding Mode Variable Structure Control is a control approach where the system structure con- tinuously switches based on a state function. The state function is determined by the deviation between the state input and the reference, ensuring that the system's state point oscillates at a high frequency with a small amplitude near a stable state plane while simultaneously moving toward the desired point. Eventually, it reaches the target and achieves stable control of the system. Figure 4.21 illustrates the control structure, where $u_1(x)$ and $u_2(x)$ are state control functions, and $s(x)$ is the switching function [10].

4.5.1.1   Sliding mode principles

Sliding Mode Variable Structure Control is a discontinuous control method, which changes the switching characteristics of the system structure according to a pre- defined rule. A switching surface is designed based on the system requirements, and the control system ensures that the state point oscillates back and forth near this switching surface—this behavior is known as sliding mode. The concept of sliding mode is introduced below using a second-order system as an example.

$$\begin{cases} \dot{x}_1 = a_1 x_1 + b_1 x_2 \\ \dot{x}_2 = a_2 x_1 + b_2 x_2 + u \end{cases} \tag{4.77}$$

A second-order system of linear time-invariant controlled objects is shown above. x_1 and x_2 are state variables; a_1, a_2, b_1, and b_2 are all constants; u represents the control function, which can be expressed as:

$$u = \begin{cases} u_1(x_1, x_2), s(x_1, x_2) < 0 \\ u_2(x_1, x_2), s(x_1, x_2) > 0 \end{cases} \tag{4.78}$$

where, $s(x_1, x_2)$ is a switching function, its expression is:

$$s(x_1, x_2) = c_1 x_1 + c_2 x_2 \tag{4.79}$$

where, both c_1 and c_2 are constants. Clearly, $s(x_1, x_2)$ is a straight line in the state variable and the state plane is composed of x_1 and x_2, and is a state-switching line in the plane. We construct an appropriate state function such that state points far from the switching line move toward it, as shown in the motion trajectory in Figure 4.22.

From Figure 4.22, we can see that there is a state point in the region where $s(x_1, x_2) > 0$, and it is far from the state switching line. Based on (4.78), the state point moves toward the state switching line under the control function $u_1(x_1, x_2)$ and will reach the state switching line within a limited amount of time. After crossing the switching line, it reaches the region where $s(x_1, x_2) < 0$. At this point, the switching function $u_2(x_1, x_2)$ takes effect, and the state point starts moving toward the switching line again, reaching the region where $s(x_1, x_2) > 0$. This represents the sliding mode of the state point, where the state point moves back and forth near the switching line and approaches the target point.

To generalize the above example to a general system, assume that the system is:

$$\dot{x} = f(x, u) \tag{4.80}$$

where x and f are n-dimensional vectors, u is an m-dimensional vector, and it exists in the range $m \leq n$. Let $s(x)$ be the state switching function, and the control

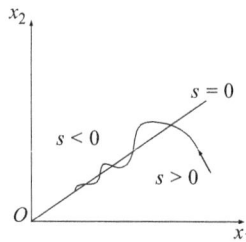

Figure 4.22   State point trajectory diagram

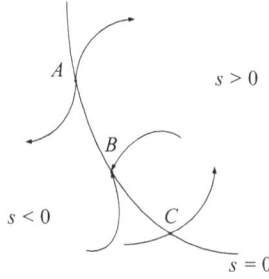

Figure 4.23 Three types of points

function can be expressed as:

$$u = \begin{cases} u_1(x), s(x) > 0 \\ u_2(x), s(x) < 0 \end{cases} \tag{4.81}$$

where, $s(x) = 0$ is an m-dimensional hyperplane, referred to as the switching sur-face. The system's state point moves across this surface through the control func-tion, and the system continues to move along the sliding mode when $s(x) = 0$, which will make the system's state point move toward the origin of the sliding mode. The state space is divided into two regions: one is the region where $s_1(x) > 0$ and the other is where $s_1(x) < 0$.

As shown in Figure 4.23, point A is the starting point, point B is the terminal point, and point C is the transition point. When the state point moves to the region adjacent to the switching surface, it starts to move toward the boundary of the sliding mode surface. When the state point moves into the region adjacent to the sliding mode surface, it starts to move toward the two terminal points of the sliding mode surface. When the state point moves from the sliding mode surface to the other side, it passes through the normal crossing to the other side of the sliding mode.

From the above, we can conclude that when the system's state point moves to a region near the starting point, it leaves the sliding mode switching surface. When the state point moves to a typical point's vicinity, the state point crosses the sliding mode switching surface and also leaves the sliding mode surface. However, when the state point moves to a region near the termination point, the state point can no longer move far from the sliding mode switching surface. This region is referred to as the sliding mode region.

The main goal of studying sliding mode variable structure control is to con-struct the control function u, set an appropriate sliding mode switching surface, and define the switching function $s(x)$ such that the control system satisfies the condi-tions for the existence of a sliding mode region. The system's state point should be able to reach or asymptotically approach the sliding mode region, and the system state should be able to stably move in the sliding mode region. The following introduces these conditions.

1. Existence of the region must satisfy:

$$\lim_{s \to 0} s \frac{ds}{dt} \leq 0 \tag{4.82}$$

2. Reachability of the region must satisfy:

$$s \frac{ds}{dt} \leq 0 \tag{4.83}$$

3. Motion stability must satisfy: We construct a positive definite Lyapunov function as follows:

$$V = \frac{s(x)^T s(x)}{2} \tag{4.84}$$

Then the stability condition is:

$$\dot{V} = s(x)^T \dot{s}(x) \leq 0 \tag{4.85}$$

4.5.1.2   Sliding mode variable structure control method

The following three basic variable structure control methods are introduced

1. Constant switch control method:

$$u = \begin{cases} k_1, s(x) > 0 \\ k_2, s(x) < 0 \end{cases} \tag{4.86}$$

where k_1 and k_2 are both constant value vectors of m dimension.
2. Function switching control method:

$$u = u^* + u_0 \text{sgn}[s(x)] \tag{4.87}$$

where, u^* is the equivalent control function when the state point of the control system moves near the switching surface of the sliding mode, and function $u_0 \text{sgn}[s(x)]$ is to ensure that the state point of the system can adjust back to the state of the sliding mode in time when it deviates from the trajectory of the sliding mode by external interference.
3. Proportional switching control method:

$$u = \sum_{i=1}^{k} f_i x_i \tag{4.88}$$

where, $k \leq n$, and the expression of the function f_i is:

$$f_i = \begin{cases} a_i, x_i s(x) > 0 \\ b_i, x_i s(x) < 0 \end{cases} \tag{4.89}$$

where a_i and b_i are constants.

4.5.1.3   Sliding mode structure control methods

The following introduces three basic types of variable structure control methods:

(1) Common Switching Control Method:

$$u = \begin{cases} k_1, s(x) > 0 \\ k_2, s(x) < 0 \end{cases} \tag{4.90}$$

where k_1 and k_2 are constant values in the m-dimensional space.

(2) Function Switching Control Method:

$$u = u^* + u_0 \text{sgn}[s(x)] \tag{4.91}$$

where, u^* is the control system state point when the state point is near the sliding mode switching surface. The function $u_0\text{sgn}[s(x)]$ ensures that the system's state point is not deviating too far from the sliding mode while ensuring the system moves toward the target state. This helps adjust the sliding mode state.

(3) Comparative Switching Control Method:

$$u = \sum_{i=1}^{k} f_i x_i \tag{4.92}$$

where $k \leq n$, and the function f_i is expressed as:

$$f_i = \begin{cases} a_i, x_i s(x) > 0 \\ b_i, x_i s(x) < 0 \end{cases} \tag{4.93}$$

where a_i and b_i are constant values.

4.5.1.4   Advantages and issues of sliding mode variable structure control

Sliding mode variable structure control has several advantages, such as achieving system decoupling, reducing the order of the object model, and providing good invariance and robustness.

(1) Achieving System Decoupling: During the sliding mode variable structure control process, the design of the sliding mode switching surface directly determines the system's state and performance. This makes the system's state and performance independent of system parameters, achieving system decoupling.

(2) Reducing the Order of the Object Model: When the system's state point enters the sliding mode, the state point is essentially constrained to move within the region near the termination point. As mentioned in Section 4.5.1.1, for an n-dimensional system, when its state point is constrained to move along an m-dimensional sliding mode switching surface, the system's sliding mode equation can be simplified and reduced to an n-m order equation. This greatly simplifies the study and analysis of the system's state, which is particularly useful for studying high-order systems.

(3) Good Invariance and Robustness: Once the system enters the sliding mode, changes in parameters or external disturbances will not affect the system's state.

Although sliding mode variable structure control has these obvious advantages, it also has an inherent problem that cannot be completely eliminated: "chattering." Since sliding mode variable structure control requires continuous switching of the state control function based on the switching function, it is a discontinuous control method. This results in inevitable "chattering" in the system. When applying sliding mode control to practical engineering systems, there are several factors that lead to "chattering" during the control process. The following lists and analyzes these factors:

1. The system inertia in actual control systems can cause a lag in the switching of control functions, leading to "chattering" in the system.
2. The time delay in the switching function causes the system state to change only after the control function switches, resulting in a lag before the control effect begins.
3. The spatial delay in the switching function causes a "dead zone" in the system variables in space.
4. Measurement errors in state observation directly affect the sliding mode switching surface, causing it to become unstable, which leads to random "chattering" in the control function switching.
5. If we apply sliding mode variable structure control to a discrete system, the discontinuity in sampling system state variables and the time delay caused by sampling will also affect the smoothness of the sliding mode control.

In response to the causes of "chattering" in sliding mode variable structure control, researchers have proposed methods to mitigate it. For example, they suggest designing more effective control methods to reduce system inertia or replacing the switching function with a smooth function. These methods have both advantages and disadvantages, and it is necessary to choose the appropriate one based on specific practical requirements.

4.5.2   Sensorless control module design based on current-type SMO

We build an SMO to estimate the back-EMF of the motor in real time and then extract the required rotor position and speed information from it, thus achieving sensorless motor control.

The mathematical model of a PMSM can be expressed as follows:

$$\begin{cases} u_\alpha = L_\alpha \dfrac{di_\alpha}{dt} + R_\alpha i_\alpha + e_\alpha \\[2mm] u_\beta = L_\beta \dfrac{di_\beta}{dt} + R_\beta i_\beta + e_\beta \end{cases} \tag{4.94}$$

$$\begin{cases} e_\alpha = -\sqrt{\dfrac{3}{2}}\psi_f \omega \sin\theta \\[4mm] e_\beta = \sqrt{\dfrac{3}{2}}\psi_f \omega \cos\theta \end{cases} \tag{4.95}$$

where, e_α and e_β are the components of the back-EMF on $\alpha\beta$-axis respectively, ω is the rotor angular velocity. It can be seen from (4.95) that the component of the motor's back-EMF e_α on α-axis is proportional to the product of the sine value of the motor rotor speed and the rotor position angle, and the component of the motor's back-EMF e_β on β-axis is proportional to the product of the cosine value of the motor rotor speed and the rotor position angle.

From (4.94), it can be obtained:

$$\begin{cases} \dfrac{di_\alpha}{dt} = -\dfrac{R_\alpha}{L_\alpha}i_\alpha - \dfrac{1}{L_\alpha}e_\alpha + \dfrac{1}{L_\alpha}u_\alpha \\ \dfrac{di_\beta}{dt} = -\dfrac{R_\beta}{L_\beta}i_\beta - \dfrac{1}{L_\beta}e_\beta + \dfrac{1}{L_\beta}u_\beta \end{cases} \tag{4.96}$$

According to the mathematical model of the stator voltage of PMSM and the theory of sliding mode variable structure, the switching function composed of the error between the estimated current signal $i*$ and the actual detected current signal i is selected:

$$s(x) = i^* - i \tag{4.97}$$

Therefore, the sliding mode switching plane is designed as:

$$s(x) = i^* - i = 0 \tag{4.98}$$

The switching method is selected as the constant switching control method mentioned in Section 4.5.1.2, and the design state control function is:

$$u = k\mathrm{sgn}(z) \tag{4.99}$$

where, k is the switching gain, a constant; $z = i^* - i$; $\mathrm{sgn}(z)$ is a symbolic function, then the graph of the function u in the plane coordinate system is shown in Figure 4.24.

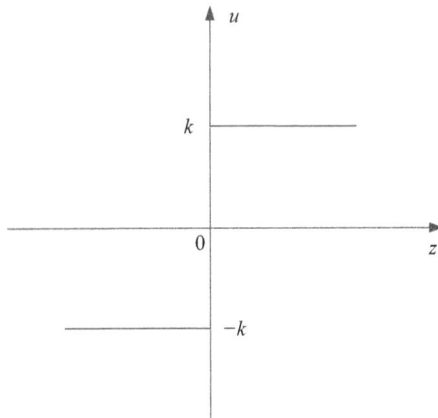

Figure 4.24   Switching function

The constant switching function (4.99) replaces the sum of the components of the back-EMF in the α-axis and β-axis in (4.96), and the state equation of the SMO of the current signal is constructed as follows:

$$
\begin{cases}
\dfrac{di_\alpha^*}{dt} = -\dfrac{R_\alpha}{L_\alpha}i_\alpha^* + \dfrac{1}{L_\alpha}u_\alpha - \dfrac{k}{L_\alpha}\mathrm{sgn}\left(i_\alpha^* - i_\alpha\right) \\[3mm]
\dfrac{di_\beta^*}{dt} = -\dfrac{R_\beta}{L_\beta}i_\beta^* + \dfrac{1}{L_\beta}u_\beta - \dfrac{k}{L_\beta}\mathrm{sgn}\left(i_\beta^* - i_\beta\right)
\end{cases}
\tag{4.100}
$$

By subtracting (4.96) from (4.100), the system error state equation of the SMO can be obtained as follows:

$$
\begin{cases}
\dfrac{d\hat{i}_\alpha}{dt} = -\dfrac{R_\alpha}{L_\alpha}\hat{i}_\alpha + \dfrac{1}{L_\alpha}e_\alpha - \dfrac{k}{L_\alpha}\mathrm{sgn}\left(\hat{i}_\alpha\right) \\[3mm]
\dfrac{d\hat{i}_\beta}{dt} = -\dfrac{R_\beta}{L_\beta}\hat{i}_\beta + \dfrac{1}{L_\beta}e_\beta - \dfrac{k}{L_\beta}\mathrm{sgn}\left(\hat{i}_\beta\right)
\end{cases}
\tag{4.101}
$$

where, $\hat{i}_\alpha = i_\alpha^* - i_\alpha$ and $\hat{i}_\beta = i_\beta^* - i_\beta$ is the observed error of the sum of the components of the stator current on the α-axis and β-axis. Next, to ensure that the system designed with the current mode of sliding mode observer can eventually achieve stability, we need to discuss the system's stability conditions. Therefore, a positive definite Lyapunov function is constructed for the sliding mode plane selected $s(x) = \left[\hat{i}_\alpha, \hat{i}_\beta\right]^T = 0$ above to determine the stability conditions of the sliding mode observer system. Build the Lyapunov function:

$$
V = \frac{1}{2}s(x)^T s(x)
\tag{4.102}
$$

Substituting $s(x)$ into the calculation yields:

$$
V = \frac{1}{2}\hat{i}_\alpha^2 + \frac{1}{2}\hat{i}_\beta^2
\tag{4.103}
$$

It can be seen from (4.103) that if $V > 0$, the Lyapunov function is positive definite. Based on the Lyapunov function derivation available:

$$
\dot{V} = \frac{1}{2}\hat{i}_\alpha \frac{d\hat{i}_\alpha}{dt} + \frac{1}{2}\hat{i}_\beta \frac{d\hat{i}_\beta}{dt}
\tag{4.104}
$$

According to the stability theory of sliding mode variable structure control system, if the current mode sliding mode observer is stable, $\dot{V} \leq 0$ must be satisfied. Substituting (4.101) into (4.104) yields:

$$
\dot{V} = -\frac{R_\alpha}{2L_\alpha}\left(\hat{i}_\alpha^2 + \hat{i}_\beta^2\right) + \frac{\hat{i}_\alpha}{2L_\alpha}\left[e_\alpha - k\mathrm{sgn}\left(\hat{i}_\alpha\right)\right] + \frac{\hat{i}_\beta}{2L_\beta}\left[e_\beta - k\mathrm{sgn}\left(\hat{i}_\beta\right)\right] \leq 0
\tag{4.105}
$$

Since $-(R_a/2L_a)\left(\hat{i}_\alpha^2 + \hat{i}_\beta^2\right)$ in formula (4.105) is less than 0, to ensure $\dot{V} \leq 0$, it is only necessary to ensure that the sum of the other two terms in formula (4.105) is less than or equal to 0, that is, to ensure that formula (4.106) is established, formula (4.106) is shown as follows:

$$k \geq \max\left(|e_\alpha|, |e_\beta|\right) \tag{4.106}$$

As long as a value of k satisfying (4.106) is selected, the system error equation of state is asymptotically stable. When the system state points after arriving at the sliding surface and stability, current observations will converge to current actual readings, has the type:

$$\begin{cases} s(x) = 0 \\ \dot{s}(x) = 0 \\ \hat{i}_\alpha = i_\alpha^* - i_\alpha = 0 \\ \hat{i}_\beta = i_\beta^* - i_\beta = 0 \\ \dfrac{d\hat{i}_\alpha}{dt} = 0 \\ \dfrac{d\hat{i}_\beta}{dt} = 0 \end{cases} \tag{4.107}$$

Substituting (4.107) into (4.101) yields:

$$\begin{cases} e_\alpha = k\mathrm{sgn}\left(i_\alpha^* - i_\alpha\right) \\ e_\beta = k\mathrm{sgn}\left(i_\beta^* - i_\beta\right) \end{cases} \tag{4.108}$$

The e_α and e_β obtained in this way will have a high discrete current error signal, so the signal is processed by low-pass filtering. High-frequency filters are used to filter the signal, and the estimated inverse EMF value obtained is shown as follows:

$$\begin{cases} \bar{e}_\alpha = \dfrac{\omega_c}{s + \omega_c} e_\alpha \\ \bar{e}_\beta = \dfrac{\omega_c}{s + \omega_c} e_\beta \end{cases} \tag{4.109}$$

where, ω_c is the cutoff frequency of the low-pass filter, $\bar{e}_\alpha$ and $\bar{e}_\beta$ the components of the motor's back-EMF on the α-axis and β-axis after the LPF processing respectively.

By getting the upper and lower formulas in (4.91) together and simplifying them, we can get:

$$\theta = -\arctan\left(\frac{e_\alpha}{e_\beta}\right) \tag{4.110}$$

Therefore, according to (4.105) and (4.110), the rotor position Angle $\bar{\theta}$ can be estimated as follows:

$$\bar{\theta} = -\arctan\left(\frac{\bar{e}_\alpha}{\bar{e}_\beta}\right) \tag{4.111}$$

Due to the use of an LPF in the previous step, the estimated $\bar{\theta}$ phase angle delay is inevitable. Therefore, an angle compensation is carried out for the estimated phase angle to obtain an accurate rotor position angle. The compensation angle $\Delta\theta$ can be calculated according to the control speed input ω_i and cut-off frequency ω_c of the motor when the motor is running:

$$\Delta\theta = \arctan\left(\frac{\omega_i}{\omega_c}\right) \tag{4.112}$$

Based on the above, the estimated value of rotor position angle θ_c and speed ω_e can be obtained as follows:

$$\begin{cases} \theta_e = \bar{\theta} + \Delta\theta \\ \omega_e = \dfrac{d\theta_e}{dt} \end{cases} \tag{4.113}$$

4.5.3   Optimization of the SMO

As discussed in Section 4.5.1.3, although sliding mode variable structure control systems can achieve system decoupling, system order reduction, and good invariance and robustness, the presence of system inertia, time delay and spatial delay in the switching function, and measurement errors in state observation can lead to the occurrence of "chattering." However, this phenomenon is an inherent issue in sliding mode control systems and cannot be completely eliminated. Therefore, it is necessary to take measures to suppress and mitigate "chattering." Several methods to reduce "chattering" have been mentioned earlier. In this section, we adopt a saturation switching function to replace the constant-value switching function, using a smoother switching function to mitigate the "chattering" issue. This approach is used to optimize the current-type sliding mode observer.

The saturation switching function designed in this chapter is *Hyperbolic Function*. With reference to the literature [10], this chapter aims to reduce the chattering phenomenon of SMO by smoothing out the control discontinuity and the step change within a boundary layer (BL) near the sliding surface. To achieve this target and obtain better performance, the switching function should satisfy the following requirements:

(a) The function is continuous.
(b) Referring to the saturation function, the upper and lower limit are 1 and -1.
(c) The slope within the BL is nonlinear.
(d) The function has no time-delay characteristic.

A hyperbolic function with the expression (4.114) is totally qualified, so it will be used in the new SMO.

$$\begin{bmatrix} F(\bar{i_\alpha}) \\ F(\bar{i_\beta}) \end{bmatrix} = \begin{bmatrix} \dfrac{e^{m\bar{i_\alpha}} - e^{-m\bar{i_\alpha}}}{e^{m\bar{i_\alpha}} + e^{-m\bar{i_\alpha}}} \\ \dfrac{e^{m\bar{i_\beta}} - e^{-m\bar{i_\beta}}}{e^{m\bar{i_\beta}} + e^{-m\bar{i_\beta}}} \end{bmatrix} \tag{4.114}$$

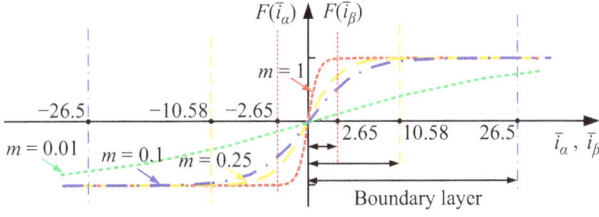

Figure 4.25   Boundary layers under different values of m

where m is a positive constant used for regulating the BL which is defined as the magnitude of the independent variable when $F = 0.99$, as in Figure 4.25. The improved SMO model used for e_α^* and e_β^* estimation can be eventually established.

Since a new switching function is adopted in the SMO, it is crucial to reappraise the stability of the observer. For the new SMO, the defined α, β-axis sliding surfaces (S_α, S_β) are $\bar{i}_\alpha$ and $\bar{i}_\beta$ which can be denoted as:

$$S = \begin{bmatrix} S_\alpha \\ S_\beta \end{bmatrix} = \begin{bmatrix} \bar{i}_\alpha \\ \bar{i}_\beta \end{bmatrix} \tag{4.115}$$

To ensure the stability of the designed observer, the following equation is necessary based on Lyapunov function:

$$V = \frac{1}{2} \cdot S^T \cdot S = \frac{1}{2}\bar{i}_\alpha^2 + \frac{1}{2}\bar{i}_\beta^2 \tag{4.116}$$

Obviously, $V > 0$. Then, according to the Lyapunov stability decision theorem, only by deducing $\frac{dV}{dt} < 0$ can we conclude that the SMO can reach a stable state. Take the time derivative of (4.116):

$$\frac{dV}{dt} = S^T \cdot \frac{dS}{dt} = \bar{i}_\alpha \frac{d\bar{i}_\alpha}{dt} + \bar{i}_\beta \frac{d\bar{i}_\beta}{dt} \tag{4.117}$$

It can be further derived as:

$$\frac{dV}{dt} = \begin{bmatrix} \bar{i}_\alpha & \bar{i}_\beta \end{bmatrix} \begin{bmatrix} -\frac{R_s}{L_s}\bar{i}_\alpha + \frac{1}{L_s}\left(e_\alpha - kF(\bar{i}_\alpha)\right) \\ -\frac{R_s}{L_s}\bar{i}_\beta + \frac{1}{L_s}\left(e_\beta - kF(\bar{i}_\beta)\right) \end{bmatrix} \tag{4.118}$$

And the polynomial expression of (4.118) is:

$$\frac{dV}{dt} = \underbrace{\left[-\left(\frac{R_s}{L_s}\bar{i}_\alpha^2 + \frac{R_s}{L_s}\bar{i}_\beta^2\right)\right.}_{\text{term1}} + \underbrace{\left.\left[\frac{1}{L_s}(e_\alpha - kF(\bar{i}_\alpha))\bar{i}_\alpha + \frac{1}{L_s}(e_\beta - kF(\bar{i}_\beta))\bar{i}_\beta\right]\right.}_{\text{term2}} \tag{4.119}$$

Obviously, term 1 is less than 0. In order to keep the SMO stable, both term 1 and term 2 are expected to be less than 0. Therefore, the observer gain can be derived to meet the following criteria:

$$k > \max \left(\left| \frac{e_\alpha}{F(\bar{i}_a)} \right|, \left| \frac{e_\beta}{F(\bar{i}_\beta)} \right| \right) \tag{4.120}$$

Clearly, the stability condition differs from another, and it can be noted that k should be much larger because $|F|$ is less than 1 over the BL range. In practice, $\bar{i}_\alpha$ and $\bar{i}_\beta$ are the estimation errors, and referring to the sliding mode theory, they will fluctuate mostly around the sliding surfaces with small variations in the range of tolerance. Define the lower tolerance as ξ, which is:

$$\min \left(|\bar{i}_\alpha|, |\bar{i}_\beta| \right) = \xi \tag{4.121}$$

And the minimum $|F|$ can be rewritten as:

$$\min |F| = \frac{e^{m\xi} - e^{-m\xi}}{e^{m\xi} + e^{-m\xi}} \tag{4.122}$$

Therefore, the observer gain can be selected as:

$$k = \left| \frac{3\sqrt{3} C_e \, e \Psi_f p \omega_{m\_max}}{2 \cdot \min |F|} \right| \tag{4.123}$$

where ω_{m_max} is the maximum speed of the machine and it can be set as the rated value. C'_e is the voltage constant relevant to the motor. The parameter k can guarantee the reachability of the observer, and during the control process, although $\bar{i}_\alpha$ and $\bar{i}_\beta$ might be smaller than ξ when the system gets to the equilibrium state, the proposed SMO can re-converge afterward. It is different from the traditional sliding form (Figure 4.26).

At the stage of state measurement, the armature currents measured by physical sensors need to pass through analog signal processing circuits (SPC). However, the SPC not only filters out the high-frequency signals that are far beyond the operating frequencies but also brings about low-pass filter delay. Taking a common SPC that

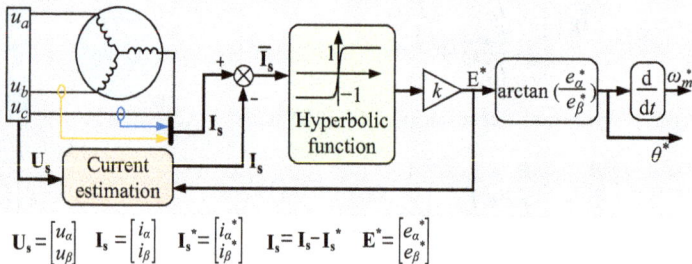

Figure 4.26   Structure of the proposed SMO

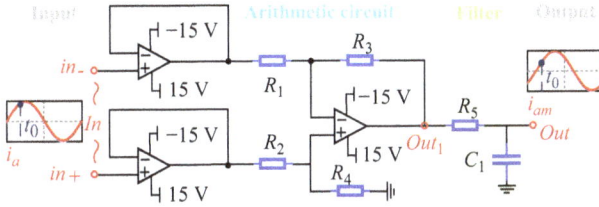

Figure 4.27   Commonly used SPC that causes phase delay to AC currents

is mainly composed of an arithmetic circuit and filter (see Figure 4.27) as an example, if a sinusoidal current signal i_a passes through it, the output signal i_{am} will lag behind. In this case, the measured current cannot accurately represent the real-time current, especially when the motor rotates over the high-speed range because the delay effect will increase as the working frequency rises.

As for the SPC in Figure 4.27 because the arithmetic circuit and the filter are cascaded, the transfer function of the overall SPC $G(s)$ can be expressed as:

$$G(s) = \underbrace{\frac{Out_1(s)}{In(s)}}_{t\_ari(s)} \cdot \underbrace{\frac{Out(s)}{Out_1(s)}}_{t\_fil(s)} \tag{4.124}$$

where $In(s)$ and $Out(s)$ are the input and output of the overall circuit in the s domain, and $Out_1(s)$ is an intermediate variable. $tf\_ari(s)$ and $tf\_fil(s)$ represent the transfer functions of the arithmetic circuit and the filter, respectively.

The relationship between the input and output of the arithmetic circuit and filter in the s domain can be written as:

$$\begin{cases} Out_1(s) = \dfrac{R_4(R_1 + R_3)}{R_1(R_2 + R_4)} \cdot in_+(s) - \dfrac{R_3}{R_1} \cdot in_-(s) \\ Out(s) = \dfrac{1}{R_5 C_1 s} \cdot Out_1(s) \end{cases} \tag{4.125}$$

where $R_{1,2,3,4,5}$ is resistance, C_1 is capacitance, $in_+(s)$ and $in_-(s)$ are the positive and negative terminals of the input, respectively.

It can be seen from (4.125) that as for the arithmetic circuit, the output equals the input when it meets the following condition:

$$\begin{cases} R_1 - R_3 = 0 \\ R_2 - R_4 = 0 \end{cases} \tag{4.126}$$

In other words, $tf\_ari(s)$ can be written as (4.127) after substituting (4.126) into (4.125):

$$tf\_ari(s) = \frac{Out_1(s)}{in_+(s) - in_-(s)} = \frac{Out_1(s)}{In(s)} = 1 \tag{4.127}$$

Moreover, based on (4.125), $tf\_fil(s)$ can be described as:

$$tf\_fil(s) = \frac{1}{R_5 C_1 s + 1} \tag{4.128}$$

Based on the aforementioned analysis, the LPF delay only comes from the filter for the SPC in Figure 4.27. As for the motor rotating at ω_m, the current phase delay caused by the filter is:

$$\Delta\varphi_{SPC} = \arctan\left(R_5 C_1 p\omega_m\right) \tag{4.129}$$

Considering the LPF delay resulting from the SPC, an effective solution to the delay issues is to compensate for the phase lag of the measured currents. On this ground, a current pre-estimation-based compensation method is proposed in this section, which is constituted of three parts: delay calculation, phase calculation, and current pre-estimation. Since the phase delays can be calculated by (4.129) analytically, the latter two parts need to be focused on mainly.

4.5.3.1   Phase calculation

Denote the measured a, b-phase currents as i_{am} and i_{bm}, respectively, and the phase currents can be adjusted to the range of -1 and 1 by dividing them by the real-time current amplitude:

$$i'_{am} = \frac{i_{am}}{\sqrt{\kappa\left(i_d^2 + i_q^2\right)}}, i'_{bm} = \frac{i_{bm}}{\sqrt{\kappa\left(i_d^2 + i_q^2\right)}} \tag{4.130}$$

where κ is a coefficient related to coordinate transformation. Namely, it is 1 and 1.5 for the equal amplitude transformation and equal power transformation, respectively. i'_{am} and i'_{bm} are the currents ranging from -1 to 1. No doubt that (4.130) is effective because the d, q-axis currents remain constant within a short period even at the stage of dynamic regulation.

Further, the real-time phases of the armature currents can be calculated by using the inverse of the sine function "arcsin," that is,

$$\begin{cases} \varphi_{am} = \arcsin\left(i'_{am}\right), \varphi_{bm} = \arcsin\left(i'_{bm}\right) \\ 0 \leq \varphi_{am} < 2\pi, 0 \leq \varphi_{bm} < 2\pi \end{cases} \tag{4.131}$$

where φ_{am} and φ_{bm} are the calculated a- and b-phase currents, respectively. However, as shown in Figure 4.5, when using (4.131) to calculate φ_{am} and φ_{bm}, there are always two possible values for each of them. Now, a new issue of how to determine which one is correct arises.

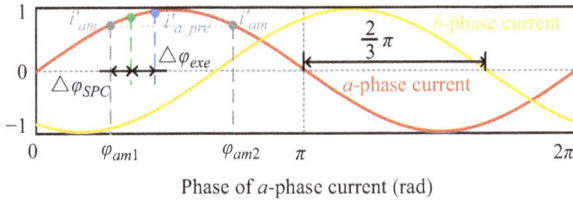

Figure 4.28   Phase calculation and current pre-estimation of a-phase current and relationship of a, b-phase currents

Because the *b*-phase current of a PMSM lags behind the *a*-phase current with $\frac{2}{3}\pi$ (see Figure 4.28), φ_{am} and φ_{bm} should satisfy the following condition:

$$\begin{cases} \varphi_{bm} = \varphi_{am} - \dfrac{2}{3}\pi + \lambda \cdot 2\pi \\ \lambda = 0, \text{if } \varphi_{am} \geq \dfrac{2}{3}\pi \\ \lambda = 1, \text{if } \varphi_{am} < \dfrac{2}{3}\pi \end{cases} \tag{4.132}$$

Denote the two calculated values of φ_{am} as φ_{am1} and φ_{am2} and φ_{bm} as φ_{bm1} and φ_{bm2}, respectively, and construct an optimization problem as follows:

$$\begin{cases} \varphi_{am\_c} = \varphi_{ami}, \varphi_{bm\_c} = \varphi_{bmj} \\ \text{s.t.min} \left| \varphi_{ami} - \dfrac{2}{3}\pi + \lambda \cdot 2\pi - \varphi_{bmj} \right|, i = 1, 2; j = 1, 2 \end{cases} \tag{4.133}$$

Theoretically, the correct current phases are φ_{am_c} and φ_{bm_c}.

4.5.3.2   Current pre-estimation

Since the phase delays as well as the real-time current phases are obtained, as shown in Figure 4.5, the currents with delays compensated can be derived as:

$$\begin{cases} i'_{a\_pre} = \sin\left(\varphi_{am\_c} + \Delta\varphi_{SPC} + \Delta\varphi_{exe}\right) \\ i'_{b\_pre} = \sin\left(\varphi_{bm\_c} + \Delta\varphi_{SPC} + \Delta\varphi_{exe}\right) \end{cases} \tag{4.134}$$

where i'_{a_pre} and i'_{b_pre} are the pre-estimated currents ranging from -1 to 1. Further, the real pre-estimated currents i_{a_pre} and i_{b_pre} can be obtained:

$$i_{a\_pre} = \sqrt{\kappa\left(i_d^2 + i_q^2\right)} \cdot i'_{a\_pre}, i_{b\_pre} = \sqrt{\kappa\left(i_d^2 + i_q^2\right)} \cdot i'_{b\_pre} \tag{4.135}$$

Using the compensated current to estimate the position and speed, delay can be avoided.

4.6   New method for full-speed ranges

The previous sensorless control algorithm cannot be applied to the full-speed domain. Second, for position and speed estimation methods used in the low-speed range, several issues exist. First, the necessary condition for implementing high-frequency injection methods is that the motor must exhibit saliency, that is, $L_d \neq L_q$. Therefore, for surface-mounted PMSMs, the saturation saliency effect of the motor must be utilized for the high-frequency injection methods. This requires the internal magnetic field of the motor to be strong enough so that the iron core operates near the region of magnetic saturation. Otherwise, the high-frequency injection method cannot be utilized. Second, the high-frequency injection method requires the injection of additional current, which cannot produce effective output electromagnetic torque. This results in lower motor efficiency, and the magnetic field generated by high-frequency signals will reduce the motor's stable operation capability. Lastly, the injected high-frequency signals generate noise, which can cause discomfort to nearby personnel. Third, back-EMF-based strategies are typically used for high speeds. However, back-EMF estimation-based methods are proven ineffective for low-speed applications because, at low speeds, the back-EMF is too small, making it difficult to obtain accurate information through arctangent functions or PLLs.

It is important to note that in practice, state variables related to rotor position information are not limited to back-EMF. For instance, motor speed, which is the most closely related variable, is usually ignored in practice. Practically, by integrating motor speed and refining estimation algorithms, there is potential to develop a more comprehensive and reliable sensorless control strategy for PMAs over full-speed range.

4.6.1   Rotor speed estimation based on SMO

In this study, the SMO is selected for speed estimation due to its robustness against system uncertainties and disturbances, which are crucial in ensuring accurate and reliable speed estimation in low-speed sensorless control of PMSMs.

4.6.1.1   Structure of SMO

Based on the sliding mode variable structure theory, it is straightforward to establish an SMO used for speed estimation. It involves substituting the motor model states (i_q and ω_e) in the math model with their estimated counterparts, where the final estimated value ω_e can be expressed using an equation that includes a switching function. On this basis, the SMO used for speed estimation can be designed as:

$$
\begin{aligned}
\frac{di_q^*}{dt} &= -\frac{1}{L_q}(L_d i_d + \psi_f)\omega_e - \frac{R_s}{L_q}i_q^* + \frac{u_q}{L_q} \\
&= -\frac{1}{L_q}(L_d i_d + \psi_f)\lambda_s F(\bar{i}_q) - \frac{R_s}{L_q}i_q^* + \frac{u_q}{L_q}
\end{aligned}
\tag{4.136}
$$

where λ_s is gain factor of the observer; i_q^* is estimated current of the observer; ω_e^* is the estimated electrical rotor speed; $\bar{i}_q$ is the error between the estimated and the real currents, namely,

$$\bar{i}_q = i_q^* - i_q \tag{4.137}$$

$F(\bar{i}_q)$ is the switching function. For the sake of low estimation errors, it is designed as a signum function, that is,

$$F(\bar{i}_q) = \begin{cases} 1 & \text{if} \quad \bar{i}_q \geq 0 \\ -1 & \text{if} \quad \bar{i}_q < 0 \end{cases} \tag{4.138}$$

Theoretically, once the SMO reaches an equilibrium state, according to the SMO's invariance property on the sliding surface, the estimated speed ω_e^* can be calculated by:

$$\omega_e^* = \lambda_s F(\bar{i}_q) \tag{4.139}$$

It is important to note that regardless of whether the motor is operating at high or low speeds, where mechanical and electrical dynamics differ, the observer remains capable of accurately estimating the speed due to its inherent sliding mode properties and robustness to frequency variations. To minimize fluctuations in the estimated speed resulting from chattering or start-up, a low-pass filter can be employed to smooth the signals. Then, the high-accuracy mechanical motor speed ω_m^* used for control is:

$$\omega_m^* = \frac{\omega_e^*}{p} \tag{4.140}$$

where p is the number of pole pairs.

4.6.1.2 Stability analysis

To analyze the stability of the SMO, a sliding surface $\mathbf{S}$ is defined as follows:

$$\mathbf{S} = \begin{bmatrix} \bar{i}_q \end{bmatrix} \tag{4.141}$$

Then, a Lyapunov function $\mathbf{V}$ can be constructed as follows:

$$\mathbf{V} = \frac{1}{2}\mathbf{S} \cdot \mathbf{S}^T = \frac{1}{2}\begin{bmatrix} \bar{i}_q^2 \end{bmatrix} \tag{4.142}$$

Further, take the derivative of (4.142) and it can be obtained that:

$$\frac{d\mathbf{V}}{dt} = \begin{bmatrix} \dfrac{d\bar{i}_q}{dt}\bar{i}_q \end{bmatrix} \tag{4.143}$$

And it can be derived that:

$$\frac{d\mathbf{V}}{dt} = \begin{bmatrix} \underbrace{-\frac{R_s}{L_q}\bar{i}_q^2}_{\text{first tem}} + \underbrace{\frac{p(L_d i_d + \psi_f)(\omega_e - \lambda_s F(\bar{i}_q))}{L_q}}_{\text{second term}} - \bar{i}_q \end{bmatrix} \tag{4.144}$$

From (4.142), it can be noticed that $\mathbf{V} > 0$, which is one of the necessary conditions for making the SMO stable. Then, as long as the derivative of $\mathbf{V}$ is less than zero, the SMO remains stable, that is:

$$\frac{d\mathbf{V}}{dt} < 0 \tag{4.145}$$

In (4.144), the first term is less than zero obviously. As for the second term, (4.143) will be satisfied constantly when it is less than zero, that is,

$$\frac{p\left(L_d i_d + \psi_f\right)\left(\omega_e - \lambda_s F\left(\bar{i}_q\right)\right)}{L_q} \bar{i}_q < 0 \tag{4.146}$$

Considering the sign of $\bar{i}_q$, (4.146) can be further derived as:

$$\begin{cases} \omega_e - \lambda_s < 0, & \text{if } \bar{i}_q \geq 0 \\ \omega_e + \lambda_s > 0, & \text{if } \bar{i}_q < 0 \end{cases} \rightarrow \lambda_s > |\omega_e| \tag{4.147}$$

Overall, λ_s should be set as a positive constant which is larger than the maximum electrical rotor speed. Considering that the proposed SMO is used for low-speed situations, there must be a value for λ_s within the working range to satisfy (4.147).

4.6.2   *Rotor position calculation*

After the speed information is obtained by using the sliding mode speed observer, the position can be obtained by using integral operation, namely,

$$\theta^* = \theta_0 + \int_0^t \omega_e^* \; dt \tag{4.148}$$

where θ^* is the estimated position; θ_0 is the initial rotor speed; t represents time. Obviously, the speed observer-based position estimation method does not require high-frequency signals and is not limited to IPMSMs.

4.6.3   *Rotor pre-positioning technique for initial position detection*

In (4.147), the initial rotor position is needed to calculate the real-time position, which is also the position when the motor is at the standstill state (zero speed). To avoid using a high-frequency signal injection-based position estimation strategy, this part presents a simple pre-positioning method to determine the initial rotor position [11].

As shown in Figure 4.29(a), according to the basic principles of vector control, the initial position of the d-axis (zero position in vector control, that is, $\theta = 0$) is defined as the rotor position when the magnetic field generated by the permanent magnet coincides with the a-phase axis ($A_1 A_2$). For the a-phase axis, it also represents the direction of the magnetic field generated after the current is applied to the a-phase winding. Theoretically, regardless of the b and c phases, if a direct

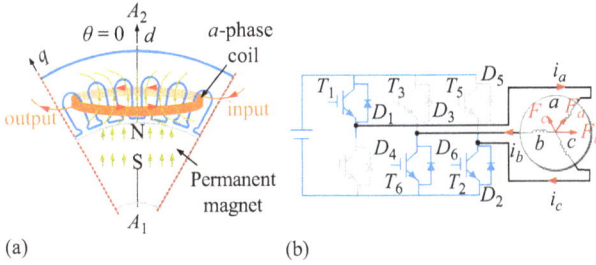

Figure 4.29 Foundations of pre-positioning method: (a) physical explanations of zero position in vector control and (b) operating mechanism of power electronics for proposed pre-positioning method

current is applied only to the a-phase winding, the rotor will be forcibly dragged and fixed to either the zero position or the reverse of the zero position, driven by the magnetic force between the stator and rotor. This paves the way for using the magnetic field generated by the stator winding to fix the rotor at a known position, which serves as θ_0 in (4.148).

However, for a three-phase motor, since the three-phase windings form a closed circuit, we cannot apply current only to the a-phase winding. Instead, when current is applied to the a-phase winding, current will also flow through the b and c-phases, generating corresponding magnetic fields. These fields may affect the rotor, causing it to be fixed at an uncertain position. To address this issue, assume that T_1, T_2, and T_6 are switched on in Figure 4.29(b). In this case, a direct current flows from the positive terminal of the power supply through the a-phase winding and returns to the negative terminal via the b and c-phase windings. Since the impedances of the b and c-phase windings are almost equal, the three-phase currents i_a, i_b, and i_c satisfy:

$$i_b = i_c = -\frac{i_a}{2} \tag{4.149}$$

Then, the relationship between the magnetic fields F_a, F_b, and F_c generated by the three-phase windings is illustrated in Figure 4.29(b). It can be seen that the direction of the resultant magnetic field remains aligned with the a-phase axis. At this time, the rotor can also be forcibly dragged to the zero position.

Based on the above analysis, when the motor is at a standstill state, we can turn on T_1, T_2, and T_6 by using pulse width modulation (PWM) signals. Then, the motor position is:

$$\theta_0 = 0 \text{ or } \pi \tag{4.150}$$

It needs to be noticed that, first, the value of θ_0, whether 0 or π, depends on the direction of the current flowing through the winding, which determines the polarity of the permanent magnet aligned with the a-phase axis. For example, as shown in Figure 4.29, if the current flows counterclockwise and the N-pole is near the winding side, then θ_0 is 0; otherwise, θ_0 is π. In practice, the value of θ_0 can be determined by identifying which of the two ends of one coil the voltage is applied to, in conjunction with the winding direction. Second, the accuracy of the above initial position detection method primarily depends on the current flowing through the motor when T_1, T_2, and T_6 are turned on. As the applied current increases, the magnetic forces exerted by the stator windings on the rotor become stronger, thereby enhancing the rotor's alignment to the zero position. By carefully controlling the current magnitude during this pre-positioning process, the method ensures that the rotor is accurately aligned to the desired initial position, which is critical for subsequent sensorless control operations. Finally, the real-time position of the motor during rotation can be further obtained.

Chapter 5

Coordination of multiple motors in robotic systems

5.1 Introduction to coordination of multiple motors

The coordination of multiple motors in robotic systems is a fundamental concept that lies at the heart of advanced robotic functionality. It involves the precise synchronization and control of two or more motors to achieve a unified goal. This is especially critical in scenarios where seamless motion, efficient energy usage, and precise task execution are required. Multi-motor coordination ensures that robotic systems function harmoniously, reducing mechanical stress, enhancing reliability, and optimizing overall performance. In an era where robotics is integral to industrial automation, healthcare, and service applications, mastering the art of coordinating multiple actuators is crucial [12].

The importance of motor coordination becomes particularly evident in tasks demanding high precision and efficiency. Dual-arm robots, for example, require precise coordination between both arms to execute complex assembly operations, such as handling delicate components or performing intricate manipulations. Similarly, mobile robots rely on synchronized motor coordination to navigate accurately, particularly in applications such as autonomous vehicles or warehouse automation. Collaborative robots, often referred to as cobots, necessitate advanced coordination strategies to safely and effectively interact with human coworkers or other robots. In these cases, coordination enables not only functional operation but also adaptability to dynamic and unpredictable environments.

When considering the broader context, the coordination of multiple motors allows robotic systems to meet increasingly stringent performance requirements. Smooth and precise operations are vital for reducing operational wear and tear, thereby extending the lifespan of the robotic components. Moreover, such coordination minimizes energy consumption by eliminating redundant or counter-productive motor movements. This efficiency contributes to the overall sustainability and cost-effectiveness of robotic solutions, particularly in industrial settings where large-scale deployments can amplify even small efficiency gains.

Coordination strategies can be broadly classified into two categories: synchronized and asynchronous coordination. These approaches reflect different methodologies for achieving the overarching objective of unified multi-motor control. Each category is suited to specific operational scenarios, depending on the level of interdependence and interaction required between the motors.

Synchronized coordination represents a scenario where all motors involved work in perfect harmony, adhering to a pre-defined trajectory or pattern. This type of coordination ensures that each motor maintains a precise positional, speed, or torque relationship with the others. It is particularly critical in robotic systems requiring high degrees of precision and uniformity. For example, dual-arm robots executing a complex task must move in perfect synchronization to maintain balance and accuracy. This is evident in assembly operations, where precise handling of components is essential. Similarly, mobile robots require synchronized coordination among their wheels to ensure smooth navigation and accurate trajectory tracking. Any discrepancy in the operation of individual wheels could result in inefficiencies or errors in motion, potentially compromising the robot's performance.

Other examples of synchronized coordination can be found in gantry systems, often employed in pick-and-place operations. Here, motors must work together to ensure accurate movement along multiple axes. Without proper synchronization, the system risks losing alignment, which could result in operational inefficiencies or damage to the objects being manipulated. These examples underscore the critical role of synchronized coordination in achieving seamless and accurate operations across a wide range of robotic applications.

On the other hand, asynchronous coordination involves motors operating independently yet in a cooperative manner. In this scenario, each motor may have its unique task or timing, but the overall system objectives are achieved without conflict. Asynchronous coordination is particularly relevant in applications where tasks are distributed across multiple actuators with varying requirements. For instance, collaborative robots often rely on asynchronous coordination to interact with humans and adapt to changing environments. These robots need to independently control each joint or actuator to perform tasks such as handing over tools or assisting in assembly operations. Unlike synchronized coordination, where uniformity is paramount, asynchronous coordination prioritizes flexibility and adaptability.

Conveyor systems offer another example of asynchronous coordination. In such systems, different motors may control distinct sections of a conveyor, each operating at different speeds or with varying timing to manage the flow of products effectively. This approach ensures that the system can handle varying workloads and product characteristics without compromising efficiency. Similarly, in heterogeneous robotic systems, which involve diverse actuators or robots working on different subtasks, asynchronous coordination plays a vital role in ensuring a seamless workflow. By allowing each component to operate independently while maintaining overall harmony, asynchronous coordination facilitates the execution of complex tasks that would be challenging to achieve through synchronized coordination alone.

The choice between synchronized and asynchronous coordination depends on several factors, including the nature of the task, the degree of precision required, and the complexity of the system. Synchronized coordination is ideal for applications where uniformity and precision are critical, while asynchronous coordination is better suited to scenarios requiring flexibility and adaptability. However, in many cases, a combination of both approaches is necessary to achieve optimal performance. For example, a robotic system may use synchronized coordination for certain components while allowing others to operate asynchronously to accommodate dynamic changes in the environment or task requirements.

5.2   Goals of multi-motor coordination

In robotics, multi-motor coordination plays a pivotal role in enabling the seamless operation of complex systems. Robots often require the precise and synchronized movement of multiple actuators to perform intricate tasks with high precision, efficiency, and robustness. This section explores the fundamental goals of multi-motor coordination, which include precision and accuracy, efficiency, robustness, scalability, and the achievement of coordinated motion in robotic systems. These goals not only ensure optimal system performance but also address challenges in dynamic environments and evolving application requirements.

5.2.1   *Precision and accuracy*

Precision and accuracy are critical in multi-motor coordination for robotic systems, as they directly impact the quality and reliability of task execution. In robotic applications such as surgical robots, assembly lines, and autonomous vehicles, even minor deviations in motor control can result in significant errors or system failures. The need for precise motor control arises from the inherent interdependence of robotic components, where the movement of one motor influences the operation of others.

Accurate coordination ensures that motors follow their intended trajectories, positions, and velocities without deviation. This involves advanced control algorithms that can handle nonlinearities, coupling effects, and disturbances in the system. For instance, field-oriented control (FOC) and model predictive control (MPC) are commonly used to achieve high levels of precision in multi-motor systems. These control methods account for dynamic interactions between motors and enable precise trajectory tracking, which is essential for applications like robotic arm manipulation or drone navigation.

Precision also extends to the synchronization of motor movements. In scenarios where multiple motors must operate concurrently—such as in multi-axis robotic arms—any mismatch in timing can lead to mechanical stress, reduced efficiency, and task failure. To address this, multi-motor coordination strategies often incorporate real-time communication protocols and distributed control systems that ensure synchronized operations. These systems leverage high-speed data exchange and feedback loops to minimize delays and discrepancies, thereby achieving the desired level of precision.

5.2.2   *Efficiency*

Efficiency is a fundamental goal of multi-motor coordination, as it directly influences the energy consumption, operational costs, and overall sustainability of robotic systems. Coordinated motor control optimizes energy usage by reducing unnecessary movements, minimizing friction and heat generation, and distributing workload evenly among actuators. This is particularly important in battery-powered robots, where energy efficiency determines the robot's operational lifespan and range.

Energy-efficient coordination requires intelligent control strategies that balance performance and power consumption. For example, in mobile robots, motors must coordinate to achieve smooth acceleration and deceleration, reducing the energy lost to sudden starts and stops. Similarly, in industrial robots, optimizing the torque and speed profiles of motors can significantly lower energy consumption while maintaining high throughput.

Another aspect of efficiency is the reduction of wear and tear on robotic components. By coordinating motors to operate within their optimal performance ranges, the system can minimize mechanical stress and extend the lifespan of actuators and other components. This not only lowers maintenance costs but also enhances the reliability of the robot.

In addition to energy savings and reduced wear, efficient coordination contributes to faster task completion. By synchronizing motor movements and optimizing motion trajectories, robots can perform tasks more quickly without sacrificing accuracy. This is particularly beneficial in time-sensitive applications, such as disaster response robots or high-speed manufacturing systems, where both speed and efficiency are paramount.

5.2.3 Robustness

Robustness is a crucial goal in multi-motor coordination, as it ensures the reliability and fault tolerance of robotic systems in unpredictable environments. Robots often operate in dynamic settings where external disturbances, sensor noise, and component failures can compromise performance. A robust coordination strategy must be able to adapt to these challenges and maintain system stability and functionality.

One approach to achieving robustness is through redundancy and fault-tolerant control. In multi-motor systems, redundancy involves having additional motors or actuators that can take over the functions of failed components. Fault-tolerant control algorithms detect and isolate faults in real time, reconfiguring the system to compensate for the loss of functionality. For instance, in a robotic arm with multiple joints, if one motor fails, the control system can redistribute the workload among the remaining motors to continue the task.

Robust coordination also involves handling uncertainties and disturbances in the system. External factors such as varying loads, environmental conditions, and interactions with unknown objects can introduce unpredictability. Advanced control techniques, such as adaptive control and robust MPC, can accommodate these uncertainties by dynamically adjusting motor commands based on real-time feedback.

In addition to fault tolerance and disturbance rejection, robustness extends to the durability of the coordination strategy itself. The control algorithms and communication protocols must be designed to withstand hardware and software limitations, ensuring consistent performance over extended periods. This is particularly important in critical applications, such as surgical robots or autonomous vehicles, where failures can have severe consequences.

5.2.4   Scalability

Scalability is an essential consideration in multi-motor coordination, as robotic systems are becoming increasingly complex with the addition of more actuators, sensors, and computational resources. A scalable coordination strategy must accommodate the growing number of components without compromising performance or introducing excessive computational overhead.

One of the challenges in scalability is managing the increased communication and computation requirements associated with larger systems. As the number of motors increases, so does the complexity of their interactions and the volume of data exchanged between components. Scalable coordination strategies address this by adopting hierarchical or distributed control architectures. In hierarchical systems, higher-level controllers oversee the overall system objectives, while lower-level controllers manage individual motors or subsystems. This reduces the computational burden on any single controller and allows the system to scale efficiently.

Distributed control architectures further enhance scalability by enabling decentralized decision-making. Each motor or subsystem operates semi-independently, communicating with its neighbors to achieve coordinated behavior. This approach reduces the reliance on centralized control and improves the system's resilience to failures or communication delays.

Scalability also involves designing control algorithms that can handle the increased dimensionality of the system. For example, in swarm robotics, where hundreds or thousands of robots must coordinate their movements, algorithms must efficiently compute and execute control commands in real time. Techniques such as multi-agent reinforcement learning and graph-based coordination methods are often employed to achieve scalable solutions.

5.2.5   Coordinated motion of robots

Coordinated motion is the ultimate goal of multi-motor coordination, as it enables robots to perform complex tasks that require the synchronized movement of multiple actuators. This includes tasks such as object manipulation, locomotion, and collaborative operations between multiple robots. Coordinated motion ensures that all motors work together harmoniously, achieving the desired outcomes without conflicts or inefficiencies.

In robotic arms, coordinated motion involves synchronizing the movements of joints to achieve smooth and accurate end-effector trajectories. This requires precise control of each joint's position, velocity, and acceleration, as well as real-time adjustment based on feedback from sensors. For example, in pick-and-place tasks, the arm must coordinate its joints to move the gripper along a specific path while avoiding obstacles and maintaining stability.

In mobile robots, coordinated motion extends to the interaction between wheels or tracks. For instance, differential-drive robots rely on the precise coordination of left and right wheel motors to achieve smooth turns and straight-line motion. Similarly, quadruped or humanoid robots require the synchronization of multiple leg joints to maintain balance and perform dynamic movements such as walking or jumping.

Coordinated motion also applies to collaborative robotics, where multiple robots work together to achieve a common goal. In these scenarios, coordination involves not only individual motor control but also the interaction between robots. This includes tasks such as carrying large objects, assembling complex structures, or performing synchronized movements in swarm robotics. Effective coordination in such cases requires advanced communication protocols, shared task planning, and conflict resolution mechanisms.

In conclusion, the goals of multi-motor coordination in robots—precision and accuracy, efficiency, robustness, scalability, and coordinated motion—are fundamental to the development of high-performance robotic systems. By addressing these goals, researchers and engineers can design robots that operate reliably and efficiently in diverse and challenging environments, paving the way for advancements in automation, healthcare, manufacturing, and beyond.

5.3   Challenges in multi-motor coordination

Apart from the challenges associated with single-motor drives, such as reliability, thermal management, and system optimization, etc., multi-motor drives also pose several unique challenges due to the increased complexity of the system. First, coordination control is essential for the successful operation of multi-motor drives, as it allows the motors to work together seamlessly to achieve the desired performance objectives. However, achieving coordination control in multi-motor drives is a complex task because it requires the design and implementation of system-level control schemes that can coordinate the operation of multiple motors. Second, mutual interference refers to the phenomenon where the operation of one motor affects the operation of another motor in the system, leading to performance degradation or even system failure. Mutual interference occurs because of electromagnetic coupling, cross-coupling of control (CCC) and loads, and physical connections of motors. In practice, developing techniques that can mitigate the effects of mutual interference is crucial. Third, communication is also an important aspect of multi-motor drives, as it enables the exchange of information between different components of the system, such as the motors, controllers, and sensors. Communication modes and protocols need to be designed to ensure reliable and secure communication between the system components. Fourth, faults in one motor can affect the operation of the entire system. However, the fault diagnosis challenges associated with multi-motor drives are substantial, primarily because of the mutual interaction between various motors and their working statuses, as well as the presence of hybrid faults. Finally, multi-motor drives may cause power quality reduction in their power systems, such as microgrids. Power quality issues such as voltage fluctuations and high reactive power can impact the performance of other connected devices or loads (DOLs) and may even lead to equipment failures.

5.4   Solutions for multi-motor coordination

5.4.1   Coordination control

Coordination control involves coordinating the movements and actions of each motor, ensuring that they work in harmony with each other and the overall system. The general goals of coordination control of multi-motor drives can vary depending on the specific application requirements. In some cases, synchronization of the motors may be necessary to achieve coordinated motion, while in other cases, asynchronization may be required to allow each motor to operate independently.

5.4.1.1   Synchronization cases

In applications where high synchronization is desired, all motors should work at the same speed in real time. In this regard, the mechanical cascaded configuration offers a significant advantage over other configurations due to its inherent properties. This structure can achieve synchronism without requiring any significant control scheme improvements. Hence, current research on improving synchronization from a control perspective mainly focuses on mechanical parallel configurations. To address this gap, we introduce system-level control schemes for two configurations: "mechanical parallel + electrical parallel (MP-EP)" and "mechanical parallel + electrical cascaded (MP-EC)," which aim to ensure or improve system synchronization.

There are several control schemes available for the MP-EP configuration, which include the parallel scheme, master-slave scheme, coupling scheme, and virtual shaft scheme. (1) Parallel scheme: The diagram in Figure 5.1 illustrates a parallel scheme where the speed reference for each motor is identical. Sensors are used to measure the actual speed of each motor, which is then discretely fed back. The difference between the feedback speed and the reference speed is regulated by various controllers, which produce pulse width modulation (PWM) signals that are applied to the inverters connected to each motor. In this scheme, achieving high synchronization depends on the control algorithms discussed in Section 5.3. If each motor exhibits marked steady-state performance, dynamics, robustness, and anti-disturbance capacity, all motors can work synchronously. However, if the operating conditions of one motor deviate from those of the other motors, synchronization will fail. (2) Master-slave scheme: Figure 5.2 depicts three different master-slave structures. In Figure 5.2(a), only the first (master) motor's speed reference is set to the desired value, while the other

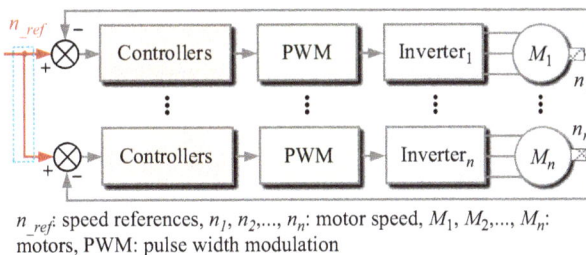

n_{ref}: speed references, $n_1, n_2, ..., n_n$: motor speed, $M_1, M_2, ..., M_n$: motors, PWM: pulse width modulation

Figure 5.1   Parallel scheme for synchronization of multiple motors

$n_{ref}, n_{ref1}, n_{ref2}, ..., n_{refn}$: speed references, $n_1, n_2,..., n_n$: motor speed, $M_1, M_2, ..., M_n$: motors, PWM: pulse width modulation

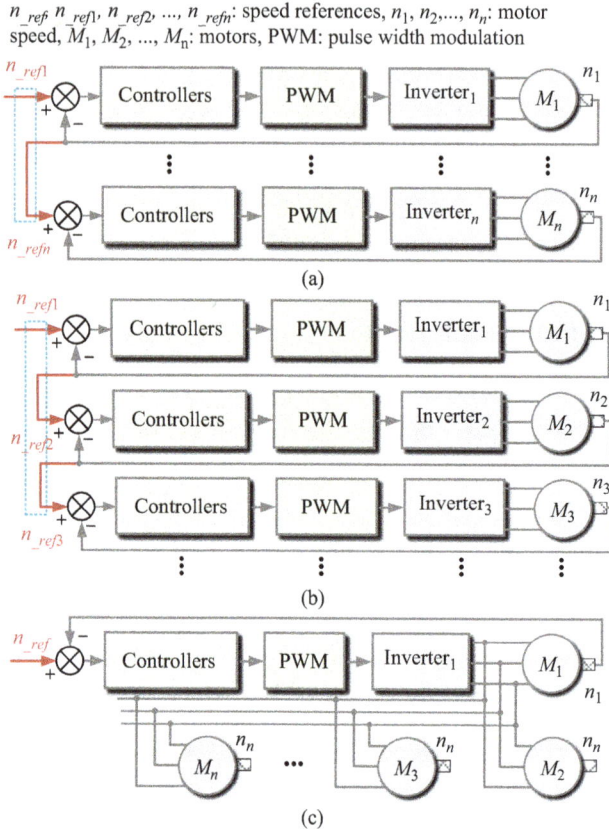

(a)

(b)

(c)

Figure 5.2   Master-slave schemes for synchronization of multiple motors: (a) the first master-slave scheme, (b) the second master-slave scheme, and (c) the third master-slave scheme

motors' speed references are equal to the real speed of the master motor. This approach tends to synchronize all slave motors with the master motor, but it may lead to delay issues because it takes time for the slave motors to track the master motor's speed. In Figure 5.2(b), all motors except the first and last ones function as both master and slave motors. This happens because the real speed of the ith ($I = 1, 2, 3, ...$) motor is set as the speed reference of the (i+1)th motor. This structure exhibits worse delay issues than the one shown in Figure 5.2(a). To minimize the impact of delay, these two schemes are suitable for motors with low inertia. Naturally, the control scheme depicted in Figure 5.2(c) is a master-slave scheme, where only one inverter drives multiple motors, and the speed of only one motor is fed back and regulated. The other motors work in open-loop control mode but with the same current/voltage frequency as the master motor. Although this scheme has no delay issues and is easy to implement, the synchronization property of the overall system depends on the tracking capacity of the slave motors. (3) Coupling scheme: Figure 5.3

n_{ref}: speed reference, n_1, n_2, ..., n_n: motor speed, $\triangle n_1$, $\triangle n_2$, ..., $\triangle n_n$: speed errors, M_1, M_2, ..., M_n: motors, PWM: pulse width modulation, K_1, K_2, ..., K_n: coefficients

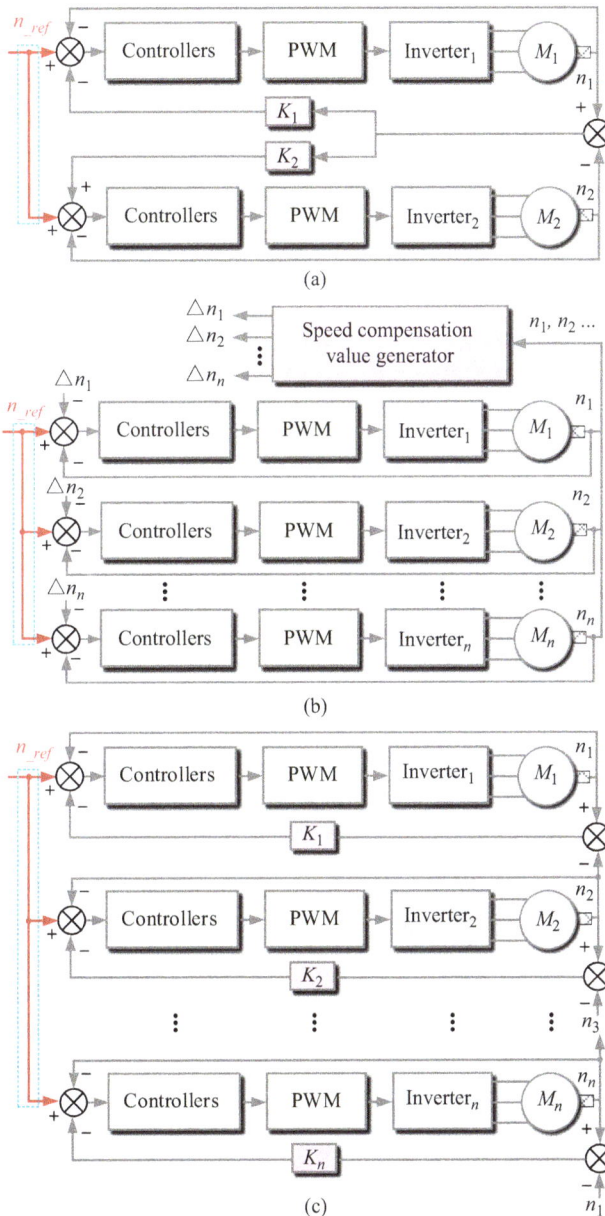

Figure 5.3   Coupling schemes for synchronization of multiple motors: (a) cross-coupling scheme, (b) improved cross-coupling scheme, and (c) ring coupling scheme

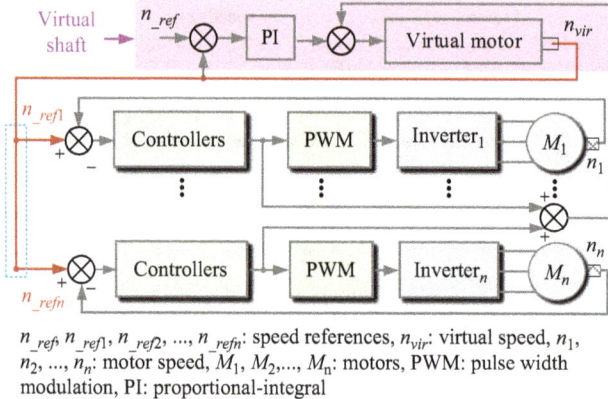

n_{ref}, n_{ref1}, n_{ref2}, ..., n_{refn}: speed references, n_{vir}: virtual speed, n_1, n_2, ..., n_n: motor speed, M_1, M_2,..., M_n: motors, PWM: pulse width modulation, PI: proportional-integral

Figure 5.4   Virtual shaft scheme for synchronization of multiple motors

illustrates three coupling schemes that use the same speed reference but different feedback information, which includes measured speed and speed errors for compensation. The structure in Figure 5.3(a), also known as the cross-coupling structure, is suitable only for dual-motor drives and uses subtraction to obtain the speed error for compensation. To extend this structure to systems with more than two motors, Figure 5.3(b) modifies the control structure by introducing a speed compensation value generator (SCVG) to calculate the speed errors to be compensated. The SCVG can be created using various methods such as numerical calculation, fuzzy logic theory, or AI theory. Figure 5.3(c) shows a ring coupling scheme that is designed for scenarios involving more than two motors. This scheme considers the error relationship between two adjacent units and utilizes it to synchronize the motors. (4) Virtual shaft: Figure 5.4 illustrates the virtual shaft scheme, which is another solution for achieving synchronous control of multiple motors. This scheme involves creating a virtual motor based on a mathematical model to simulate the properties of the real motors. The output speed of the virtual motor is set as the reference for the real motors. This scheme is similar to the first master-slave scheme in Figure 5.2, but it eliminates the delay issue. However, the real motors' speeds may not precisely track the reference given in the virtual motor control process.

The MP-EC configuration has the potential to achieve high synchronization due to the consistent current phase, amplitude, and frequency flowing through all motors. This makes it a competitive solution for coordination control. However, similar to the structure in Figure 5.2(c), the limited number of one or two inverters used in the system can solely be controlled by the PWM signals generated from the feedback speed information of one motor. This leaves the remaining motors to work in open-loop mode, resulting in a natural master-slave scheme. To ensure satisfactory synchronization, the slave motors must possess a high tracking capacity. Thus, this scheme is typically applied in systems equipped with DC motors or low-inertia AC motors.

5.4.1.2   Asynchronization cases

In the case of multi-motor drives operating in asynchronous mode, the motors are commonly connected in both mechanical and electrical parallel configurations since each motor may have a distinct operating speed. Figure 5.5 shows the control diagram, where speed references are independent of each other. This is a different parallel scheme than the one shown in Figure 5.1. In this case, high-performance motor-targeted control algorithms that are presented in Section 5.3 play a dominant role in achieving coordinated operation of the entire system. Specifically, if each motor in the system can be precisely and rapidly controlled regardless of external and internal impacts, the entire system can meet the application requirements.

5.4.2   Mutual interference

As shown in Table 5.1, mutual interference issues of multi-motor drives reflect in the following aspects: electromagnetic interference (EMI), CCC, cross-coupling of

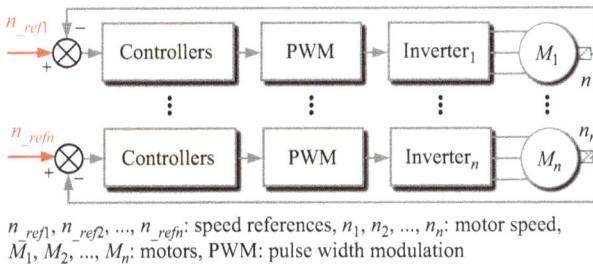

$n_{ref1}, n_{ref2}, ..., n_{refn}$: speed references, $n_1, n_2, ..., n_n$: motor speed, $M_1, M_2, ..., M_n$: motors, PWM: pulse width modulation

Figure 5.5   Parallel scheme for asynchronization of multiple motors

Table 5.1   Considerations of mutual interference

Aspects	FEATURES AND IMPACTS	Available solutions
EMI	• Sources: drive, motor, and wiring • Affect electronic components • Result in performance reduction and low reliability	• Shielding • Filtering • Grounding • Isolation
CCC	• Sources: control algorithms, signal interference, etc. • Either beneficial or harmful	• Harmful effect reduction • Advanced control method • Decoupling method
CCL	• Sources: motors and loads • Result in performance reduction and low reliability	• Load-sharing algorithm • Feedback control method • Decoupling method
DC-link interactions	• Sources: motors share the same DC-link • Voltage ripples and low stability	• Filters (e.g., capacitors and inductors) • Isolation (optocouplers)

loads (CCL), and DC-link interactions. Clear explanations for each of them are detailed as follows.

EMI is a common issue in both single-motor and multi-motor drives. However, multi-motor systems pose unique challenges. In such systems, multiple motors are typically controlled by a single drive, leading to potential EMI issues as each motor generates its own electromagnetic fields that can interfere with each other and other electronics. Moreover, the wiring and cabling in multi-motor systems are more complex, increasing the risk of EMI problems. EMI can affect the components in the system, including sensors, communication modules, and control circuits, leading to instability, malfunctions, or system shutdown. To address EMI in multi-motor drives, engineers can employ various solutions, including shielding, filtering, grounding, and isolation. Shielding involves enclosing the motors and their associated electronics in a conductive material, such as a Faraday cage, to block electromagnetic waves. Filtering involves inserting components such as capacitors, inductors, or resistors into the circuit to attenuate unwanted signals. Grounding and isolation techniques can also be used to separate the motors electrically and mechanically, reducing the impact of EMI. Besides, proper motor and load placement, as well as careful design of the electrical system, can also help reduce EMI and improve the overall reliability of the multi-motor drive.

CCC refers to the influence that one motor's control has on another motor's behavior. It is important to emphasize that this effect can either be beneficial or harmful to the system, depending on specific circumstances. Specifically, while intentional signal coupling techniques in the master-slave and coupling schemes described in Section 5.4.1 can enhance synchronization, unintended CCC caused by signal interference and other unexpected factors can result in instability and system failure. On the other hand, to mitigate harmful cross-coupling effects in multi-motor drives, several solutions are available. One approach is to utilize advanced control algorithms that account for interactions between motors. These algorithms can help ensure appropriate control signals are sent to each motor, minimizing unintended effects on other motors. Another solution is to implement decoupling techniques that modify control signals to eliminate cross-coupling effects. However, the decoupling techniques may require additional sensors and computational resources. Therefore, it is important to carefully consider the specific circumstances and requirements of each multi-motor drive to determine the most appropriate solution for addressing CCL.

In multi-motor drives, CCL is a complex issue due to the interactions between the different motors and their load. As the load on one motor changes, it can cause changes in the operating conditions of the other motors, leading to cross-coupling effects. The extent of the cross-coupling effects depends on factors such as the mechanical coupling between the motors and the type of load. CCL in multi-motor drives can lead to performance reduction. On this ground, several solutions have been developed. First, advanced load-sharing algorithms that can dynamically distribute the load among the motors based on their capacity and operating conditions are effective in reducing the impact of CCL. Second, feedback control systems that adjust the load on each motor in real time based on the measured load and operating conditions can help to mitigate CCL. Third, similar to CCC, decoupling techniques, such as modifying the control signals that take the interactions between the motors into account, can also be employed.

In a multi-motor drive, DC-link interactions can occur when multiple motors share the same DC-link, as all motors cannot operate synchronously. These interactions can cause unwanted voltage fluctuations, voltage sag, and voltage swell, which can in turn affect the performance of the other motors. To mitigate this issue, advanced control algorithms such as MPC based on direct ripple control can be implemented to simultaneously regulate the DC-link voltage fluctuations and improve motor performance. Another effective solution is to install filters such as DC-link capacitors or inductors, which can reduce voltage fluctuations and sudden changes so as to improve the overall system performance. Besides, some motor drives employ isolation techniques such as optocouplers to minimize crosstalk between different drives and improve the isolation between the DC links. These techniques can help to minimize the impact of DC-link interactions and improve the performance and reliability of multi-motor drives.

5.4.3   Intercommunication

Intercommunication is a necessary aspect of multi-motor drives as it enables the exchange of information between the different components of the system, including motors, controllers, and sensors. The reasons are as follows. First, it enables coordinated operation between the different motors in the system. In many industrial applications, multiple motors may be used to drive a single DOL, with each motor performing a specific function. In this case, effective communication between the motors is crucial to ensure that they operate in coordination with each other, which can help to prevent overloading and ensure smooth operation of the system. Second, intercommunication facilitates fault detection. The multiple motors used in an industrial application may be subjected to harsh operating conditions, which can cause wear and tear over time. In this case, it is essential to monitor the performance of the motors to detect any potential faults before they lead to system failures. By implementing effective communication protocols, the system can send real-time data from the motors to the controllers or sensors, enabling the early detection and diagnosis of faults. Third, intercommunication is also necessary for system performance optimization. By exchanging information between different components of the system, the controllers can adjust the operating parameters of the motors to optimize their performance. For instance, the controller can adjust the motor speed or torque based on the load or operating conditions to improve the efficiency and performance of the system. Finally, intercommunication helps ensure safety and compliance with industry regulations. Effective communication between the different components of the system can help to prevent safety hazards and ensure that the system operates within the prescribed safety limits. This is particularly important in aerospace, automotive, manufacturing, etc., where safety regulations are stringent. So far, high-reliability intercommunication technologies have been developed for multi-motor drives, including communication modes and communication protocols.

As depicted in Figure 5.6, there are three commonly used wired communication modes for multi-motor drives: (1) a host connected with distributed chips (H-DC), (2) a single central chip (CC), and (3) network-on-chip (NC). The H-DC mode consists of a host and a group of distributed chips (C_1, C_2, C_3, and C_4) connected through communication wires. The host in the system executes control algorithms, synthesizes all information, and sends control commands to the distributed chips, serving as the core of intercommunication architecture. This mode is well-suited for large-scale distributed systems, but with the increase in distributed chips and communication demands, clock synchronization and communication bandwidth become the bottleneck for high-speed communication. As for the CC mode, all control algorithms are implemented in a single digital signal processor (DSP), field programmable gate array (FPGA), or in the cloud. In this case, intercommunication can be achieved within the CC (see Figure 5.7), which is now focused on by the authors. The CC mode is highly

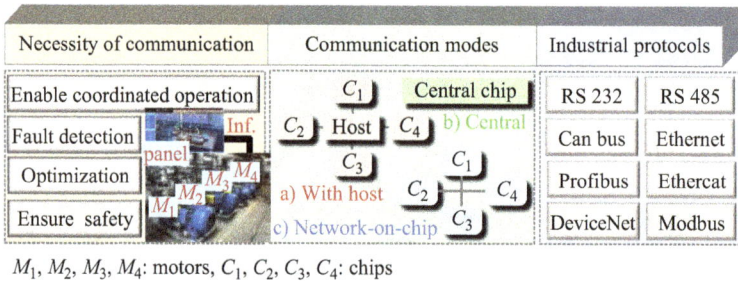

M_1, M_2, M_3, M_4: motors, C_1, C_2, C_3, C_4: chips

Figure 5.6   Intercommunication of multiple motors

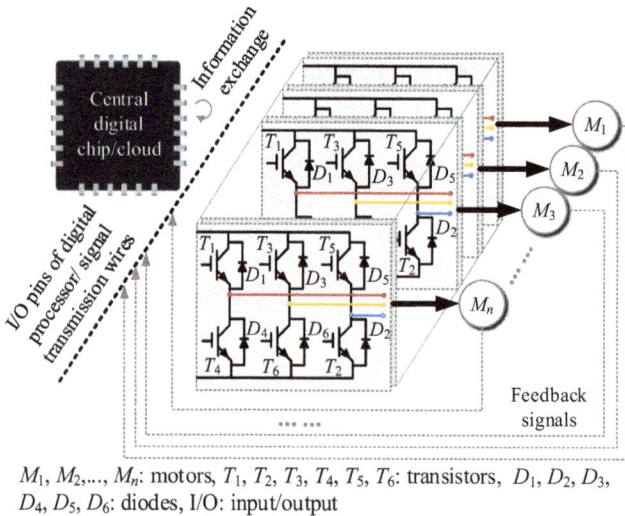

M_1, M_2,..., M_n: motors, T_1, T_2, T_3, T_4, T_5, T_6: transistors, D_1, D_2, D_3, D_4, D_5, D_6: diodes, I/O: input/output

Figure 5.7   Communication mode using one central chip

integrated, but when the number of motors is large, the computing performance of a single digital processor cannot meet the requirements, but cloud computing becomes suitable. Finally, similar to the H-DC, the NC is achieved by relying on multiple distributed chips. But the host is no longer employed. Instead, the distributed chips can share information with each other. Compared to the other two communication modes, the NC mode endows the multi-motor control drives with high integration and sufficient computing resources, but the disadvantage is that it requires a more complex design and programming. Therefore, selecting an appropriate communication code is essential for achieving efficient communication between multiple motors while considering the trade-offs between cost and complexity.

Intercommunication between multiple motors or chips can be achieved efficiently by using a variety of communication protocols. A multi-motor control circuit is proposed that employs the controller area network (CAN) bus protocol. Similarly, the CANopen protocol is utilized to control six brushless direct current (BLDC) motors. Another approach is presented, where a LAN-based system is developed using a full-duplex communication method that combines the advantages of RS-422 and CAN. The verification results demonstrate high communication precision in the system. Multi-motor control algorithms are implemented on an ABB programmable logical controller AC500 PM 571, which is connected to the drives via Profibus. Additionally, various standard communication protocols are compared for speed-controlled belt conveyors, such as RS232, RS485, ProfiBus, Interbus, DeviceNet, Modbus, and Ethernet. The advantages and disadvantages of each protocol are discussed. After further comparison, the authors conclude that the CANopen and DeviceNet protocols are superior when considering the telecommunication interaction arrangement of microprocessor devices of different types in the composition of multi-motor electric drives. Overall, the studies discussed above highlight the significance of selecting the appropriate communication protocol for multi-motor systems. It is evident that the choice of protocol plays a crucial role in achieving efficient and reliable intercommunication between motors.

In addition to wired communication methods, wireless communication technologies such as Wifi, Bluetooth, Zigbee, or cellular networks can also be employed for communication between devices without the need for physical wires. These wireless communication methods offer advantages such as flexibility, scalability, and remote accessibility. However, they may introduce challenges related to reliability, security, and latency compared to wired communication.

5.4.4 Interdependent fault diagnosis

From a faulty component perspective, faults in a single-motor drive can be divided into four main categories: motor faults, inverter faults, sensor faults, and mixed faults involving multiple components. However, multi-motor drives face not only the same types of faults but also more complex fault categories, such as simultaneous multi-motor failures, simultaneous multi-inverter failures, simultaneous multi-motor and multi-inverter failures, sensor failures coupled with multiple motors and/or inverters, and more complicated mixed faults of multiple components that include couplers between motors.

Single-motor drives exhibit corresponding and evident fault characteristics for analysis and diagnosis due to the independence of each component. Therefore, various fault methods are currently available, including model-based, signal-based, and data-based methods. However, extending these fault diagnosis methods, especially the model-based and signal-based methods, to multi-motor drives is challenging due to mutual influences in hardware, feedback signals, and control software. These challenges can be summarized in the following five aspects. (1) System modeling: Models of the multi-motor drives are more complex and difficult to accurately describe, which affects the accuracy of model-based methods. (2) Feature signal selection: One fault feature may correspond to multiple fault types, or a fault feature may not be evident under system interaction, which affects signal-based fault diagnosis. (3) Signal analysis: A sensor may measure the combined workload of multiple motors, such as the bus current sensor, making it extremely difficult to distinguish fault information in the signal and correspond it to the fault information of which component. (4) Feature changes: Due to the mutual influence among multiple motor systems, the fault characteristics exhibited by a particular fault in a single-motor drive cannot be reproduced in a multi-motor drive. (5) Fault location: For instance, it is difficult to identify the faulty motor when multiple motors fail simultaneously.

On the one hand, since it is difficult to implement model-based fault and signal-based methods in multi-motor drives, only a few studies are focusing on developing the relative techniques. Researchers developed a model-based approach to address the power loss fault in a large-scale multi-motor web-winding system, where the fault influence can be treated as a disturbance. The approach establishes a dynamic model of the web-winding system first and then, develops disturbance observation-based strategies to detect the fault. However, there is still a lack of targeted fault diagnosis methods for the aforesaid specific faults in multi-motor drives.

On the other hand, by contrast with the model-based and signal-based methods, the data-based methods without the need for a precise model of the system have been widely studied. They can directly extract features from data and achieve fault classification, which makes them more adaptable to multi-motor drives with unclear fault features. In recent years, a variety of data-based methods have been proposed for multi-motor fault diagnosis, which includes machine learning-based methods, statistical analysis-based methods, and deep learning-based methods. Machine learning-based methods are one of the most common strategies for fault diagnosis or classification of multi-motor drives. They can use historical data to build fault detection models, which can accurately distinguish normal and abnormal system states, and then identify the fault types. For example, a decision tree algorithm is given for fault diagnosis of the coal transportation system, and a support vector machine algorithm is used for fault diagnosis of unmanned aerial vehicle motors. As for statistical analysis-based methods, principal component analysis (PCA) and independent component analysis (ICA) are commonly used statistical analysis methods. They can extract fault features from high-dimensional data and reduce the dimensionality of data, facilitating fault diagnosis. The methodology of using PCA for diagnosing faults in induction motors (IMs) is introduced, and the literature clarifies the possibility of using ICA to detect faults in a multi-motor drive but does not introduce the implementation procedures at length. Deep learning-based methods have become popular in recent years due to their excellent performance in pattern recognition and fault diagnosis. Now, convolutional neural networks (CNNs) and recurrent neural networks (RNNs) are commonly used deep learning models. CNNs can automatically learn the fault features from raw data and achieve high accuracy in fault diagnosis. They are suitable for fault diagnosis of multi-motor drives with image data, such as thermal images and vibration images. A multi-stream CNN for feature extraction from and fusion of motor vibration and stator current maps is presented for IM fault classification, and CNN is used for fault diagnostics of motors, where the input is the time-frequency map of the vibration signal. In terms of RNNs, they are suitable for fault diagnosis of multi-motor drives with time-series data, such as current signals and vibration signals, etc. A series arc fault diagnosis and a line selection method based on RNN for a multi-load system are introduced.

It is worth noting that data-based methods have many advantages in fault diagnosis of multi-motor drives. First, they can deal with the non-linear and non-stationary features of multi-motor drives. Second, they can effectively diagnose multiple faults in multi-motor drives, including simultaneous motor and inverter faults, sensor faults, and mixed faults involving multiple components. Third, they can handle the large amounts of data generated by multi-motor drives and provide real-time fault diagnosis results. These advantages make data-based methods the future direction of multi-motor fault diagnosis in real-world applications. Although data-based methods offer promising benefits, they require a significant amount of training data to achieve optimal accuracy. Failure to meet this data requirement may compromise the accuracy of the results. In addition, the intricate execution complexity and high hardware resource demands exceed those of traditional methods, emphasizing the urgent need for further enhancement.

5.4.5   Power quality

Section 5.4.4 discusses the mutual effects of multiple motors that share a DC-link. However, since a multi-motor system may only be one of several electric equipment in a power system, such as a microgrid, parallel-connected equipment may not function properly due to the multi-motor system's negative impacts on the power quality of the microgrid. Therefore, researchers are currently striving to minimize the multi-motor system's influence on the power system as well. As shown in Table 5.2, existing methods to accomplish this include hardware filtering, current phase-shift control, direct voltage control, and current modulation. These methods aim to improve the power quality of a multi-motor system and reduce its adverse impacts on the power system, ensuring that other parallel-connected devices work normally.

Hardware filtering is a commonly used technique in multi-motor systems to mitigate harmonic distortion and improve power quality. This technique adopts passive or active filters, which can be designed to filter out specific harmonics or a

Table 5.2   Features of different power quality improvement methods

Methods		Features
Hardware filtering	Passive filtering	• Simple implementation and low-cost • Reduced efficiency and only for specific harmonics
	Active filtering	• Complex and expensive compared to passive filter • Reduced efficiency and for a broad range of harmonics
Current phase shift		• Require no additional filtering equipment and low cost • Complex algorithm and possible efficiency reduction • Possible instability issue
Direct voltage control		• Simple compared to current phase-shift control • Require no additional filtering equipment but the sensor
Current modulation		• Require no additional filtering equipment and low cost • Complex implementation

range of harmonics. Passive filters use inductive and capacitive components (e.g., LC and LCL filters) to filter out harmonic frequencies, which are simple and low-cost, but only capable of filtering out one or a few specific harmonic frequencies. Active filters use power electronic devices such as inverters to generate an opposing harmonic waveform that cancels out the harmonic distortion. For instance, this chapter focuses on upgrading power quality in DC electrical drives by introducing a shunt active power filter using active-reactive power (P-Q) theory to optimize the total harmonic distortion (THD) and power factor. Compared to the passive filters, the active ones are more complex and expensive, but can filter out a broader range of harmonic frequencies and compensate for reactive power. Hence, the choice of filter type depends on the specific requirements of the multi-motor system and the level of harmonic distortion that needs to be mitigated. Moreover, there are also several typical features that need to be addressed for the hardware filtering methods. First, although they are effective in mitigating harmonic distortion, they can lead to increased losses and reduced efficiency due to the use of more components. Second, the selection of the filter design must take into account the power rating, load characteristics, and operating conditions of the multi-motor system.

Current phase-shift control is a method used to improve the power quality of multi-motor systems by removing harmonic distortion. This technique takes full advantage of phase-shifted current waveforms to cancel out the harmonic current generated by the motor. A control strategy must be developed to calculate the optimal phase shift for each motor. For example, the effectiveness of phase-shifted current control for mitigating harmonics is investigated. A current phase-shift control method that incorporates a particle swarm optimization (PSO) algorithm is proposed to optimize power levels and firing angles for multi-drive systems with a fixed number of drives. The objective of the optimization is to minimize the THD level of the current at the point of common coupling. Current phase-shift control is advantageous because it does not require additional filtering equipment, which can reduce costs. However, it can also lead to reduced efficiency, as the motors may not operate at their optimal energy efficiency state. Another challenge of current phase-shift control is the potential for instability if the phase shift is not maintained correctly. Despite these challenges, current phase-shift control is a powerful method for removing harmonic distortion and improving power quality, particularly in multi-motor systems with motors operating at different loads and speeds.

Direct voltage control is a technique that regulates the DC-link voltage to improve power quality in multi-motor systems. This method involves feedback and regulation of the DC-link voltage, which helps maintain stable and consistent power levels. Direct voltage control is easier to implement than current phase-shift control, which requires developing a control strategy to calculate the optimal phase shift for each motor. This happens because only automatic voltage regulators (AVR) are needed. Additionally, direct voltage control eliminates the need for additional hardware filtering equipment, reducing the complexity of the physical system. Overall, direct voltage control offers several advantages over current phase-shift control and hardware filtering, making it an effective technique for improving power quality in multi-motor systems. However, an extra DC-voltage sensor needs to be installed for measurement, and the cost will only decrease slightly compared to the active filtering strategy.

Compared to the direct voltage control strategy, the current modulation method is a current-targeted control technique for power quality improvement. It controls the current drawn from the DC-link and modifies it to remove harmonic distortion. As introduced, the current modulation method utilizes a reference current generator that produces a pre-programmed modulation signal to generate the switching signals for the inverter. The modulation signal is synchronized with the three-phase input current through a phase-locked loop (PLL) system, and the switching angles are compared to a sinusoidal waveform to produce the required current waveform. The current modulation method optimizes the current waveform to reduce specific harmonics to meet the maximum allowable harmonic levels. This technique has been proven to effectively reduce THD by utilizing the inherent physical components of the system, such as the inverter and sensors, making it a cost-effective solution for improving the power quality of multi-motor drives. However, the algorithm used for current modulation is relatively complex.

5.5   Case studies and applications

5.5.1   Dual-arm robotic systems

Dual-arm robotic systems represent a significant advancement in robotics, designed to replicate and extend the capabilities of human bimanual operations. These systems are increasingly utilized in industrial, healthcare, and service domains, where tasks demand complex manipulation, precision, and adaptability. Collaborative manipulation, enabled by dual-arm robots, involves coordinated motion and task execution across two robotic arms, which often emulate human hands in their functionality and interaction capabilities. Effective coordination strategies are critical in ensuring that both arms operate in unison or complementarily to achieve the desired outcomes, particularly in environments with high precision requirements, dynamic conditions, or the need for close human-robot collaboration. The fundamental challenge in dual-arm coordination lies in achieving seamless synchronization between the two robotic arms while respecting constraints such as physical coupling, kinematic limitations, and dynamic interactions with the environment. To address this challenge, control strategies are designed to manage the motion trajectories, forces, and interaction dynamics of the arms in real time. One widely adopted approach is motion synchronization, wherein both arms follow pre-defined or dynamically generated trajectories to ensure harmonious operation. For example, in assembly tasks, the robotic arms may coordinate to hold and manipulate different parts of an object, such as aligning components for insertion or fastening. Synchronization ensures that each arm's motion complements the other's, avoiding collisions or misalignments that could compromise the task's success.

To enable motion synchronization, many systems rely on master-slave control architectures. In this configuration, one arm (the master) drives the overall motion trajectory, while the other arm (the slave) adjusts its movements to follow or complement the master arm's actions. This approach is particularly effective in tasks requiring precise coordination, such as gripping and assembling fragile components. For example, in tasks where one arm holds a delicate item while the other arm applies force to assemble it, the master-slave strategy ensures that the force applied does not exceed the tolerances of the object being manipulated. However, the master-slave approach has limitations, particularly when tasks demand higher autonomy or adaptability for each arm. To overcome these challenges, researchers have developed distributed control architectures, where both arms are treated as semi-independent agents capable of cooperative decision-making. These systems utilize shared data, such as object position, force feedback, and task progress, to dynamically adjust each arm's behavior for optimal performance.

Force control plays a crucial role in collaborative manipulation, especially when the robot interacts with objects or the environment. Unlike purely motion-based strategies, force control considers the interaction forces between the robotic arms and the objects they manipulate. Impedance control and admittance control are widely used techniques in this context. Impedance control focuses on regulating the relationship between force and displacement, allowing the arms to adapt their movements in response to external forces. This is particularly important in tasks such as polishing, sanding, or surface cleaning, where maintaining consistent contact force is essential for quality outcomes. On the other hand, admittance control modifies the robot's motion trajectory based on the measured interaction forces, enabling compliant behavior. For example, when one arm encounters resistance while pushing an object, admittance control allows the other arm to adjust its position to maintain the intended task trajectory.

In addition to force control, the integration of sensor-based feedback systems enhances the dual-arm robot's ability to coordinate effectively. Sensors such as cameras, tactile sensors, and force-torque sensors provide real-time data on object position, contact forces, and environmental conditions. Vision-based systems, in particular, enable robots to detect and track objects, recognize task-relevant features, and adjust their actions accordingly. For instance, in tasks involving object handovers, one arm may use visual input to locate and orient the object before passing it to the other arm, which adjusts its grasp based on tactile feedback. The combination of multiple sensory modalities ensures that the robot can adapt to dynamic and unpredictable environments, enhancing its versatility and reliability.

Human–robot interaction (HRI) introduces additional complexity and opportunities in dual-arm coordination. Collaborative robots, or cobots, are designed to work alongside humans, often requiring close physical and functional interaction. In such scenarios, dual-arm robots must not only coordinate their internal actions but also align with human actions and intentions. For example, in an industrial setting, a worker and a dual-arm robot may collaborate on an assembly line, where the robot assists by holding a component while the worker fastens it. To ensure safety and efficiency, the robot must detect and predict the worker's movements, adjust its posture to avoid interference, and synchronize its actions with the worker's pace. Advanced HRI systems employ machine learning algorithms to model human behavior and predict task requirements, enabling more intuitive and seamless collaboration.

One notable application of dual-arm coordination is in medical robotics, where precision and adaptability are paramount. Dual-arm robots are used in surgical procedures to manipulate instruments, hold tissues, or perform suturing with high accuracy. In these applications, the coordination strategies must account for the delicate nature of human tissues, the need for precise force application, and the dynamic movements of the surgical team. For example, in robotic-assisted minimally invasive surgery, one arm may hold a camera to provide visual feedback, while the other arm manipulates surgical tools. The robot's control system must synchronize these actions to maintain a stable field of view while executing intricate surgical maneuvers. The integration of real-time imaging, force feedback, and haptic interfaces further enhances the robot's capability to perform complex medical tasks.

In industrial applications, dual-arm robots are revolutionizing processes such as automated assembly, material handling, and packaging. For instance, in automotive manufacturing, dual-arm robots assemble components such as engines, doors, and dashboards, requiring precise alignment and fastening. The robots' ability to handle multiple tools simultaneously and switch between tasks reduces production time and increases efficiency. Similarly, in the electronics industry, dual-arm robots assemble delicate components such as circuit boards, leveraging their precise coordination capabilities to place and solder tiny parts without damaging them.

Artificial intelligence (AI) and machine learning (ML) are driving significant advancements in dual-arm coordination strategies. By leveraging AI algorithms, robots can learn from past experiences, optimize their control parameters, and adapt to new tasks or environments. For example, reinforcement learning enables robots to develop coordination strategies through trial and error, gradually improving their performance in collaborative manipulation tasks. In addition, neural networks can model complex relationships between sensory inputs and motor commands, enabling robots to handle non-linear and dynamic interactions. These AI-driven approaches are particularly valuable in unstructured environments, where pre-programmed strategies may fall short.

Despite the remarkable progress in dual-arm coordination, several challenges remain. One key challenge is the computational complexity associated with real-time control and decision-making for two arms operating in a shared workspace. Advanced control algorithms, while effective, often require significant computational resources, limiting their scalability. Additionally, ensuring the safety and reliability of dual-arm systems in dynamic and human-centric environments remains a priority. Researchers are exploring innovative solutions such as probabilistic planning, predictive control, and shared autonomy to address these challenges.

In conclusion, dual-arm robotic systems represent a transformative technology with immense potential across diverse applications. The development and implementation of coordination strategies for collaborative manipulation have enabled these robots to achieve levels of precision, adaptability, and efficiency that were once unattainable. As advancements in control systems, sensing technologies, and artificial intelligence continue to evolve, dual-arm robots are poised to play an increasingly vital role in shaping the future of robotics and automation.

5.5.2 Mobile robots

Synchronization of motors in mobile robots is a cornerstone of their functionality, directly influencing their ability to navigate, maneuver, and perform tasks with precision and reliability. In mobile robotic systems, motor synchronization is primarily essential for two fundamental subsystems: traction and steering. These systems ensure smooth motion, accurate trajectory tracking, and efficient energy use, which are vital for applications ranging from autonomous vehicles and warehouse automation to outdoor exploration and disaster response. Achieving synchronization in these subsystems requires advanced control algorithms, robust communication mechanisms, and precise feedback systems to address dynamic operating conditions and environmental challenges.

Motor synchronization in the context of traction refers to the coordinated control of multiple motors responsible for propelling a mobile robot. In many mobile platforms, particularly those with differential drive or skid-steer configurations, independent motors drive each wheel or track. Synchronization ensures that all propulsion units operate at the correct speed, torque, and direction to achieve the desired movement. For example, to drive in a straight line, the left and right wheels of a differential-drive robot must rotate at identical speeds. Any discrepancy in motor operation could cause the robot to veer off course, leading to inefficiencies or task failures.

Achieving synchronized traction involves several control strategies, with proportional-integral-derivative (PID) controllers being among the most commonly employed methods. PID controllers continuously adjust the motor inputs based on the difference between the desired and actual velocities of the wheels. By using feedback from encoders or other sensors, the system compensates for disturbances such as uneven terrain, wheel slip, or varying loads. For instance, in a differential-drive robot operating on a slope, the PID controller adjusts the motor torque to ensure that both wheels maintain the same speed, countering the effects of gravity.

In more advanced systems, MPC is utilized to optimize motor synchronization for traction. Unlike PID controllers, which operate reactively, MPC predicts future states of the system based on a mathematical model and optimizes control actions accordingly. This predictive capability allows the system to account for dynamic changes in the environment, such as obstacles or surface irregularities, and maintain synchronized motion. For example, an autonomous delivery robot navigating a crowded warehouse may use MPC to adjust wheel speeds dynamically, ensuring smooth and efficient movement even in the presence of obstacles.

Synchronization also plays a crucial role in steering, where precise control of motor angles and speeds determines the robot's turning radius, trajectory, and overall maneuverability. In mobile robots with Ackermann steering configurations, commonly used in autonomous vehicles, synchronization ensures that the front wheels align correctly to achieve the desired steering angle. Any misalignment between the left and right steering motors could result in increased tire wear, reduced stability, or inaccurate trajectory tracking. Synchronization is equally important in robots with omnidirectional wheels, such as Mecanum or omni-wheels, where each wheel must rotate at a specific speed and direction to achieve complex motions like lateral sliding or diagonal movement.

One common approach to motor synchronization in steering systems is the use of kinematic modeling and real-time feedback. Kinematic models define the relationships between wheel angles, velocities, and the robot's overall motion, enabling precise control of steering motors. For example, in an autonomous vehicle, the kinematic model calculates the required wheel angles for a given turning radius and speed. Feedback from sensors such as rotary encoders or steering angle sensors ensures that the actual motor positions match the desired values, maintaining synchronization during steering maneuvers.

In robots with holonomic drive systems, where the motion is not constrained to specific directions, synchronization becomes even more complex. These systems require precise coordination of all motors to execute smooth and accurate motions. For instance, a robot with Mecanum wheels uses independent motors to control each wheel, and the synchronization of these motors determines the robot's ability to move in any direction without changing its orientation. Advanced control algorithms, such as inverse kinematics and force distribution methods, are used to calculate and synchronize the motor commands needed for holonomic motion.

The integration of sensor fusion techniques further enhances motor synchronization in traction and steering systems. By combining data from multiple sensors, such as inertial measurement units (IMUs), GPS, and wheel encoders, sensor fusion provides a comprehensive understanding of the robot's position, orientation, and motion. This information enables more accurate synchronization by compensating for sensor errors, noise, and environmental disturbances. For example, an outdoor exploration robot navigating rough terrain may use sensor fusion to detect and correct deviations in its trajectory caused by uneven surfaces or wheel slips.

Motor synchronization in mobile robots is also critical for energy efficiency and system longevity. Coordinated control of traction and steering motors reduces unnecessary power consumption, minimizes wear and tear on mechanical components, and ensures smooth operation. For instance, in an electric autonomous vehicle, synchronized motor control prevents the overloading of individual motors, optimizing energy usage and extending battery life. Similarly, in warehouse robots that operate continuously for long periods, synchronized motion reduces mechanical stress, lowering maintenance requirements, and increasing reliability.

Decentralized control architectures are increasingly being adopted to enhance motor synchronization in mobile robots, particularly in systems with a large number of actuators. In decentralized architectures, each motor controller operates semi-independently, communicating with neighboring controllers to achieve coordinated behavior. This approach reduces the computational burden on a central controller and improves the system's scalability and fault tolerance. For example, in a swarm of mobile robots performing collaborative tasks, decentralized synchronization ensures that each robot maintains its position and alignment relative to the group, enabling seamless collective motion.

In addition to traditional control strategies, machine learning and artificial intelligence (AI) are emerging as powerful tools for motor synchronization in mobile robots. Reinforcement learning algorithms, for instance, enable robots to learn synchronization strategies through trial and error, optimizing their performance over time. In one application, an autonomous vehicle may use reinforcement learning to adapt its steering synchronization to different road conditions, such as wet or icy surfaces. Similarly, neural networks can model the complex relationships between motor commands, environmental factors, and system dynamics, enabling more accurate and robust synchronization in unstructured environments.

Despite these advancements, challenges remain in achieving perfect synchronization of motors in mobile robots, particularly in dynamic and unpredictable environments. Factors such as communication delays, sensor inaccuracies, and mechanical imperfections can introduce synchronization errors, affecting the robot's performance. To address these challenges, researchers are exploring innovative solutions such as real-time adaptive control, predictive synchronization algorithms, and fault-tolerant control strategies. These approaches aim to enhance the robustness and reliability of motor synchronization, ensuring consistent performance even in the face of uncertainties.

Synchronization of motors for traction and steering is a critical aspect of mobile robot design and operation. By leveraging advanced control algorithms, sensor feedback systems, and emerging technologies like AI, researchers and engineers continue to push the boundaries of what mobile robots can achieve. These innovations not only improve the efficiency, precision, and reliability of robotic systems but also pave the way for their deployment in increasingly complex and demanding applications. Whether in autonomous vehicles, warehouse automation, or outdoor exploration, synchronized motor control remains at the heart of mobile robotics, enabling robots to navigate the world with confidence and precision.

Chapter 6
Fault tolerance in permanent magnet actuator systems

Despite the advantages, permanent magnet actuators (PMAs) are inherently subject to a range of mechanical, electrical, and thermal stresses. These stresses arise from factors such as fluctuating operational loads, environmental conditions, and prolonged usage, making fault detection and management a vital area of research and development. The faults in robotic joint PMAs can broadly be categorized into motor faults, reducer faults, and controller faults. Each type presents unique challenges and impacts on system performance. Motor faults, for instance, may arise from issues like demagnetization of the rotor magnets, winding insulation breakdown, or bearing wear. Reducer faults often stem from gear wear, lubrication failure, or structural deformation, leading to increased vibration and reduced transmission accuracy. Controller faults, on the other hand, involve issues in sensors, control algorithms, or power electronics, potentially resulting in unstable or imprecise actuator behavior. Understanding these fault mechanisms is essential for designing effective diagnostic and prognostic systems. The interdependence between motor, reducer, and controller components further complicates fault identification, as a fault in one subsystem can cascade into others, exacerbating system-level malfunctions.

Fault diagnosis plays a pivotal role in ensuring the reliability and safety of robotic systems. Timely detection of faults in PMAs not only prevents catastrophic failures but also minimizes downtime and maintenance costs. In applications such as autonomous manufacturing and robotic surgery, undetected faults can lead to significant operational and financial risks. Hence, robust diagnostic techniques that combine mechanical, electrical, and computational insights are indispensable. In recent years, advancements in sensing technologies and data-driven algorithms have enabled significant progress in fault diagnosis. Techniques such as vibration analysis, thermal imaging, and current signature analysis provide valuable insights into the health of PMAs. Meanwhile, artificial intelligence (AI) and machine learning (ML) methods have revolutionized fault prediction and classification, offering the potential to identify subtle anomalies that traditional methods might overlook.

The dynamic and uncertain environments in which robotic systems operate present unique challenges for fault management. For example, industrial robots often function in high-load and high-speed scenarios, where transient faults might not be immediately apparent. Similarly, service robots deployed in public or residential spaces must cope with unstructured environments, making fault diagnosis even more challenging. Another significant challenge lies in the scarcity of fault data. Since robotic systems are typically designed to be highly reliable, the occurrence of faults is relatively rare. This paucity of data makes it difficult to train and validate fault diagnosis models, particularly those based on deep learning or other data-intensive approaches. Small-sample fault diagnosis methodologies, therefore, represent an important area of ongoing research.

Fault tolerance refers to the ability of a system to continue functioning, at least partially, despite the presence of faults. In robotic joint PMAs, achieving fault tolerance is critical for ensuring uninterrupted operation in mission-critical applications. Fault-tolerant strategies involve a combination of hardware redundancy, software reconfiguration, and advanced control algorithms. Hardware redundancy, such as using multiple sensors or actuators, allows the system to switch to backup components in case of a failure. This approach ensures that the robot can maintain functionality while minimizing performance degradation. Reconfigurable control systems adapt to faults by redistributing tasks among functional components. For example, if a specific actuator experiences reduced performance, the control system can compensate by adjusting the workload of other actuators. Advanced control algorithms enable real-time adjustments to system parameters based on fault conditions. These methods enhance the robustness and reliability of robotic joint PMAs under varying operational scenarios. Integrated health monitoring systems provide real-time diagnostics and prognostics, allowing operators to predict potential failures and implement corrective actions before they occur. These systems often leverage machine learning and data analytics to detect subtle fault indicators. By integrating these fault-tolerant strategies, robotic systems can achieve higher levels of resilience and reliability. This not only reduces downtime and maintenance costs but also enhances the overall safety and performance of the robot in dynamic environments.

The ultimate goal of fault diagnosis and management in robotic joint PMAs is to achieve intelligent and resilient robotic systems. This involves not only detecting and diagnosing faults but also implementing fault-tolerant control strategies that allow robots to continue functioning, albeit at reduced performance levels, until repairs can be made. Techniques such as redundancy, reconfiguration, and adaptive control are critical in this regard. Emerging technologies, including graph neural networks, extended state observers, and physics-informed neural networks, are poised to play a transformative role in this domain. By leveraging the inherent relationships between different components of the PMA system, these approaches enable more accurate and efficient fault diagnosis, even under complex operating conditions.

6.1   Typical faults in permanent magnet actuator systems

6.1.1   Motor faults

PMAs are integral components in modern robotic systems, valued for their high efficiency, precise control, and excellent performance characteristics. The motor fault mechanisms within PMAs can significantly affect their operation, leading to performance degradation, increased maintenance costs, and, in extreme cases, total system failure. These faults can originate from various sources, including electrical, mechanical, and magnetic issues. Understanding these motor faults, their causes, detection methods, and impacts on the actuator's operation is essential for maintaining the long-term reliability of PMAs in robotic applications.

The classification of motor faults in PMAs typically involves electrical, mechanical, and magnetic failures, as shown in Figure 6.1. Electrical faults are primarily associated with the stator and rotor windings, as well as the power supply system. These faults include short circuits, open circuits, and ground faults. A short circuit in the motor windings occurs when the insulation between the winding turns breaks down, causing a low-resistance path that results in an excessive current flow. This can lead to overheating, insulation breakdown, and eventually failure of the motor. In contrast, an open circuit occurs when a winding connection is broken, resulting in a loss of current and torque, which compromises the motor's efficiency. Ground faults occur when the winding insulation deteriorates, creating an unintended path between the windings and the motor housing, leading to erratic motor behavior, potential damage, and even electrical hazards. Mechanical faults, on the other hand, are typically related to the physical components of the motor, such as bearings, shafts, and rotor alignment. A common mechanical fault is bearing wear, which occurs over time due to friction and can lead to misalignment, increased vibration, and noise. Rotor imbalance, which results from uneven mass distribution in the rotor, can also cause vibration and stress on other motor components, reducing the actuator's overall efficiency. Misalignment, which occurs when the motor shaft or components are not properly aligned during installation or due to wear, can cause additional mechanical stress and contribute to accelerated wear of critical components. Finally, magnetic faults are specific to the magnetic components of the motor. Demagnetization of the permanent magnets can occur due to external factors such as high temperatures, overloading, or electrical disturbances. Demagnetization reduces the motor's torque capacity and efficiency. Additionally, magnetic saturation can occur when excessive current or external magnetic fields cause the motor's magnetic materials to reach a point where they no longer efficiently contribute to torque production, further impairing motor performance.

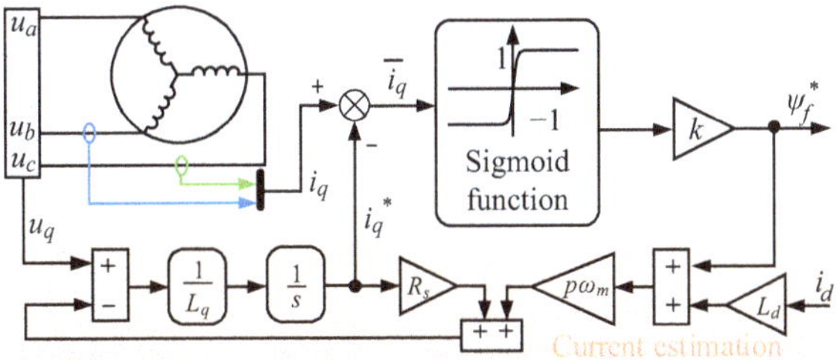

Figure 6.1   Structure of the proposed SM-RFO

The causes of motor faults in PMAs are varied and often interrelated. One of the most common causes is thermal stress, which can result from prolonged overloading or insufficient cooling. High temperatures can degrade insulation materials, leading to insulation failures and, in some cases, the demagnetization of the permanent magnets. Electrical overloads are another significant cause of motor faults. Operating the motor above its rated current can cause the windings to overheat, compromising their insulation and, in severe cases, leading to short circuits. Environmental factors, such as high humidity, dust, and corrosive gases, can accelerate the degradation of motor components. Corrosive environments, in particular, can cause deterioration of the motor's housing and bearings, while dust accumulation can affect the cooling efficiency and contribute to overheating. Manufacturing defects such as poor material quality, improper assembly, or inadequate testing can also lead to motor faults. For example, if the rotor is improperly balanced or the windings are poorly insulated, the motor may exhibit signs of failure earlier than expected. Mechanical overload, which occurs when the motor is subjected to forces beyond its design limits, can result in excessive wear and tear on both electrical and mechanical components. Aging and wear are inevitable in all mechanical systems, and PMAs are no exception. As the motor ages, the materials in the bearings, windings, and magnets can degrade, leading to a gradual decrease in performance and an increased likelihood of failure. Improper installation can also contribute to motor faults. Incorrect alignment or insufficient lubrication during installation can lead to early motor failure, as misalignment can cause undue stress on both the rotor and the bearings.

Motor faults can have a range of impacts on the performance of a PMA system. Performance degradation is perhaps the most immediate consequence of a fault. Whether the fault is electrical, mechanical, or magnetic, it can reduce the torque output, decrease efficiency, and affect speed regulation. For example, a demagnetized motor will produce less torque for the same current input, resulting in reduced performance. Faults can also increase energy consumption. As a motor experiences mechanical or electrical degradation, it often requires more energy to perform the same task, leading to higher operational costs. Faulty motors can also generate vibration and noise, especially when bearings are worn or the rotor is unbalanced. This not only affects the comfort and safety of operating personnel but also leads to further wear on system components. Increased wear on other system components is another significant concern. Faulty motors can place excessive loads on other parts of the system, such as the gearbox, controller, or structural elements, leading to cascading failures. Safety concerns are particularly critical when motor faults progress to more severe stages. Electrical faults, such as short circuits, or mechanical failures, such as bearing or rotor failures, can result in catastrophic consequences like fires or mechanical breakdowns, endangering both personnel and equipment. Finally, the cost of repairs and downtime is a significant consideration. Repairing or replacing a faulty motor involves direct costs (e.g., parts and labor) and indirect costs (e.g., system downtime, productivity loss), which can be minimized with early fault detection and proactive maintenance.

Several strategies can help mitigate motor faults and extend the lifespan of PMAs. Design considerations play a critical role in preventing faults from occurring in the first place. For instance, using high-quality materials, redundant protection systems, and robust motor housings can help prevent failures caused by environmental or thermal stress. Preventive maintenance is another essential strategy. Regular inspections of the motor's bearings, windings, and cooling system can help identify potential issues before they escalate into more severe problems. Thermal management is particularly important in preventing overheating. Cooling systems, such as fans, heat sinks, or liquid cooling, can help regulate the motor's temperature and prevent thermal damage. Fault-tolerant control systems can be implemented to ensure that the motor continues to function at reduced capacity in the event of a fault. These systems can detect faults, isolate faulty components, and adjust the motor's operation to minimize performance degradation while maintaining functionality. Condition monitoring systems that track vibration, temperature, and electrical signatures provide real-time data that can be analyzed for early fault detection. By analyzing these signals over time, operators can predict when a motor is likely to fail and take corrective actions, such as scheduling maintenance or replacing worn components before a catastrophic failure occurs.

In conclusion, motor faults in PMA systems pose significant challenges to the reliability and performance of robotic systems. Understanding the types of faults, their causes, and the impact they can have on the system is essential for effective maintenance and fault management. With advancements in fault detection technologies, such as vibration analysis, thermography, and electrical signature analysis, along with proactive maintenance strategies and fault-tolerant control systems, the risks associated with motor faults can be significantly reduced, ensuring the long-term operation of PMA-based robotic systems. As motor technology and fault management practices continue to evolve, future research may offer even more efficient and effective methods for dealing with motor faults, contributing to the continued advancement of robotics and automation systems.

6.1.2 Faults of reducer

Reducers, also known as gearboxes, are critical components in many PMA systems, particularly in robotic applications, where precise motion control and torque multiplication are often required. These mechanical components are used to reduce the speed of the motor's output while simultaneously increasing the torque, allowing the actuator to perform more effectively in high-load scenarios. While reducers play a vital role in optimizing system performance, they are also susceptible to a variety of faults that can significantly impact the overall functionality of PMA systems. Understanding the different types of reducer faults, their causes, effects, and methods of diagnosis is essential for ensuring the longevity and reliability of robotic actuators.

Reducer faults can generally be categorized into mechanical and lubrication-related issues, with each having distinct causes and effects. Mechanical faults in reducers are often related to wear and tear of the gears, bearings, or shafts, which are fundamental components in the gear mechanism. These faults can occur due to excessive load, inadequate maintenance, or manufacturing defects. Lubrication faults, on the other hand, result from the improper application or degradation of lubricating oils or greases used to reduce friction and prevent overheating of the moving parts inside the reducer.

One of the most common mechanical faults is gear tooth wear or damage. This occurs when the gears experience excessive friction or stress, leading to the gradual loss of material on the teeth. Over time, this can result in incomplete meshing of the gears, producing excessive noise, vibration, and reduced torque transmission efficiency. The gears may also experience tooth pitting, where small pits form on the surface due to repeated cyclic loading, leading to a loss of surface integrity and a further decrease in the effectiveness of the gear system.

Another critical mechanical fault is bearing failure, which is often caused by poor lubrication, misalignment, or excessive load. Bearings are essential for supporting rotating shafts and reducing friction between moving parts. When bearings become worn or damaged, they can result in increased friction, leading to overheating, vibration, and potentially catastrophic failure of the reducer. Shaft misalignment can also be a significant contributor to reducer faults. If the shafts in the reducer are not properly aligned, it can cause uneven loading on the gears and bearings, leading to rapid wear and potential failure. Misalignment may occur due to incorrect assembly, thermal expansion, or deformation of structural components over time.

Overheating is another prevalent fault in reducers. Excessive temperatures can cause the lubricant inside the reducer to break down, leading to inadequate lubrication, which further accelerates wear and tear on the mechanical components. Overheating can also deform critical parts of the reducer, such as gears and bearings, causing misalignment and loss of efficiency. In some cases, severe overheating may result in the complete failure of the reducer, rendering it inoperable.

Lubrication faults are often the result of inadequate lubrication or contaminated lubricant. Insufficient lubrication can cause increased friction between the gears and bearings, leading to overheating and accelerated wear. Contaminants such as dust, metal particles, or moisture in the lubricant can also cause damage to the internal components, leading to erosion, pitting, or corrosion. Incorrect lubrication—such as the use of the wrong type of lubricant or the incorrect viscosity—can also lead to similar issues, reducing the efficiency of the reducer and contributing to premature failure.

6.1.2.1 Causes of reducer faults

The causes of reducer faults are multifaceted, involving both external and internal factors. Excessive load is one of the most common contributors to reducer failure. Reducers are designed to operate under specific load conditions, and exceeding these limits can cause significant stress on the gears, bearings, and shafts, leading to rapid wear and eventual failure. For example, operating a reducer in an application where the torque demands exceed the system's rated capacity can result in gear tooth damage, bearing wear, or shaft deformation. Similarly, shock loads—brief, high-intensity loads that exceed the rated operating conditions—can cause immediate damage to the reducer components.

Another key cause of faults is poor maintenance practices. Regular inspection, lubrication, and cleaning of the reducer are essential for maintaining optimal performance. If these practices are neglected or performed incorrectly, it can lead to the accumulation of contaminants in the lubricant, which, in turn, can damage the gears and bearings. Additionally, improper alignment during installation or over time due to structural deformations can increase the likelihood of misalignment, resulting in uneven load distribution and premature wear of the mechanical components. Manufacturing defects in the reducer components can also contribute to faults. For example, gears may have defects in the material structure, such as cracks or voids, that weaken their load-bearing capacity. Similarly, poor heat treatment during manufacturing can lead to hardened spots or imbalanced gears, which can affect the reducer's performance and durability.

Environmental factors can also exacerbate reducer faults. For instance, temperature fluctuations can affect both the viscosity of the lubricant and the dimensional stability of the mechanical components. Extreme cold can cause the lubricant to thicken, reducing its ability to lubricate the components effectively, while excessive heat can cause the lubricant to break down, leading to increased friction and wear. Similarly, dust, dirt, and moisture can infiltrate the reducer housing, contaminating the lubricant and causing damage to the internal components. Operating in corrosive environments—such as in the presence of salts, chemicals, or gases—can cause the reducer components to degrade over time, leading to rusting, pitting, and eventual failure.

Improper assembly or installation is another cause of reducer faults. If the reducer is not assembled correctly, misalignments, improper tightening of bolts, or incorrect positioning of gears and bearings can result in immediate and long-term issues. This is particularly crucial during the installation of large PMAs, where the reducer must be carefully aligned with the motor and the load to ensure smooth operation. Even during routine maintenance, errors such as over-tightening or under-tightening bolts, improper cleaning methods, or failure to replace worn seals can lead to internal damage or leaks in the lubrication system.

The impacts of reducer faults on a PMA system can be severe and far-reaching. The most immediate effect is the reduction in torque transmission efficiency, as worn or damaged gears may not mesh properly. This can cause a loss of power output and make it difficult for the actuator to meet the required load demands. In some cases, it may lead to an inability to achieve the desired motion control, affecting the overall performance of the robotic system. Another consequence is the increase in vibration. Faulty reducers, especially those with misaligned shafts, worn gears, or damaged bearings, often produce excessive vibrations during operation. These vibrations can propagate throughout the actuator and even into the rest of the robotic system, causing further damage to other components, such as the motor or structural parts. Prolonged exposure to these vibrations can lead to structural degradation and eventual failure of other system components. Noise generation is another significant issue. Gearboxes that suffer from excessive wear or misalignment tend to produce louder and more irregular sounds. These noises are not just indicative of mechanical inefficiencies, but they can also create an uncomfortable or unsafe operating environment for personnel. Over time, noise can also cause hearing damage, especially in industrial settings where noise levels are already high. Energy inefficiency is a less obvious, but still significant, impact of reducer faults. A malfunctioning reducer often requires the motor to work harder to achieve the same torque output, resulting in increased energy consumption. This not only drives up operating costs but can also reduce the overall efficiency of the robotic system. In some cases, the increased energy demand can lead to motor overheating or electrical overload, which may cause additional system faults.

The most catastrophic consequence of a reducer fault is total failure, which can lead to system downtime, costly repairs, and even safety hazards. A seized bearing, for instance, can cause the entire reducer to lock up, stopping the actuator from functioning entirely. Similarly, gear failures may result in the loss of motion control or torque output, potentially rendering the actuator useless. In some cases, failure of the reducer can cause a cascade of system faults, affecting the motor, controller, and other interconnected parts of the PMA system. This can lead to costly repairs and extended downtime, negatively impacting system availability and performance.

6.1.3   Faults of controller

The controller in a PMA system is responsible for interpreting the control signals and adjusting the motor's behavior, ensuring that the actuator functions within desired operational parameters. The controller receives data from sensors (e.g., position, speed, and torque sensors), processes this data through various algorithms, and determines the control commands for the inverter to drive the PMSM. The inverter, in turn, uses semiconductor devices like metal-oxide-semiconductor field-effect transistors (MOSFETs) to convert the input DC voltage into AC voltage and regulate motor operation. As a crucial component in this system, any fault within the controller or the inverter's MOSFETs can lead to significant performance degradation or failure.

Controller faults in PMA systems can arise from hardware or software issues. Hardware faults include failures in the controller components (e.g., the micro-controller, MOSFETs, sensors, and communication interfaces), while software faults refer to programming errors or algorithmic issues within the control logic. Additionally, sensor faults and communication faults between the controller and other system components can exacerbate the issues, leading to unreliable actuator operation. Hardware faults within the controller typically involve component fail-ures that disrupt the normal functioning of the system. A microcontroller failure, for example, may occur due to electrical overstress, improper handling, or manu-facturing defects, causing the controller to fail in executing control algorithms or processing sensor data. Another critical hardware component is the inverter, which uses MOSFETs to regulate the current supplied to the motor. MOSFETs in the inverter play a pivotal role in switching the DC voltage from the power supply to the AC voltage used by the permanent magnet synchronous motor (PMSM), based on the control signals from the controller. A failure in the MOSFETs, due to factors like overheating, overcurrent, or electromagnetic interference (EMI), can cause malfunction in the inverter, leading to poor motor performance, motor stalling, or even damage to the motor itself.

When MOSFETs in the inverter fail, the controller may no longer be able to precisely control the motor's speed or torque. These failures can be caused by factors like thermal stress due to excessive current or inadequate cooling, voltage spikes, or EMI from the surrounding environment. Overvoltage conditions, often caused by sudden spikes in the power supply, can damage the MOSFETs and prevent the inverter from supplying the proper AC waveform to the motor, leading to instability or motor shutdown. Overcurrent conditions, on the other hand, can result from excessive motor load or incorrect control signals, causing the MOSFETs to overheat and fail. EMI from other devices can also disrupt the functioning of MOSFETs by introducing noise into the switching circuit, leading to the misoperation of the inverter and imprecise control of the motor. In addition to MOSFETs, other critical components in the inverter, such as capacitors, inductors, and gate drivers, can also fail, causing instability in the power conversion process. For example, gate driver faults can prevent the MOSFETs from switching correctly, leading to erroneous current output to the motor, resulting in erratic performance.

Software faults are another important source of controller failures. These faults can manifest as errors in the algorithms, incorrect logic, or programming bugs that prevent the controller from functioning as intended. For example, if the control loop within the software is poorly designed or inadequately tuned, it may lead to instability in motor performance, such as excessive oscillations, slow response times, or failure to reach the desired position or speed. One example of a software fault is a control loop instability resulting from improperly configured parameters in a proportional integral-derivative (PID) controller, which might lead to an uncontrolled motor response, excessive overshoot, or underperformance. Sensor faults also play a significant role in controller failures. Sensors provide critical feedback data, such as motor position, speed, and temperature, which are essential for adjusting control signals to the inverter. If a sensor malfunctions or fails to provide accurate data, the controller may make incorrect decisions that could result in inaccurate motor control. For instance, if a position sensor is faulty, the controller might not be able to determine the motor's position correctly, leading to issues like overshooting or undershooting the desired position. Similarly, a temperature sensor failure might prevent the controller from detecting overheating conditions, risking damage to the MOSFETs or the motor.

Finally, communication faults between the controller and other components (such as sensors, the inverter, or the power supply) can disrupt the control process. If the controller does not receive timely or accurate sensor data due to faulty wiring, signal interference, or protocol mismatches, it may not respond appropriately to changes in system conditions, causing delays, instability, or motor failure.

The causes of controller faults can be attributed to both internal factors (e.g., component degradation, design flaws, or coding errors) and external factors (e.g., environmental conditions, power supply issues, or interference). The most common causes of hardware faults are electrical stress, overheating, and component aging. For example, MOSFET failures in the inverter can occur due to electrical overstress from overvoltage or overcurrent conditions. Thermal stress is another major factor leading to MOSFET failure. As MOSFETs are subjected to high currents and switching frequencies, they can overheat, especially in systems without adequate cooling or thermal management. Overvoltage conditions can occur when there is a sudden spike in the power supply, which can exceed the voltage rating of the MOSFETs, leading to breakdown and failure of the switching components.

Aging and wear of electronic components, including MOSFETs, are also common causes of hardware faults. Over time, capacitors and semiconductor materials degrade, reducing the reliability of the inverter circuit. In particular, MOSFETs can suffer from latch-up, where the MOSFET remains in an irreversible state of conduction, potentially leading to circuit failure. Additionally, manufacturing defects in the MOSFETs, such as improper doping or poor material quality, can make them prone to premature failure. Software faults arise from coding errors, incorrect algorithms, or incomplete system modeling. For instance, errors in the control logic or algorithm can lead to incorrect calculations for the inverter switching sequence or control commands, which can result in erratic motor operation. These errors could be as simple as an improperly tuned PID controller, which would lead to instability in motor speed or position control. Overflow or underflow errors in software, caused by improper handling of data types or incorrect calculations, could also lead to incorrect control signals being sent to the inverter. Sensor faults can arise from environmental factors, such as temperature changes, vibrations, or exposure to electromagnetic fields, which can affect sensor accuracy. EMI can disrupt the signals from the sensors, leading to inaccurate or delayed sensor readings, which the controller interprets incorrectly. Moreover, sensor drift over time can cause the sensors to provide inaccurate feedback, leading to incorrect control adjustments. Similarly, sensor miscalibration during installation or system setup can also result in poor performance, even if the sensors themselves are functioning properly. Communication faults are typically caused by issues in the wiring, signal integrity, or data transmission protocols. For instance, loose connections, damaged cables, or signal noise can degrade the data quality being communicated between the controller, sensors, and inverter. Communication protocol mismatches or time synchronization errors between different system components can also lead to incorrect or delayed control actions.

The impacts of controller faults are wide-ranging, affecting the performance, safety, and longevity of the PMA system. Performance degradation is one of the most immediate consequences of controller faults. When MOSFETs in the inverter fail or software algorithms are not functioning correctly, the motor may not operate as intended. This could manifest as poor motor control, where the desired position or speed is not achieved, or the system may exhibit oscillations or delays due to unstable control loops. Loss of stability is another significant impact, especially in systems where multiple actuators need to work together. Faults in the controller or inverter can lead to desynchronization between actuators, causing the system to behave unpredictably and potentially fail to maintain equilibrium. Safety risks also arise from controller faults. For instance, a failure in the MOSFETs or control logic could result in uncontrolled motor movement, which might lead to mechanical damage, injury, or destruction of surrounding equipment.

6.1.4 Fault diagnosis and tolerance

The development of fault diagnosis systems for PMAs has grown significantly due to their critical role in modern electromechanical systems, especially in robotics. Fault diagnosis, as a discipline, began to take shape in the mid-20th century and has since evolved in response to advancements in computer science, sensor technology, and data processing methods. With the increasing complexity of systems relying on PMAs, such as industrial robots, electric vehicles, and automation systems, fault detection and diagnosis have become crucial for ensuring operational reliability and safety. PMAs, as key components in many applications, are vulnerable to a variety of faults, which can significantly affect system performance, energy efficiency, and safety.

A PMA's operational state refers to the overall condition of the actuator during its functioning, encompassing factors such as mechanical performance, electrical parameters, and environmental influences. These states can be classified into three broad categories: normal, abnormal, and fault states. Monitoring the actuator's condition involves continuously measuring key parameters such as current, voltage, temperature, vibration, and torque. By analyzing these measurements, engineers can determine whether the actuator is functioning within expected limits or whether there is a deviation that indicates a potential failure. Such monitoring is critical not only for maintaining the actuator's performance but also for planning predictive maintenance and preventing costly downtime.

The state of the actuator is determined by various performance parameters, such as the torque generated, the power consumption, and the mechanical vibrations it produces. These parameters can change over time due to various factors such as wear and tear, electrical or mechanical faults, or environmental conditions. Diagnosing faults in PMAs involves identifying when these parameters deviate from their expected values, and more importantly, understanding the underlying causes of these deviations. Fault diagnosis plays a crucial role in minimizing the risks of actuator failure and extending the life of the equipment.

6.1.4.1 Fault diagnosis

Fault diagnosis methods for PMAs can be divided into several categories based on the techniques and approaches used. These methods are designed to detect faults in various components of the actuator, including the rotor, stator, bearings, and control systems. Traditional fault diagnosis techniques often involve the use of mathematical models of the system, which are then compared with real-time data to detect discrepancies. However, the increasing complexity of PMA systems has led to the adoption of more advanced methods, including AI-based approaches, which offer promising results in terms of accuracy and reliability.

One of the traditional methods for fault diagnosis in PMAs is the mathematical model-based approach, which uses a theoretical model to simulate the behavior of the actuator under normal and faulty conditions. This method typically requires a deep understanding of the actuator's physical characteristics and the ability to model its dynamic behavior accurately. However, this approach can be difficult to implement in practice due to the complexity of the system and the challenges associated with accurately modeling nonlinearities and uncertainties. Despite these challenges, mathematical model-based methods remain a valuable tool in fault diagnosis, especially when combined with other techniques.

Signal processing-based methods are another widely used diagnostic tool for PMAs. These methods involve analyzing the input and output signals from the actuator to identify any deviations from normal operation. By monitoring parameters such as current, voltage, and power, these methods can detect faults in the system by identifying irregularities in the signal waveforms. Techniques such as Fast Fourier Transform (FFT) are often employed to analyze the frequency content of the signals, as different types of faults often result in distinct frequency patterns. Although signal processing techniques are effective for detecting faults such as rotor imbalance or winding faults, they can be less reliable when dealing with complex or multi-fault scenarios.

State estimation methods, such as Kalman filters, are another important tool in PMA fault diagnosis. These methods aim to estimate the internal state of the actuator from external observations, such as sensor data, in order to detect discrepancies between the actual and expected performance. By continuously updating the system's state estimation, these methods can detect faults in real time and provide valuable information about the actuator's condition. One advantage of state estimation methods is their ability to detect incipient faults, which may not be immediately obvious from raw sensor data.

Artificial intelligence-based methods, particularly those involving machine learning, have become increasingly popular in the diagnosis of faults in PMAs. These methods are capable of handling complex, nonlinear, and uncertain fault patterns, making them well-suited for modern actuator systems. For example, neural networks can be trained on large datasets of sensor measurements to recognize fault patterns and predict the likelihood of failure. These methods can learn from experience and improve over time, making them a powerful tool for fault detection and diagnosis. Additionally, fuzzy logic systems can be used to handle uncertainty and imprecision in sensor data, allowing for more robust fault diagnosis in real-world applications.

6.1.4.2   State monitoring

Effective monitoring of PMAs is essential for early fault detection and ensuring optimal performance. Several monitoring techniques are commonly used to assess the condition of PMAs, each with its advantages and limitations. These methods typically involve the use of sensors to measure key performance indicators such as temperature, current, voltage, and vibration, with the data being analyzed to detect deviations from normal operation.

Vibration monitoring is one of the most commonly used techniques for detecting faults in PMAs. By placing vibration sensors on critical components of the actuator, such as the bearings or housing, engineers can measure the amplitude and frequency of vibrations produced during operation. Different types of faults, such as rotor imbalance, bearing wear, and misalignment, produce distinct vibration patterns that can be analyzed using spectral analysis techniques. This allows engineers to detect potential issues early and take corrective action before the faults lead to actuator failure.

Current monitoring is another important technique for diagnosing faults in PMAs. By measuring the current flowing through the actuator, it is possible to detect abnormalities such as current imbalance or excessive current draw, both of which can indicate underlying faults. Techniques such as Park's vector approach, which transforms the measured current into a rotating coordinate system, are often used to analyze the current signals in greater detail. This method helps to identify subtle issues such as winding faults or rotor asymmetry that may not be immediately apparent in the raw sensor data.

Temperature monitoring is also critical in assessing the health of PMAs. Sensors embedded in key components such as the stator, rotor, and bearings can provide real-time temperature data, allowing for the detection of overheating issues. Overheating is a common cause of failure in PMAs, and monitoring temperature levels can help prevent damage caused by excessive heat. If the temperature exceeds predefined thresholds, it can trigger an alarm, allowing operators to take corrective measures such as reducing the load or improving the cooling system.

Instantaneous power monitoring is another valuable technique for detecting faults in PMAs. By measuring the voltage and current in real time, engineers can calculate the instantaneous power consumed by the actuator and monitor for any irregularities. Fluctuations in power consumption can be indicative of underlying issues such as mechanical resistance, electrical faults, or load imbalances. This method provides a more comprehensive understanding of the actuator's performance, allowing for the detection of faults that may not be apparent from a single parameter alone.

In recent years, AI-based monitoring techniques have gained popularity for fault diagnosis in PMAs. AI algorithms, particularly machine learning models, can process large volumes of sensor data and identify patterns associated with specific fault types. These models can be trained on historical data to recognize fault signatures and predict future failures with high accuracy. AI-based methods are particularly useful in complex systems where traditional diagnostic methods may struggle to handle the volume and complexity of data. Additionally, machine learning models can improve over time, enhancing their fault detection capabilities as they are exposed to more data.

6.1.4.3 Fault tolerance

Fault tolerance is a vital aspect of ensuring that PMA systems can continue to operate safely and effectively in the event of component failures. Given their critical role in applications like robotics, automotive, aerospace, and industrial machinery, PMA systems need to maintain functionality, even when faults occur. Fault tolerance in PMA systems involves strategies and mechanisms that allow the system to detect faults, isolate them, and continue operating—often with degraded performance—without causing catastrophic system failure. These strategies address the unique operational characteristics of PMAs, such as high precision, rapid response times, and the need for high reliability under varying load conditions.

One of the primary methods of achieving fault tolerance in PMA systems is through redundancy. This approach involves duplicating critical components so that if one fails, a backup can take over seamlessly. In robotic applications, for instance, redundancy can be applied to the actuators themselves. A dual-actuator setup, where two PMAs are installed to manage a single robotic joint or task, ensures that if one actuator fails, the remaining actuator can compensate for the loss, maintaining the system's function. This redundancy is particularly valuable in high-risk environments, where actuator failure can lead to significant operational disruptions. Similarly, redundant control systems or distributed controllers can be used, where multiple control units are responsible for managing the performance of different parts of the actuator system. In the case of controller failure, other controllers can take over the task, preventing total system failure. Additionally, the use of redundant sensors helps maintain accurate performance monitoring. If one sensor fails, backup sensors can continue providing feedback on critical parameters such as position, speed, temperature, and torque, ensuring that the PMA system can adjust in real time to maintain proper operation.

Another crucial strategy for ensuring fault tolerance in PMA systems is the use of fault-tolerant control. Fault-tolerant control techniques involve adaptive control systems that adjust the control law in response to faults, ensuring the actuator can continue to perform its task even if there is a malfunction. In PMA systems, adaptive control techniques like model reference adaptive control or Gain Scheduling are particularly effective. When a fault, such as demagnetization or winding damage, occurs, these adaptive control methods allow the system to modify its control parameters—such as increasing the current or adjusting the voltage applied to the motor—to compensate for the change in the actuator's performance. This adaptability helps the system continue functioning despite the fault, albeit sometimes with reduced performance. Another widely used fault-tolerant technique is model predictive control, which allows the control system to optimize its inputs by predicting the future states of the system based on its current conditions. This is particularly useful in PMA systems where faults can lead to sudden and unpredictable changes in the motor's behavior. Model predictive control can dynamically adjust control signals to mitigate the effects of faults and ensure that the PMA continues operating within its safe limits.

Fault isolation and reconfiguration are also critical components of fault tolerance in PMA systems. Fault isolation involves identifying the source of a fault so that corrective actions can be taken quickly. In PMA systems, fault detection often relies on advanced diagnostic techniques like motor current signature analysis, vibration monitoring, or thermal analysis. These methods allow the system to pinpoint issues such as bearing wear, rotor misalignment, or overheating. Once a fault is detected and isolated, the system can undergo reconfiguration to mitigate the impact of the fault. For example, if a PMA experiences a fault in one of its windings, the control system can redistribute the current to the remaining windings, ensuring that the actuator can still produce torque, though potentially at a lower capacity. In more complex systems, reconfiguration might involve switching to a backup actuator or modifying the load distribution in a multi-actuator setup. This reconfiguration ensures that the PMA system continues to operate reliably, even when individual components fail.

In recent years, self-healing and predictive maintenance techniques have become increasingly important in enhancing fault tolerance for PMA systems. These approaches use data-driven models to predict potential failures before they occur, enabling the system to take corrective actions proactively. Self-healing algorithms allow the actuator to autonomously adjust its behavior in response to detected faults. For instance, if an increase in temperature is detected due to excessive loading, the system can automatically reduce the current or adjust its speed to prevent thermal damage. Similarly, if a fault is detected in a component like a bearing, the system can switch to another actuator or modify the control inputs to reduce the load on the failing actuator. These self-healing actions help prevent minor issues from escalating into major failures, thereby improving system longevity and reliability.

Moreover, predictive maintenance algorithms can be employed to forecast when certain components, such as bearings, windings, or sensors, are likely to fail based on historical data and real-time performance metrics. By continuously monitoring parameters like vibration, torque fluctuations, and temperature, these algorithms use machine learning and statistical models to predict failure events. This allows the system to schedule maintenance or component replacement before a failure occurs, reducing the risk of unplanned downtime and extending the life-span of the PMA system. Predictive maintenance is especially valuable in industrial settings where the cost of failure is high and operational continuity is critical.

Lastly, distributed control is a strategy that can significantly enhance fault tolerance in PMA systems, particularly in complex multi-actuator configurations. In a distributed control system, each actuator has its own local controller that communicates with other controllers and the central control unit. This distributed architecture ensures that a fault in one actuator or controller does not disrupt the operation of other actuators. For example, in a robotic system with multiple PMAs, if one actuator fails, the remaining actuators can continue to operate, and the overall system can compensate for the loss by redistributing tasks. Distributed control systems also allow for faster fault detection and response times since each actuator's local controller can independently detect faults and adjust its behavior without waiting for instructions from a central controller.

In conclusion, fault tolerance is an essential feature of PMA systems, ensuring that they continue to operate reliably even in the presence of faults. Techniques like redundancy, fault-tolerant control, fault isolation and reconfiguration, self-healing, predictive maintenance, and distributed control help mitigate the impact of component failures. These strategies allow PMA systems to remain operational, albeit with potentially reduced performance, while avoiding catastrophic failure. As PMA systems are used in increasingly critical applications, the importance of fault-tolerance mechanisms will continue to grow, enabling these systems to meet the high reliability and safety standards required in modern industrial and robotic systems.

6.2 Implementation of model-based demagnetization fault diagnosis

The demagnetization faults can be divided into uniform demagnetization and partial demagnetization. Uniform demagnetization refers to that all permanent magnets installed in the motor are demagnetized to the same level. This happens when the PMs experience a pretty similar operating environment. For instance, the internal temperature of the motor rises significantly in a uniform pattern. As for partial demagnetization, it occurs due to local variations, and local heating is such a typical factor. Although there exist differences between the uniform and partial demagnetization faults, they are consistent in reducing motor performance. Hence, detecting either of them is crucial and has attracted much attention from both academia and industry.

Based on the signals used for demagnetization fault diagnosis, the existing strategies can be grouped into four categories: magnetic flux-based methods, back electromotive force (EMF)-based methods, current-based methods, and vibration/noise-based methods. *Magnetic flux-based methods*: Considering that the demagnetization fault can reduce or distort the air-gap magnetic field of the PMSM, monitoring the magnetic flux precisely is the most direct way for fault evaluation. Generally, physical sensors such as Gaussmeter and Hall effect sensors can be used for flux measurement, but this scheme sacrifices the size and performance of the motor. In order to discard the flux sensors, flux observers are developed to estimate the magnetic field in real time. *Back EMF-based methods*: The back-EMF of the PMSM can reflect flux variations because they are linearly dependent. However, this method requires the machine to work as a generator so as to be only applicable offline. *Current-based methods*: It is well known for motor current signature analysis. The rationale behind it is that the demagnetization fault can generate more or less harmonic components in the faulty motors compared to the healthy ones. However, it is difficult to extract and distinguish the specific current harmonic components corresponding to the demagnetization faults. This happens because first, different motors have different properties, and second, other kinds of faults, such as eccentricity, may occur simultaneously, influencing the current characteristics. *Vibration/noise-based methods*: Magnetic field variations could generate unbalanced radial forces, which can further influence the vibration and acoustic properties of the motor. The implementations of this method are complicated and high-cost because both modal analysis and vibrating/audio sensors are required. Among the aforesaid fault diagnosis methods, only the first one can detect both the uniform and the partial demagnetization faults online, while the others are mainly suitable for either offline scenarios or partial demagnetization faults.

The existing flux observers include an extended Kalman filter, an integral flux observer, a model reference adaptive system, and a sliding mode rotor flux observer (SM-RFO). Although these observers have been successfully applied in practice, there are still several issues deserving further investigation. First, the precision of the traditional extended Kalman filter algorithm is low, especially in nonlinear situations, which does not satisfy high-performance applications. As for the integral flux observer and model reference adaptive system algorithms, the initial setting values that are difficult to determine have a marked impact on the estimation results. Second, although the sliding mode flux observers have the advantages of simple structure, fast response, and strong anti-interference capability, few studies discuss the robustness of the observers. Specifically, because an SM observer must be designed on the basis of the machine model, it must be parameter-dependent. Hence, the flux estimation precision will be influenced by the machine parameters, such as the stator inductances that will shift when the fault occurs. To solve the issue, this part introduces a precise demagnetization fault diagnosis strategy based on multiple robust sliding mode observers under variable conditions [13].

6.2.1   Sliding mode rotor flux observer

6.2.1.1   Structure of the SM-RFO

The rotor flux information is contained in (2.13). Therefore, based on the SM variable structure theory, the SM-RFO can be designed as follows (see Figure 6.1):

$$\frac{di_q^*}{dt} = \frac{L_d}{L_q}p\omega_m i_d - \frac{R_s}{L_q}i_q^* + \frac{u_q}{L_q} - \frac{kF(\bar{i_q})}{L_q}p\omega_m \tag{6.1}$$

where i_q^* is the estimated current of the observer, k is gain coefficient, $\bar{i_q}$ is the error between the estimated current and the real current, namely, $\bar{i_q} = i_q^* - i_q$. $F(\bar{i_q})$ is the switching function (SF), and for the sake of low chattering, a sigmoid function instead of the traditional one can be adopted as the SF, which can be described as:

$$F(\bar{i_q}) = \frac{1 - e^{-\bar{i_q}}}{1 + e^{-\bar{i_q}}} \tag{6.2}$$

When the observer gets stable, the estimated flux linkage ψ_f^* is:

$$\psi_f^* = kF(\bar{i_q}) \tag{6.3}$$

Then, the error $err_{\_est-ori}$ between the estimated flux linkage and the original one ψ_{f_ori} detected before the PMSM is put into use can be adopted to determine whether the demagnetization fault occurs, that is,

$$\begin{cases} err_{\_est-ori} = \psi_f^* - \psi_{f\_ori} \\ err_{\_est-ori} < \psi_{thr}, fault\ occurs \\ err_{\_est-ori} \geq \psi_{thr}, no\ fault \end{cases} \tag{6.4}$$

where ψ_{thr} is a negative threshold value.

6.2.1.2   Stability analysis

For the SM-RFO, define the sliding surface S as:

$$S = \tilde{i}_q \tag{6.5}$$

To analyze the stability of the novel flux observer, a Lyapunov function is established as:

$$V = \frac{1}{2} \cdot S^T \cdot S = \frac{1}{2} \tilde{i}_q^2 \tag{6.6}$$

Apparently, V is greater than zero. Further, according to the Lyapunov stability decision theorem, only when the time derivative of V is less than zero will it be concluded that the SM-RFO can reach a stable state, which is derived as:

$$\frac{dV}{dt} = S^T \cdot \frac{dS}{dt} = \tilde{i}_q \frac{d\tilde{i}_q}{dt} \tag{6.7}$$

Subtract (2.13) from (6.1), and substitute the result into (6.7):

$$\frac{dV}{dt} = \underbrace{-\frac{R_s}{L_q} \tilde{i}_q^2}_{term1} \underbrace{- \frac{p\omega_m \tilde{i}_q}{L_q} (kF(\tilde{i}_q) - \psi_f)}_{term2} \tag{6.8}$$

From (6.2) to (6.8), it can be seen that *term*1 is less than 0. In order to keep the SM observer stable, *term*2 is supposed to be less than 0 as well, that is,

$$-\frac{p\omega_m \tilde{i}_q}{L_q} (kF(\tilde{i}_q) - \psi_f) < 0 \tag{6.9}$$

Then, taking the sign of $\tilde{i}_q$ into account, (6.9) can be further deduced as:

$$\begin{cases} kF(\tilde{i}_q) - \psi_f > 0, \; if \tilde{i}_q > 0 \\ kF(\tilde{i}_q) - \psi_f < 0, \; if \tilde{i}_q < 0 \end{cases} \to k > \frac{\psi_f}{|F(\tilde{i}_q)|} \tag{6.10}$$

Considering that $\tilde{i}_q$ is the current estimation error, ideally, its lower bound is expected to be zero. However, the estimated current cannot always track the real one precisely in practice because of external disturbances, chattering and delays, and so on. Hence, we can assume that the magnitude of the current estimation error ranges from $|\xi_{min}|$ to $|\xi_{max}|$ when the observer reaches the equilibrium state. In this case, the minimum value of $|F(\tilde{i}_q)|$ is:

$$\min(|F(\tilde{i}_q)|) = \frac{1 - e^{-|\xi_{min}|}}{1 + e^{-|\xi_{min}|}} \tag{6.11}$$

Further, to ensure (6.10) is satisfied, k can be set as:

$$k = \frac{\psi_f}{\min(|F(\tilde{i}_q)|)} \tag{6.12}$$

It can be found that there exists a constant that is much larger than the real flux linkage, maintaining the SM-RFO to be stable.

6.2.1.3 Impacts of inductance mismatch on flux observation and fault diagnosis

Practically, the measured d, q-axis inductances need to be used for flux estimation when implementing the proposed SM-RFO based on (6.1). However, it is not easy to measure the inductances accurately with the magnetic saturation effect considered in real applications, thereby affecting the precision of the flux estimation and fault diagnosis results.

Denote the variations of the d, q-axis inductance as ΔL_d and ΔL_q, respectively, namely,

$$L_{d\_mea} = L_d - \Delta L_d, L_{q\_mea} = L_q - \Delta L_q \tag{6.13}$$

where L_{d_mea} and L_{q_mea} are the measured inductances. And then substituting them into (6.1), which is then subtracted from (2.13), the error between the real flux linkage and the estimated one $err_{\_rea\text{-}est}$ can be derived as:

$$err_{\_rea\text{-}est} = \psi_f - \psi_f{}^* = \frac{L_q}{p\omega_m}\frac{d\bar{i_q}}{dt} - \frac{\Delta L_q}{p\omega_m}\frac{d i_q{}^*}{dt} - \Delta L_d i_d - \frac{R_s}{p\omega_m}\bar{i_q} \tag{6.14}$$

Considering that when the SM-RFO reaches the stable state, the following conditions are satisfied:

$$\bar{i_q} \approx 0, \frac{d\bar{i_q}}{dt} \approx 0, i_q{}^* \approx i_q \tag{6.15}$$

Hence, (6.14) can be further simplified as:

$$err_{\_rea\text{-}est} = -\frac{\Delta L_q}{p\omega_m}\frac{d i_q}{dt} - \Delta L_d i_d \tag{6.16}$$

It can be seen from (6.16) that, first, when using inaccurate inductances to calculate the flux linkage, $err_{\_rea\text{-}est}$ cannot remain zero. Second, the motor working status influences the magnitude of the flux estimation error if the measured inductances do not comply with the real ones. To intuitively illustrate the impacts of inductance mismatch, take a surface-mounted PMSM prototype of which parameters are given in Table 6.1 as an example, and Figure 6.2 depicts the relationship between $err_{\_rea\text{-}est}$ and ΔL_d, ΔL_q under different conditions, from which four aspects need to be addressed. First, regardless of the inductance variations and motor speed, when the zero-d-axis current control strategy ($i_d = 0$) is used to control the motor to rotate stably ($d i_q/dt = 0$), the flux estimation results are accurate. Second, the larger the inductance deviations are, the larger the flux estimation errors become. Third, the accuracy of the flux estimation results decreases as the machine speed rises. Fourth, in comparison with the derivative of the q-axis current, the magnitude of the d-axis current has a larger impact on the flux estimation errors. As long as the d-axis current is non-zero and even

Table 6.1　Parameters of PMSM prototypes

Parameter	SPMSM	IPMSM	Unit
Stator winding resistance R_s	0.05	0.605	Ω
Measured d-axis inductance L_{d_mea}	3	12.65	mH
Measured q-axis inductance L_{q_mea}	3	13.5	mH
Number of pole pairs p	3	2	–
Moment of inertia J	0.1	0.012	kg·m^2
Original flux linkage ψ_{f_ori}	0.083	0.6873	Wb
Flux linkage threshold ψ_{thr}	−0.01	−0.08	Wb

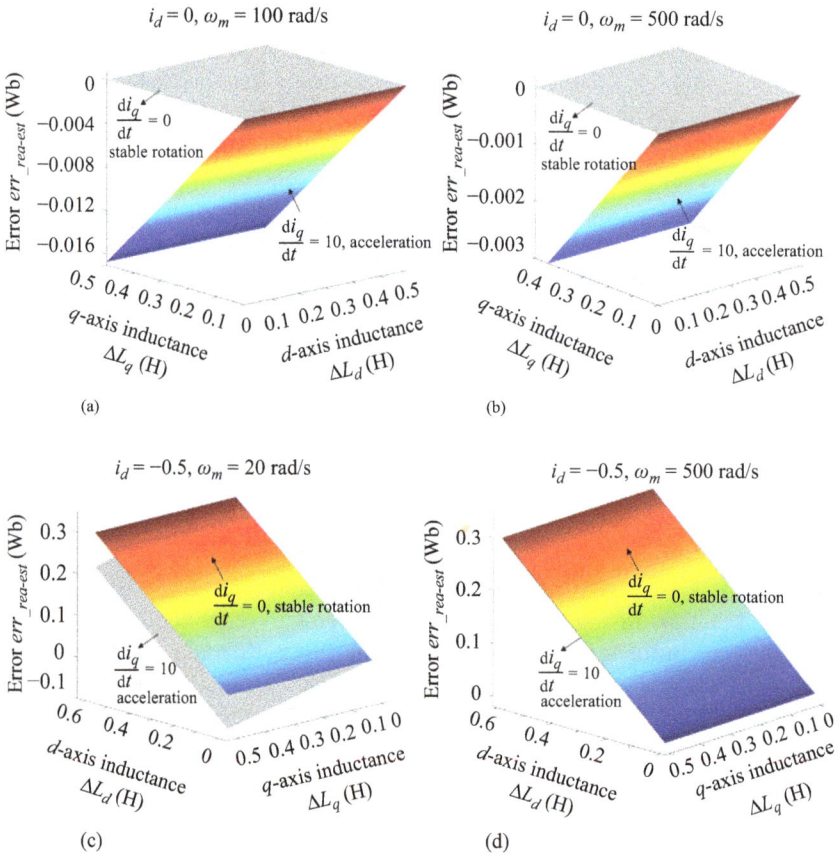

Figure 6.2　*Relationship between err_rea-est and ΔL_d, ΔL_q under different conditions: (a) speed maintains at 100 rad/s and d-axis current is 0, (b) speed maintains at 500 rad/s and d-axis current is 0, (c) speed maintains at 20 rad/s and d-axis current is −0.5 A, and (d) speed maintains at 500 rad/s and d-axis current is −0.5 A*

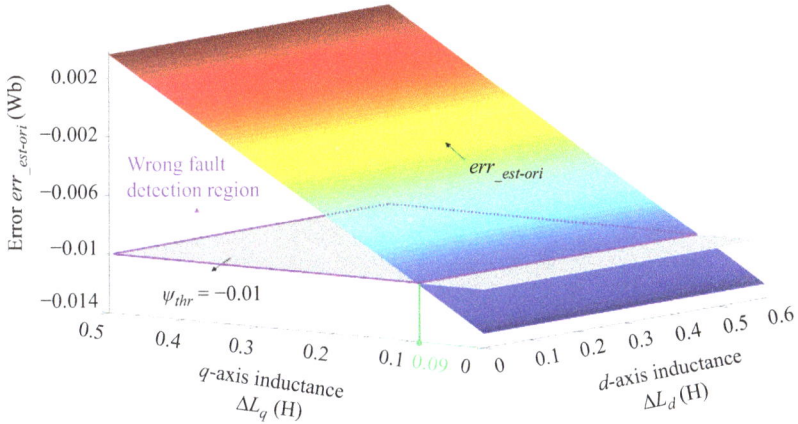

Figure 6.3   Relationship between err_est-ori and ΔL_d, ΔL_q when d-axis current is 0, the derivative of q-axis current is 10, and the motor speed is maintained at 100 rad/s

when it is just small (e.g., −0.5 A), there are large estimation errors when inductance mismatch occurs.

To clearly demonstrate the impacts of the inductance variations on the demagnetization fault detection process, assume that the real flux linkage of the motor is 0.7 Wb (fault occurs because ψ_{thr} is −0.01 Wb). Then, (6.16) and (6.4) can be employed to calculate ψ_f^* and $err_{\_est\_ori}$ once the flux estimation error is obtained, respectively. Figure 6.3 shows the relationship between $err_{\_est\_ori}$ and ΔL_d, ΔL_q when $i_d = 0$, $di_q/dt = 10$ and $\omega_m = 100$ rad/s. It can be seen that when the inductance mismatch phenomenon occurs, it is probable that the demagnetization fault is diagnosed incorrectly. For instance, when $\Delta L_d = 0$ while ΔL_q is larger than 0.09 H, the calculated results of $err_{\_est\_ori}$ are over −0.01 Wb. In this case, no fault will be detected when using (6.4) for fault judgment. Overall, the inductance mismatch issue brings about the difficulty in precisely identifying the flux linkage and diagnosing the demagnetization fault.

Overall, it is necessary to develop effective techniques to eliminate the impacts of inductance mismatch on the SM-RFO to improve the fault diagnosis accuracy.

6.2.1.4   Applicable conditions of proposed observer and fault diagnosis method

Based on (6.16), the flux linkage estimation accuracy varies under different working conditions when the inductances are mismatched, and the comparative results are shown in Table 6.2 at length. If the proposed SM-RFO is applied to the particular working condition ($i_d = 0$, $di_q/dt = 0$), the flux linkage estimation and fault detection accuracy can be guaranteed. Hence, for the sake of high accuracy, when using the proposed SM-RFO and fault diagnosis method, the zero-*d*-axis

Table 6.2   Comparison of accuracy under different conditions

Status		Accuracy	Descriptions of working conditions
i_d	di_q/dt		
$= 0$	$= 0$	High	d-axis current remains 0, steady state
	$\neq 0$	Pretty low	d-axis current remains 0, dynamic state
$\neq 0$	$= 0$	Low	d-axis current is not 0, steady state
	$\neq 0$	Low	d-axis current is not 0, dynamic state

current control strategy should be employed to regulate the motor to run stably, which is defined as "applicable conditions."

Undoubtedly, the above applicable conditions limit the scope of the proposed SM-RFO and fault diagnosis strategy. In terms of the other conditions dissatisfying the applicable ones (defined as "non-applicable conditions"), the aforesaid flux estimation and fault diagnosis methods might not be effective. Hence, it is meaningful to develop effective techniques to eliminate the impacts of the inductance mismatch, extending the applicability of the proposed strategies.

6.2.2   Accurate demagnetization fault diagnosis

In this section, robust d, q-axis sliding mode inductance observers (SM-IOs) are developed first. Then, the estimated parameters are incorporated into the SM-RFO to enhance the flux estimation and fault detection accuracy.

6.2.2.1   Structure of sliding mode inductance observers

Theoretically, considering that the differential equations (2.12), (2.13), and (2.19) contain the information of L_d and L_q, each of them can be used for inductance estimation by using the SM variable structure theory. However, because the flux linkage in (2.13) is uncertain as well, it is better not to use it to construct an IO. Besides, L_q is a numerator while L_d is a denominator in (2.12), making it more suitable to estimate the q-axis inductance, and (2-19) will be used to calculate the d-axis inductance.

As depicted in Figure 6.4, two steps are included in the inductance estimation process. First, the measured d-axis inductance L_{d_mea}, together with the resistance, is substituted into (2.13) to construct the q-axis SM-IO and obtain the estimated q-axis inductance L_q^*. Second, the pre-estimated inductance L_q^* and the other accurate parameters are substituted into (2.19) to obtain the estimated d-axis inductance L_d^*.

According to the SM variable structure theory, the q-axis SM-IO based on (2.12) can be represented as:

$$\frac{di_d^*}{dt} = -\frac{R_s}{L_{d\_mea}} i_d^* + \frac{k_{\_Lq} F(\bar{i_d})}{L_{d\_mea}} p\omega_m i_q + \frac{u_d}{L_{d\_mea}} \tag{6.17}$$

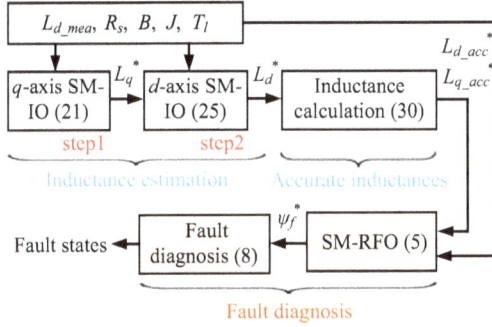

Figure 6.4   Execution procedures of the proposed inductance observers, flux observer, and fault diagnosis strategy.

where $i_d{}^*$ is the estimated current, k_{Lq} is the gain coefficient, $\bar{i}_d$ is the error between $i_d{}^*$ and the real current i_d. $F(\bar{i}_d)$ is the switching function that satisfies (6.2). When the system arrives at the equilibrium state, the real-time q-axis inductance equals the estimated value, that is,

$$L_q = L_q{}^* = k\_{Lq}F(\bar{i}_d) \tag{6.18}$$

where $L_q{}^*$ represents the estimated q-axis inductance. What needs to be mentioned is that first, when using the q-axis SM-IO, the inductance values on the nameplate L_{d_mea} are required to be adopted in (6.18). Whereas, considering that it is prone to encountering the mismatch issue, the accuracy of the estimation results might be influenced, and when subtracting (6.17) from (2.12), the q-axis inductance estimation error err_{Lq} can be derived as:

$$err\_{Lq} = L_q{}^* - L_q = \underbrace{\frac{L_{d\_mea}}{p\omega_m i_q}\frac{di_d{}^*}{dt} - \frac{L_d}{p\omega_m i_q}\frac{di_d}{dt}}_{term1} + \underbrace{\frac{R_s}{p\omega_m i_q}\bar{i}_d}_{term2} \tag{6.19}$$

Further, considering that $\bar{i}_d = 0$ and $di_d{}^*/dt = di_d/dt$ when the observer operates stably, (6.19) can be rewritten as:

$$err\_{Lq} = L_q{}^* - L_q = \frac{(L_{d\_mea} - L_d)}{p\omega_m i_q}\frac{di_d{}^*}{dt} \tag{6.20}$$

Referring to the q-axis SM-IO derivation process, the d-axis SM-IO can be constructed as (6.21) after substituting $L_q{}^*$ into (2.19):

$$\frac{d\omega_m{}^*}{dt} = \frac{1.5p}{J}\left(\psi_f i_q + (k\_{Ld}F(\overline{\omega_m}) - L_q{}^*)i_d i_q\right) - \frac{B\omega_m{}^*}{J} - \frac{T_l}{J} \tag{6.21}$$

where $\omega_m{}^*$ is the estimated speed, k_{Ld} is the gain coefficient, $\overline{\omega_m}$ is the speed estimation error. The flux information in (6.21) arises from the SM-RFO. The estimated d-axis inductance is:

$$L_d^* = k_{\_Ld}F(\overline{\omega_m}) \tag{6.22}$$

Differing from (6.17), the d-axis SM-IO does not require the measured inductance information. But if the pre-estimated L_q^* is inaccurate, the estimation result of (6.21) cannot reflect the real inductance either. And the d-axis inductance estimation error $err_{\_Ld}$ is:

$$err_{\_Ld} = L_d^* - L_d = \underbrace{\frac{J}{1.5pi_di_q}\frac{d\overline{\omega_m}}{dt}}_{term1} + \underbrace{\frac{B}{1.5pi_di_q}\overline{\omega_m}}_{term2} + \underbrace{L_q^* - L_q}_{term3} \tag{6.23}$$

In (6.23), because $\overline{\omega_m}$ equals zero, $term1$ and $term2$ are zero. Then, it can be simplified as:

$$err_{\_Ld} = L_d^* - L_d = L_q^* - L_q \tag{6.24}$$

Based on the aforementioned analysis, the accuracy of the proposed inductance observers can also be affected by the measured d-axis inductance. This is similar to the SM-RFO. To avoid estimation errors, combine (6.20) and (6.24) to form a system of equations as:

$$\begin{cases} L_q^* - L_q - \dfrac{(L_{d\_mea} - L_d)}{p\omega_m i_q}\dfrac{di_d}{dt} = 0 \\ L_d^* - L_d - L_q^* + L_q = 0 \end{cases} \tag{6.25}$$

In (6.25), L_d and L_q represent the accurate inductance values to be calculated (two unknown variables), while the other parameters, including L_d^*, L_q^*, and L_{d_mea} are known. Hence, it is a univariate quadratic equation. Denote the solutions to the equation as $L_{d\_acc}^*$ and $L_{q\_acc}^*$, which are the accurately estimated d, q-axis inductances, respectively, and they can be derived as:

$$\begin{cases} L_{d\_acc}^* = \dfrac{p\omega_m i_q L_d^* - L_{d\_mea}\dfrac{di_d}{dt}}{\dfrac{di_d}{dt} - p\omega_m i_q} \\ L_{q\_acc}^* = \dfrac{(L_d^* - L_{d\_mea} - L_q^*)\dfrac{di_d}{dt} + L_q^* p\omega_m i_q}{\dfrac{di_d}{dt} - p\omega_m i_q} \end{cases} \tag{6.26}$$

6.2.2.2   Stability analysis

The prerequisite for using the proposed SM-IOs to obtain accurate inductances is that they can reach a stable state. To analyze their stability, two independent Lyapunov functions are needed. First, define the sliding surfaces $\mathbf{S}_{dq}$ as:

$$\mathbf{S}_{dq} = [S_{id}, S_{\omega m}]^T = [\bar{i}_d, \overline{\omega_m}]^T \tag{6.27}$$

Second, the Lyapunov functions $\mathbf{V}_{dq}$ can be established as:

$$\mathbf{V}_{dq} = diag\left(\frac{1}{2}\mathbf{S}_{dq} \cdot \mathbf{S}_{dq}{}^T\right) = \frac{1}{2}\left[\bar{i}_d{}^2, \overline{\omega}_m{}^2\right]^T \qquad (6.28)$$

where *diag*( ) is the function to extract diagonal elements of a matrix, and the derivative of $\mathbf{V}_{dq}$ is:

$$\frac{d\mathbf{V}_{dq}}{dt} = \left[\frac{\bar{i}_d\,\frac{d\bar{i}_d}{dt}, \overline{\omega}_m\,\frac{d\overline{\omega}_m}{dt}}{dt}\right]^T$$

$$= \left[\begin{array}{c} \underbrace{-\frac{R_s}{L_{d\_mea}}\bar{i}_d{}^2}_{term1} + \underbrace{\frac{(k_{\_Lq}F(\bar{i}_d) - L_q)}{L_{d\_mea}}p\omega_m i_q \bar{i}_d}_{term2} \\ \underbrace{-\frac{B\overline{\omega}_m{}^2}{J}}_{term1} + \underbrace{\frac{1.5pi_d i_q \overline{\omega}_m}{J}(k_{\_Ld}F(\overline{\omega}_m) - L_d)}_{term2} \end{array}\right] \qquad (6.29)$$

Then, to make the observers stable, the following conditions (*term2* is less than zero) should be satisfied:

$$\begin{cases} \dfrac{(k_{\_Lq}F(\bar{i}_d) - L_q)}{L_{d\_mea}}p\omega_m i_q \bar{i}_d < 0 \\ \dfrac{1.5pi_d i_q \omega_m}{J}(k_{\_Ld}F(\overline{\omega}_m) - L_d) < 0 \end{cases} \qquad (6.30)$$

Interestingly, unlike (6.9), apart from the errors between the estimated and the real values ($\bar{i}_d$ and $\overline{\omega}_m$), the working states (i_d and i_q) of the motor influence the stability of the observers. Table 6.3 comprehensively shows the stability conditions of each observer. Being similar to (6.10)–(6.12), it can be deduced that there must exist constants making both SM-IOs stable. In practice, once $k_{\_Lq}$ and $k_{\_Ld}$ are designed and selected based on the real-time working status, following Table 6.3, both observers will remain stable.

Table 6.3   Stability conditions of SM-IOs

Status		Stability conditions	
		$k_{\_Lq}$	$k_{\_Ld}$
$i_q > 0$	$i_d > 0$	$< -L_q/|F(\bar{i}_d)|$	$< -L_d/|F(\overline{\omega}_m)|$
	$i_d < 0$	$< -L_q/|F(\bar{i}_d)|$	$> -L_d/|F(\overline{\omega}_m)|$
$i_q < 0$	$i_d > 0$	$> -L_q/|F(\bar{i}_d)|$	$> -L_d/|F(\overline{\omega}_m)|$
	$i_d < 0$	$> -L_q/|F(\bar{i}_d)|$	$< -L_d/|F(\overline{\omega}_m)|$

6.2.2.3   Accurate fault diagnosis strategy

By using the estimated inductances in (6.26) to construct the SM-RFO in (6.1), the negative impacts of inductance mismatch on the flux identification accuracy will be rejected. Then, the accurate flux information can be substituted into (6.4) to diagnose whether a PMSM encounters the demagnetization fault (see Figure 6.4). It can be found that the proposed IOs contribute to enhancing the robustness of both the SM-RFO and the fault diagnosis strategy when $i_d \neq 0$ or $di_q/dt \neq 0$, so their applicability is extended.

6.2.2.4   Test results

In order to comprehensively verify that the proposed demagnetization fault diagnosis strategy based on the SM-RFO is effective and accurate under different working situations, a simulation is conducted on an surface-mounted permanent magnet synchronous motor (SPMSM). For the SPMSM, the d-axis current usually remains at zero. The simulation model is established in MATLAB®/Simulink® 2018b. The simulation setups include that: (1) For the purpose of comprehensiveness, the simulation is divided into three parts, the first of which compares the flux estimation and fault diagnosis results under the applicable conditions. The second part will prove that if the SM-RFO experiences inductance mismatch issues, the flux estimation errors will appear when the working conditions change ($i_d \neq 0$ or $di_q/dt \neq 0$), addressing the necessity of using inductance observers to improve the system's robustness. In the third part, the effectiveness of the proposed fault diagnosis method based on robust SM inductance observers is verified. (2) k is set as 0.5. (3) The motor is controlled by the traditional vector control strategy in which the d-axis current reference can be manually designed. (4) The load is 5 Nm.

Results without SM-IOs incorporated under applicable conditions: Figure 6.6 shows the simulation results of the SM-RFO without incorporating SM-IOs under the applicable conditions, that is, $i_d = 0$ and $di_q/dt = 0$. First, in Figure 6.5(a) and Figure 6.5(b), the flux linkage of the motor equals the original value 0.083 Wb between 0 and 2.5 s, and after that, it is set as 0.7 Wb (demagnetization fault occurs). In addition, the d, q-axis inductances used for constructing the SM-RFO are consistent with the real ones ($\Delta L_d = 0$, $\Delta L_q = 0$). It can be seen that the SM-RFO is able to detect the flux linkage quickly and accurately, without estimation errors. Meanwhile, the fault diagnosis result is correct. Second, in Figure 6.5(c) and Figure 6.5(d), the flux linkage is set as 0.7 Wb during the whole test process, and an inductance mismatch is considered. In comparison with Figure 6.5(a) and Figure 6.5(b), it can be noted that the flux estimation results can still track the real ones, but small fluctuations appear when the inductance mismatch problem occurs. This can be tackled by using low-pass filters in practice. Obviously, the fault diagnosis accuracy is still high even if the inductances are mismatched. Overall, these illustrate that the proposed fault diagnosis method based on the SM-RFO is effective and accurate as long as the working conditions satisfy those introduced in Section 6.2.2.2. Simultaneously, the analytical results presented in (6.16) can be partially verified.

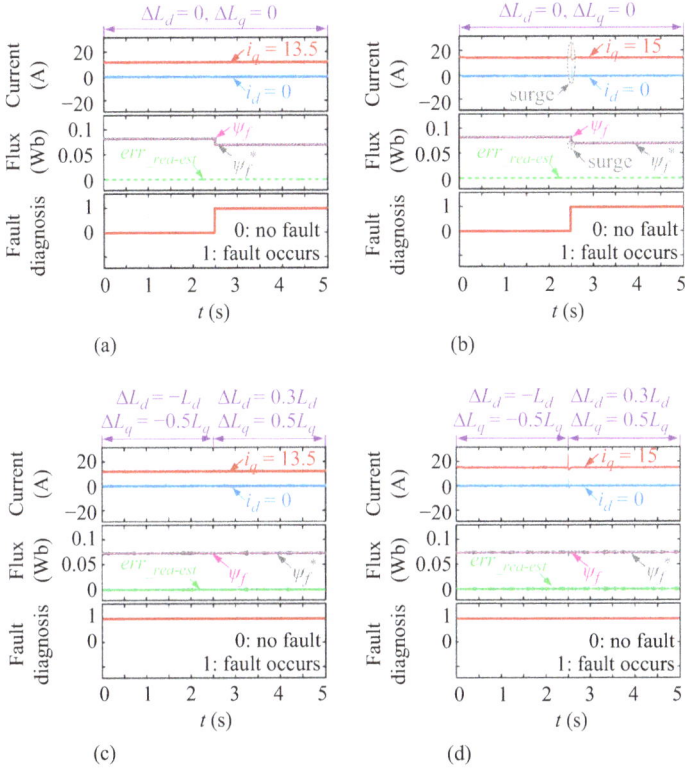

Figure 6.5   *Results of SM-RFO without incorporating SM-IOs ($i_d = 0$, $di_q/dt = 0$):*
(a) speed is 20 rad/s, (b) speed is 500 rad/s, (c) speed is 20 rad/s, and
(d) speed is 500 rad/s

Results without SM-IOs incorporated under non-applicable conditions: To further illustrate that the flux estimation errors comply with (6.20), Figure 6.6 shows the simulation results under different working conditions when the d, q-axis inductances satisfy that $\Delta L_d = -L_d$ and $\Delta L_q = -0.5L_q$. In detail, di_q/dt equals 0 and i_d is -10 A in Figure 6.6(a), di_q/dt equals 0 and i_d is -5 A in Figure 6.6(b). The motor speed is 20 rad/s. When the d-axis current is maintained at -10 A and -5 A, the flux estimation errors are -0.03 Wb and -0.015 W, respectively. This complies with the theoretical results, which can also be indicated by Figure 6.6(c).

Results with SM-IOs incorporated under non-applicable conditions: When the d, q-axis inductances used for constructing the SM-RFO are inaccurate, that is, $\Delta L_d = -L_d$ and $\Delta L_q = -0.5L_q$, the simulation results under non-applicable working conditions are shown in Figure 6.7. During the test, between 0 s and 2.5 s, the motor rotates stably at 20 rad/s and the d-axis current is set as -10 A ($i_d = -10$ A, $di_q/dt = 0$), after which it rises to 50 rad/s (i_d and di_q/dt vary in the process). First, the estimation results of the proposed d, q-axis SM-IOs are 3 mH, indicating that they are

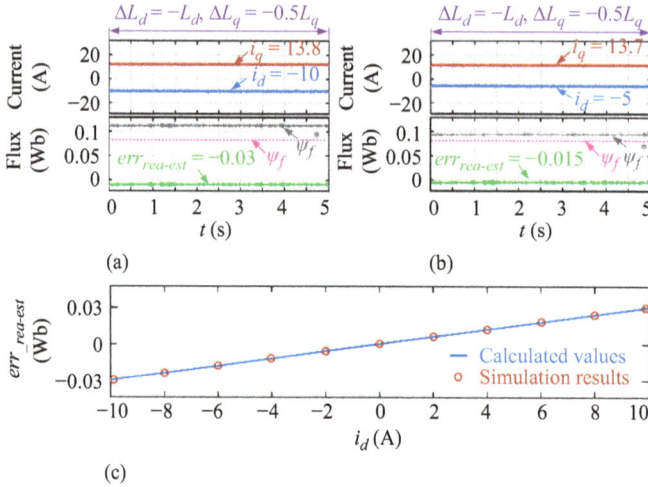

Figure 6.6   *Results of SM-RFO without incorporating the SM-IOs under non-applicable conditions (ω_m=20 rad/s, ΔL_d= $-L_d$ and ΔL_q= $-0.5L_q$): (a) $di_q/dt = 0$ and $i_d = -10$ A, (b) $di_q/dt = 0$ and $i_d = -5$ A, and (c) comparative results when d-axis current changes*

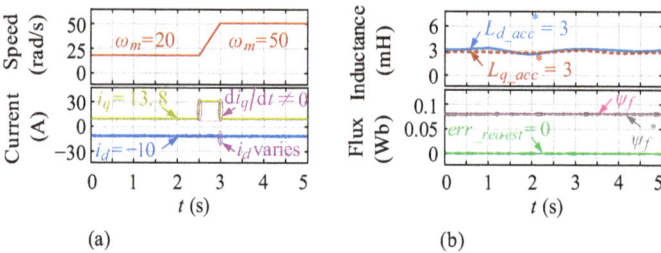

Figure 6.7   *Results with SM-IOs incorporated under non-applicable conditions: (a) speed and current performance and (b) inductance and flux estimation results*

highly robust against inductance variations. Even though the SM-IOs encounter inductance mismatch issues, they can still estimate the inductance values precisely. This lays the ground for accurate flux estimation and fault detection. Second, after using the estimated inductances to construct the SM-RFO, the flux estimation results show high accuracy. Specifically, the flux estimation errors are nearly zero, which are completely different from those in Figure 6.7(a). Further, when using the estimated flux for demagnetization diagnosis, the fault can be precisely determined. Overall, both the proposed observers and fault diagnosis methods are effective.

Chapter 7

Future trends and emerging technologies in permanent magnet actuator control

7.1 Next-generation actuation paradigms

7.1.1 Bio-inspired actuators

Bio-inspired actuators are emerging as a transformative technology in robotics, combining principles derived from the natural world with cutting-edge engineering innovations. These actuators are designed to mimic the functionality, performance, and adaptability of biological systems, seeking to overcome the limitations of traditional actuators such as motors and hydraulic systems. Their unique characteristics, including flexibility, efficiency, and scalability, have made them a subject of intense research and development (R&D). As robotics moves toward more flexible, adaptable, and efficient systems, bio-inspired actuators are poised to play a key role in revolutionizing the field.

Bio-inspiration, or biomimicry, is the practice of drawing inspiration from nature's designs to solve complex engineering problems. In the context of actuators, this involves studying how natural systems—ranging from muscles and tendons to the movement of plants and animals—generate motion and applying these mechanisms to synthetic systems. Biological organisms have evolved intricate mechanisms to perform highly efficient, robust, and adaptable motions. Whether it is the way a cheetah accelerates or the way a Venus flytrap captures prey, nature offers a wealth of inspiration for the design of actuators in robotics.

For instance, human muscles, which operate through the contraction of fibers using chemical energy, inspire actuators that rely on similar principles of contraction and relaxation. Insects, known for their ability to move rapidly and with extreme precision, inspire miniature, lightweight actuators that could be employed in small-scale robotics. The adaptability of bio-inspired actuators, allowing them to function in variable environments and under a range of conditions, is a significant advantage over traditional rigid actuators.

The concept of bio-inspired actuation transcends simple mimicry. It seeks not only to replicate biological mechanisms but to improve upon them, optimizing for efficiency, scalability, and versatility. Actuators designed with these principles in mind can perform a wider range of tasks, exhibit improved energy efficiency, and provide smoother motion control than traditional solutions.

The design principles behind bio-inspired actuators are rooted in the underlying mechanics of natural movement. Biological systems exhibit remarkable efficiency in converting energy into movement, often with minimal waste and highly adaptable responses. To capture these advantages, bio-inspired actuators focus on several key principles.

One fundamental principle is the use of compliance and flexibility. Unlike traditional rigid actuators, bio-inspired actuators often incorporate flexible materials or structures that allow for smooth, elastic deformations. This design enables them to handle a wider range of movements and forces, similar to how soft tissues and muscles in the human body function. By mimicking the deformable structures found in biological organisms, bio-inspired actuators can achieve more fluid and nuanced motions. These actuators also offer significant safety improvements, as their soft nature reduces the likelihood of damaging delicate objects or human operators in applications requiring close interaction.

Another key feature of bio-inspired actuators is their energy efficiency. Biological systems are highly optimized for low-energy operation. For example, the human body uses minimal energy to maintain posture and generate movement. In bio-inspired actuators, this efficiency is often realized through the use of materials that can store and release energy like biological muscles or tendons. For instance, elastic actuators store energy during one phase of movement and then release it during another, enhancing the actuator's efficiency and reducing the need for continuous energy input.

The use of distributed control systems is another principle inspired by biology. Many biological organisms, particularly invertebrates, rely on decentralized control to generate complex motions. Rather than having a central brain directing every movement, the control is distributed across multiple sensors and actuators, each making local decisions based on environmental input. In bio-inspired actuators, this concept can be applied through the use of modular, decentralized actuator units that communicate with each other to achieve coordinated movement. This approach offers robustness and flexibility in complex or unpredictable environments.

Bio-inspired actuators can be broadly classified into several categories, each based on a biological mechanism or system that they emulate. These categories include soft actuators, muscle-like actuators, and biohybrids, among others. Each type brings unique advantages, which make them suitable for different applications in robotics.

7.1.1.1   Soft actuators

Soft actuators are a class of actuators designed to mimic the soft and deformable characteristics of biological tissues. These actuators are typically made from elastic or flexible materials, such as silicone or rubber, which enable them to bend, twist, stretch, and compress in ways that traditional actuators cannot. Soft actuators take inspiration from biological organisms such as octopuses, worms, and other soft-bodied animals that rely on the flexibility of their bodies for movement.

Soft actuators operate through mechanisms such as pneumatic or hydraulic pressure, shape-memory alloys (SMAs), or electroactive polymers (EAPs). For example, pneumatic soft actuators inflate or deflate in response to changes in air pressure, mimicking the way certain animals use air or liquid to inflate body parts for movement. These actuators are highly flexible and can change shape to adapt to their environment, making them ideal for tasks that require precise, adaptable movements.

7.1.1.2 Muscle-like actuators

Muscle-like actuators are designed to mimic the contraction and relaxation of natural muscles, which convert chemical energy into mechanical work. These actuators use various mechanisms to replicate the functionality of biological muscles, including EAPs, artificial muscles, and SMAs.

EAPs, for example, expand or contract when an electric field is applied, similar to the way muscles contract in response to neural stimulation. These materials can be used to create actuators that not only mimic the movement of muscles but also replicate their high force-to-weight ratio, making them ideal for applications requiring both strength and compactness. Artificial muscles based on polymers or other flexible materials are also being developed to offer similar benefits in terms of efficiency and adaptability.

7.1.1.3 Biohybrid actuators

Biohybrids combine living tissues with synthetic materials to create actuators that leverage the unique capabilities of biological systems. For example, a biohybrid actuator might integrate muscle tissue into a robotic structure, allowing for natural muscle contraction and movement alongside mechanical elements. This approach is particularly promising for applications where the flexibility and responsiveness of biological systems are essential, such as in soft robotics or bio-robotics.

A notable example of biohybrids is the use of biological muscles grown in the lab and incorporated into actuators for movement. These muscle tissues can contract and relax just like natural muscle, providing a high degree of natural movement. Although biohybrids are still in the experimental stages, they represent a cutting-edge approach to actuator design.

Bio-inspired actuators hold significant potential across a wide range of applications. From healthcare to space exploration, these actuators could revolutionize the way robots interact with their environment, offering solutions that are more adaptable, efficient, and safe than current technologies.

7.1.1.4 Soft robotics

Soft robotics is one of the most promising fields for bio-inspired actuators. Soft robots, which rely on flexible materials and actuators, can perform delicate tasks in unstructured environments. These robots are particularly useful in medical applications, where they can assist with minimally invasive surgeries or provide rehabilitation support. For instance, soft actuators based on the principles of muscle contraction can be used in robotic prosthetics, allowing for more natural and flexible movements than traditional rigid prostheses. Additionally, soft actuators in grippers and manipulators can safely handle delicate objects, making them ideal for tasks in industries such as food processing or agriculture.

7.1.1.5   Exoskeletons and wearable robotics

Bio-inspired actuators also play a crucial role in the development of exoskeletons and wearable robots designed to assist people with disabilities or enhance human capabilities. By mimicking the function of human muscles, these actuators can provide smooth, coordinated movements that support or augment the wearer's own muscles. These devices can be used for rehabilitation, providing physical therapy by mimicking natural movements, or they can help individuals with mobility impairments regain function.

7.1.1.6   Robotics for space exploration

In space exploration, bio-inspired actuators offer the potential to create highly adaptable robotic systems capable of navigating the extreme environments of space. Bio-inspired actuators, such as soft robots that can deform and adapt to obstacles, may be used to explore surfaces like those on the Moon or Mars. Their ability to squeeze into tight spaces, adapt to uneven terrain, and interact safely with delicate objects could make them invaluable for tasks such as equipment maintenance, sample collection, or even search and rescue operations in extraterrestrial environments.

7.1.1.7   Underwater and marine robotics

Marine robotics is another field where bio-inspired actuators are making an impact. Bio-inspired designs that replicate the propulsion mechanisms of fish and other aquatic animals are being explored to create more efficient and maneuverable underwater robots. Soft actuators that mimic the undulating motion of fish tails or the oscillation of jellyfish tentacles offer the potential for highly efficient, silent movement in aquatic environments. These actuators are capable of making precise movements in tight or hazardous spaces, such as within coral reefs or submerged structures.

7.1.1.8   Humanoid robotics

Humanoid robots, which are designed to replicate human form and function, can greatly benefit from bio-inspired actuators. The complex movement required for a humanoid robot to walk, jump, or manipulate objects with dexterity and precision can be achieved more effectively through muscle-like actuators that replicate natural human motion. This not only improves the robot's performance but also ensures that the movements are smoother, more adaptable, and energy-efficient compared to traditional motors or hydraulic systems.

While bio-inspired actuators offer numerous advantages, they are not without their challenges. One of the major hurdles is the material limitations. Despite advances in materials science, there is still a gap between the properties of biological materials and their synthetic counterparts. For example, while muscle fibers exhibit exceptional efficiency and strength, replicating these properties in synthetic actuators requires ongoing research into novel materials and manufacturing techniques. Durability is another challenge, as bio-inspired actuators often rely on soft or flexible materials that may degrade over time, especially in demanding environments. Addressing this issue requires the development of materials with higher resilience and longer lifespans. Additionally, control mechanisms for bio-inspired actuators remain complex. Developing effective control algorithms that can handle the highly nonlinear and often unpredictable behavior of bio-inspired actuators is an ongoing area of research.

Despite these challenges, the future of bio-inspired actuators is bright. With advancements in material science, control algorithms, and manufacturing technologies, these actuators are likely to become a critical component of next-generation robots. As robots become more integrated into our daily lives and industries, bio-inspired actuators will enable more adaptable, efficient, and human-like robots that can perform a diverse array of tasks.

7.1.2   Quantum and nano-technology applications

Quantum and nano-technologies are rapidly reshaping the landscape of actuator systems, including permanent magnet actuators (PMAs). The integration of quantum and nano-scale phenomena into actuator design promises unprecedented improvements in efficiency, performance, and functionality. While still in the early stages of research and application, quantum and nano-technologies hold vast potential to unlock capabilities far beyond those possible with traditional materials and designs. In the context of bio-inspired actuators and robotics in general, these technologies are poised to offer transformative advantages. This section explores the potential of quantum and nano-technologies in actuator development, their current status, and the future directions they are likely to take.

To fully appreciate the potential applications of quantum and nano-technologies in actuators, it is essential to first understand the fundamental principles behind these technologies. At the quantum level, particles like electrons and photons behave according to the laws of quantum mechanics, exhibiting phenomena such as superposition, entanglement, and tunneling. These behaviors allow for the development of highly sensitive and efficient devices that exploit quantum properties to achieve outcomes impossible with classical systems.

Nanotechnology, on the other hand, deals with the manipulation of matter on an atomic or molecular scale, typically between 1 nanometer and 100 nanometers. At this scale, the properties of materials often differ significantly from those at the macro-scale. For example, the mechanical, electrical, and optical properties of materials can change drastically at the nanoscale, providing new opportunities for actuator development. Nano-materials like carbon nanotubes (CNTs), graphene, and quantum dots exhibit remarkable strength, conductivity, and flexibility, offering significant advantages in various applications.

One of the most exciting potential applications of quantum technology in actuator systems is the creation of quantum actuators. Quantum actuators could revolutionize the way we approach motion control by exploiting quantum mechanical phenomena to achieve extreme precision and efficiency in actuator systems. Unlike traditional actuators, which rely on classical mechanics, quantum actuators would operate on principles such as quantum superposition and entanglement, allowing them to perform complex tasks with minimal energy expenditure and maximal accuracy.

The idea behind quantum actuators is rooted in the concept of quantum coherence, where quantum systems can exist in multiple states simultaneously until measured. This principle has been shown to provide extraordinary advantages in fields such as quantum computing and sensing, where information is processed in parallel, allowing for faster and more efficient operations. When applied to actuators, quantum coherence could enable actuators to perform multiple tasks or generate movement in multiple directions simultaneously, effectively boosting their performance while reducing energy consumption.

One specific example of a quantum-based actuator involves the use of quantum dots. These are semiconductor nanoparticles that have quantum mechanical properties, particularly in terms of their electronic characteristics. Quantum dots can be used in actuators to create systems that can respond to external stimuli with extreme sensitivity and precision. For instance, quantum dots embedded in a material could be used to create a smart material that changes shape or properties when exposed to an electric field, offering new possibilities for actuators that require a highly responsive and adaptable performance.

In addition to quantum dots, superconducting materials are another candidate for quantum actuators. Superconductors, which exhibit zero electrical resistance at low temperatures, could be used to create actuators that experience no energy loss during movement. The potential for zero-loss operation, combined with the quantum mechanical properties of superconducting materials, would result in actuators that are not only highly efficient but also capable of maintaining their performance over extended periods without degradation. Superconducting actuators could find applications in space exploration, high-performance computing, and other fields requiring high precision and minimal energy expenditure.

Nanotechnology is already making significant strides in actuator design, and its application to PMAs and bio-inspired actuators is generating exciting new possibilities. The unique properties of nano-materials, such as CNTs, graphene, and piezoelectric materials, can provide actuators with unprecedented strength, flexibility, and responsiveness.

CNTs are cylindrical structures made of carbon atoms, with diameters on the nanometer scale. CNTs have remarkable mechanical properties, including exceptional strength-to-weight ratios and extraordinary flexibility. These properties make them ideal for use in actuators, where both strength and flexibility are essential for high-performance applications. CNTs can be used to create actuators that mimic the action of biological muscles, offering the ability to contract and expand under an applied voltage or other stimulus. The high tensile strength of CNTs also allows for the creation of actuators capable of generating significant force while maintaining a lightweight structure, a characteristic that is particularly valuable in mobile robots and drones.

CNT-based actuators could be used in a wide range of applications, from medical robotics to aerospace technology. For example, in prosthetics, CNTs could be integrated into the actuators of artificial limbs to provide more natural, fluid movements. In space exploration, CNT-based actuators could enable precise and flexible robotic arms to perform delicate tasks, such as sample collection or equipment maintenance, in the harsh environment of space.

Graphene, a single layer of carbon atoms arranged in a two-dimensional honeycomb structure, is another revolutionary material that has the potential to significantly enhance actuator performance. Graphene is known for its remarkable electrical, thermal, and mechanical properties, making it an ideal candidate for use in high-performance actuators. It is incredibly strong, lightweight, and flexible, with electrical conductivity and heat dissipation capabilities that far exceed traditional materials. Additionally, graphene has an extraordinary ability to conduct strain, which can be leveraged in actuators that require high precision and responsiveness.

Graphene-based actuators could have significant applications in fields ranging from robotics to energy harvesting. For example, in wearable robotics, graphene could be used to create soft, flexible actuators that conform to the human body's movements while providing enhanced strength and durability. In robotic manipulation, graphene-based actuators could allow for rapid, adaptive movements, enabling robots to handle fragile objects with high precision.

Piezoelectric materials generate an electrical charge in response to mechanical stress, making them ideal candidates for use in actuators. At the nanoscale, piezoelectric materials exhibit enhanced properties that can be harnessed for extremely fine control of motion. Nano-piezoelectric actuators can generate small deformations with a high degree of accuracy, making them useful in applications where minute movements are required. For example, in microrobotics or medical applications, nano-piezoelectric actuators could be used to control the precise positioning of surgical instruments or to provide feedback in haptic systems.

The combination of piezoelectric materials with other nanostructured materials, such as graphene or CNTs, is a promising approach for creating hybrid actuators with superior performance. These hybrid systems could exploit the strengths of each material to create actuators that are both highly sensitive and capable of generating significant force in response to minimal input.

One of the key advantages of quantum and nano-technologies in actuator systems is their potential to dramatically improve energy efficiency. Traditional actuators, such as electric motors, typically suffer from energy losses due to friction, heat dissipation, and other factors. In contrast, quantum and nano-scale actuators could operate with significantly reduced energy consumption, thanks to their unique properties. For example, nano-scale actuators made from materials like CNTs or graphene could minimize energy losses by reducing the amount of heat generated during operation. In addition, quantum-based actuators could exploit quantum tunneling and superposition to reduce the energy required to achieve specific movements, potentially allowing actuators to function with much lower power consumption than conventional systems.

Moreover, quantum and nano-technologies can enable self-powered actuators through the development of energy harvesting systems. For instance, piezoelectric nano-materials could be used to create actuators that not only perform work but also generate electrical energy from their motion, making them self-sustaining. These types of actuators could be used in remote or inaccessible environments, such as deep-sea or space exploration, where access to external power sources is limited.

While the potential applications of quantum and nano-technologies in actuator design are vast, there are significant challenges that must be addressed before these technologies can be widely adopted. One of the main challenges is the scalability of quantum and nano-materials. Many of these materials and technologies are still in the early stages of development, and it remains to be seen whether they can be manufactured at the scale required for practical actuator systems. Additionally, the integration of quantum and nano-technologies into existing actuator systems poses significant engineering challenges, particularly in terms of ensuring compatibility with traditional components and systems.

Another challenge is the cost of producing quantum and nano-materials. While the potential benefits of these materials are clear, their production can be expensive, and scaling up manufacturing processes for widespread use in actuators will require significant investment and technological advancements.

Despite these challenges, the future prospects of quantum and nano-technologies in actuator systems are highly promising. As research progresses, breakthroughs in material science, quantum computing, and nano-fabrication techniques are likely to overcome many of the current obstacles. In the coming decades, quantum and nano-based actuators could become a cornerstone of next-generation robotics, offering exceptional performance, energy efficiency, and versatility in a wide range of applications.

The ongoing development of quantum and nano-technologies promises to expand the possibilities of actuator design and robotics in ways that are difficult to predict today. By exploiting the unique properties of materials at the atomic and molecular scale, we are entering a new era of actuator technology that will likely redefine what is possible in automation, robotics, and beyond.

7.1.3 *Morphing and programmable actuators*

In the realm of advanced robotics, the demand for flexible, adaptive, and highly responsive actuator systems is growing. Traditional actuators, such as electric motors and hydraulic systems, are typically limited in terms of their shape, size, and the range of motion they can provide. This is where morphing and programmable actuators come into play. These innovative actuators allow robots to dynamically alter their shape, stiffness, and functionality based on changing environmental conditions or task requirements. As a result, they offer unprecedented levels of versatility and adaptability, making them ideal candidates for complex, real-world applications ranging from soft robotics to aerospace engineering.

This section delves into the fundamental principles, types, applications, and challenges associated with morphing and programmable actuators, exploring how these technologies are reshaping the landscape of robotics.

To begin with, it is essential to define what is meant by morphing and programmable actuators. Morphing actuators are a class of actuators that enable a system to undergo significant changes in shape or geometry in response to external stimuli, such as electric fields, magnetic fields, or thermal gradients. Unlike traditional actuators, which have fixed geometries and are limited in their ability to modify shape, morphing actuators can continuously adapt their physical form to suit specific tasks or environmental conditions.

Programmable actuators, on the other hand, combine the ability to morph with the capability to perform a range of motions or functions in a highly controlled and reversible manner. These actuators can be programmed to exhibit a series of movements or behaviors that are predefined or adapted in real-time according to sensory inputs. The combination of morphing and programmability makes these actuators ideal for creating systems that require high levels of flexibility, such as soft robots, bio-inspired robots, and other advanced robotic systems.

The key characteristics of morphing and programmable actuators include:

1. Shape adaptability: These actuators can change their form to suit different functions or environments, providing greater versatility than traditional fixed actuators.
2. Programmability: The ability to control the actuator's behavior allows for more precise and dynamic responses to external stimuli, enabling a wide range of applications.
3. Energy efficiency: Morphing actuators often require less energy to function than traditional systems because they can optimize their structure based on the task at hand.
4. Softness and flexibility: Many morphing actuators are soft, allowing for compliance with delicate or irregularly shaped objects, ideal for tasks like handling fragile items or interacting with humans.

The development of morphing and programmable actuators has seen significant advances in recent years, leading to a variety of actuator designs that leverage different principles and materials. These actuators can be broadly categorized into soft actuators, shape-memory actuators, EAPs, and others.

Soft actuators are often made from flexible, deformable materials that enable them to change shape and size in response to external forces. Unlike rigid actuators, which rely on mechanical components such as gears and motors, soft actuators deform continuously to produce movement. These actuators are particularly useful in applications requiring a high degree of compliance, such as human-robot interaction, medical devices, and bio-inspired robotics.

One popular type of soft actuator is the pneumatic artificial muscle (PAM). PAMs operate by inflating or deflating in response to changes in pressure, allowing them to mimic the contraction and expansion of biological muscles. These actuators are lightweight, energy-efficient, and highly adaptable, making them suitable for soft robotic applications, such as prosthetics and exoskeletons.

Another category of soft actuators is hydraulic actuators, which use pressurized fluids to produce movement. These actuators can deliver significant force in a compact form, and they have been used in large-scale robotics, such as in the construction or agricultural sectors.

SMAs are materials that can "remember" and return to a predetermined shape when heated above a certain threshold temperature. When an SMA is subjected to heat, it undergoes a phase change that causes it to expand or contract. The ability to return to a predetermined shape after a change in temperature makes SMAs ideal for use in programmable actuators. For instance, nickel-titanium alloys (Nitinol) are widely used in applications where precise motion control is required. The actuators made from Nitinol can change their form when an electrical current is applied, causing a rapid expansion or contraction. These actuators have been used in a range of robotic applications, including medical devices such as stents and artificial muscles for soft robots.

EAPs are another class of materials that exhibit a change in size or shape when subjected to an electric field. EAPs can be classified into two types: ionic EAPs and electronic EAPs. Ionic EAPs, also known as ionomers, undergo deformation in response to an electric field that causes ions within the material to migrate, creating a mechanical strain. Electronic EAPs, such as dielectric elastomers, generate deformation when subjected to an electric field that causes a change in the material's volume. These materials are highly versatile and can produce large deformations with relatively low voltages, making them ideal for soft robotics and actuators requiring fine control over their movements. The primary advantage of EAPs is their ability to provide high strain, meaning they can produce significant deformation with a small amount of applied energy.

Magnetostrictive actuators are materials that change their shape or size in response to an applied magnetic field. These actuators can be used to create programmable systems that alter their shape in response to changes in the external environment. They are particularly useful in applications requiring high precision and fast response times. Terfenol-D, a magnetostrictive alloy, is one of the most widely used materials in magnetostrictive actuators due to its high sensitivity and efficient energy conversion capabilities.

Electrostatic actuators use the force generated by an electric field between two charged plates to produce movement. These actuators are lightweight, highly energy-efficient, and can be fabricated at the micro and nano scales, making them suitable for micro-robots and nano-robots. Electrostatic actuators operate on the principle that oppositely charged objects experience a force of attraction, while objects with like charges repel each other. By controlling the magnitude and direction of the electric field, electrostatic actuators can be precisely controlled for various tasks.

7.2   Convergence of PMAs with broader technological ecosystems

7.2.1   Actuators in augmented reality (AR) and virtual reality (VR)

The integration of actuators into augmented reality (AR) and virtual reality (VR) systems has become a key area of innovation, significantly enhancing the immersive experience. While AR and VR have traditionally been associated with visual and auditory stimulation, the inclusion of haptic feedback, motion tracking, and physical interaction through actuators brings these technologies to life in a much more tactile and dynamic way. Actuators provide a crucial bridge between the virtual world and the user's physical experience, offering feedback that mimics real-world sensations such as touch, texture, pressure, and motion. This section explores the role of actuators in AR and VR, their impact on user experience, the different types of actuators used, and the challenges and future directions of this integration.

AR and VR technologies immerse users in virtual environments, either by overlaying digital content onto the real world (AR) or by completely replacing the real world with a simulated one (VR). While the visual and auditory components of these experiences are crucial for immersion, the addition of haptic feedback through actuators can deepen the sense of presence. When users can feel the virtual environment, it enhances their ability to interact with it in a natural, intuitive way. This tactile feedback is essential for various AR and VR applications, such as gaming, training simulations, therapeutic exercises, and remote collaboration.

In VR, where users are entirely detached from their physical surroundings, actuators provide critical feedback that helps the brain interpret and navigate the virtual space. For example, when a user interacts with a virtual object, an actuator might simulate the sensation of resistance, texture, or weight, providing a sense of "realness" to the virtual interaction.

In AR, where digital objects are overlaid on the real world, actuators can simulate interactions with both physical and virtual objects. For instance, in an industrial AR application, a technician might use an AR headset to view maintenance instructions overlaid on machinery while simultaneously feeling tactile feedback from the system through actuators embedded in a wearable suit or gloves.

The ability to simulate touch, movement, and force with actuators creates a more holistic and immersive interaction with the virtual world, significantly enhancing the user experience.

Several different types of actuators are employed in AR and VR systems to provide haptic feedback and motion sensations. These actuators vary in their underlying mechanisms and the types of feedback they offer, but all share the goal of enriching the virtual experience by adding a tactile dimension. Below are the most commonly used actuators in AR and VR environments.

Vibration is one of the most straightforward forms of haptic feedback used in AR and VR systems. Vibrational actuators are commonly found in controllers, gloves, and suits, providing users with feedback through small vibrations. This feedback can be used to simulate various sensations, such as the feeling of an object being touched, an impact, or the presence of an obstacle in the virtual environment. While simple, vibrational feedback is highly effective for many VR and AR applications, particularly in gaming and training simulations. The vibrations can be adjusted in frequency, intensity, and duration to represent different types of inter-actions. For example, a high-frequency vibration might simulate the sensation of a soft touch, while a low-frequency, high-intensity vibration could represent the feeling of a collision or a powerful impact.

Force feedback actuators go beyond simple vibrations and provide users with the sensation of resistance or force as they interact with the virtual world. These actuators are typically found in specialized devices like force-feedback joysticks, VR gloves, and haptic suits. Force feedback can simulate a range of sensations, including the resistance of pushing against an object, the feeling of holding or gripping something, or the force of an impact. A common example of a force feedback actuator is the linear actuator used in robotic arms or VR gloves. These actuators work by moving along a linear path in response to user input or virtual interactions, providing real-time resistance that mimics real-world physical sensa-tions. For example, if a user attempts to pick up a virtual object, the actuator might simulate the weight of the object by providing resistance during the motion, enhancing the realism of the interaction.

EAPs are materials that change shape or size when an electrical voltage is applied. These materials are increasingly being explored for use in AR and VR actuators due to their flexibility, responsiveness, and ability to provide continuous deformation. EAPs are used in soft actuators that can simulate more natural, fluid movements in VR and AR interfaces. This technology is particularly useful in wearables like gloves or suits that need to conform to the body and produce various tactile sensations, such as stretching, contraction, or vibration.

EAP-based actuators offer several advantages, including low energy con-sumption, lightweight design, and the ability to be embedded into wearable systems without adding significant bulk. In VR gloves, for instance, EAP actuators could simulate the sensation of grasping different materials or textures, providing a more immersive haptic experience.

Piezoelectric actuators generate motion in response to an applied electric field, allowing them to produce high-precision, small-scale movements. These actuators are well-suited for applications in AR and VR that require fine, subtle feedback, such as simulating the texture or shape of an object. Piezoelectric actuators can generate microscopic vibrations, enabling the simulation of fine textures and the sensation of surface details that might otherwise be missed with other actuator technologies. For example, a piezoelectric actuator could be used in a haptic glove to simulate the feeling of touching a rough surface, like sandpaper or fabric. The actuator would produce high-frequency vibrations that mimic the small changes in texture felt when interacting with such surfaces, enhancing the realism of the user's experience.

Magnetorheological (MR) and electrorheological (ER) actuators are unique in that they change the viscosity of a fluid in response to magnetic or electric fields, respectively. This change in viscosity allows these actuators to provide programmable resistance and simulate a wide variety of tactile sensations. MR and ER actuators can be used to simulate soft or hard surfaces, varying stiffness, or even simulate the sensation of an object moving through a medium, like water or air. These actuators are often used in haptic suits or wearable devices that need to provide high-fidelity feedback across a wide range of sensations. For example, in a VR training simulation for industrial workers, these actuators could simulate the feeling of manipulating various tools, providing varying levels of resistance based on the virtual interaction.

The integration of actuators into AR and VR systems has led to a wide range of innovative applications that span gaming, healthcare, education, industrial training, and even remote work. Below are some of the most notable applications of actuators in these immersive technologies. One of the most popular uses of AR and VR actuators is in the realm of gaming and interactive entertainment. VR gaming platforms, such as Oculus Rift, HTC Vive, and PlayStation VR, are integrating advanced haptic feedback to immerse players in virtual worlds. Actuators in controllers, gloves, and body suits provide real-time feedback that simulates a variety of in-game actions. Whether it is the sensation of firing a weapon, swinging a sword, or feeling the impact of an explosion, actuators help create a more engaging and lifelike experience for the user.

In the healthcare sector, actuators integrated into AR and VR systems are playing a significant role in medical training and rehabilitation. For example, medical students can use VR simulations to practice surgery or other medical procedures, with haptic feedback provided by actuators to simulate the feeling of cutting tissue or applying pressure. This feedback helps build muscle memory and enhances the realism of the training process.

In rehabilitation, actuators are used in VR-based physical therapy systems, where patients can interact with virtual environments while receiving haptic feedback to simulate physical movements. For example, patients recovering from strokes or injuries can use VR gloves embedded with actuators to practice hand movements, with the actuators providing feedback on the success or failure of each motion. This use of actuators aids in recovery by improving motor skills and coordination.

AR and VR are increasingly being used for industrial training and safety simulations, where actuators provide tactile feedback to simulate real-world scenarios. Workers in fields such as manufacturing, construction, or mining can use VR simulations to learn how to operate machinery, handle hazardous materials, or respond to emergencies. Actuators in haptic suits or gloves can simulate the feeling of handling tools, equipment, or materials, giving workers a more realistic sense of the tasks they will perform in the field.

In remote collaboration, actuators can be used to enhance teleoperation and virtual collaboration by allowing users to interact with physical objects and systems remotely. Through the use of haptic gloves, suits, or exoskeletons, operators can control robotic arms, drones, or other machines and receive real-time tactile feedback, enhancing their sense of control and awareness over the operation. This technology has applications in various industries, including manufacturing, exploration, and even remote surgery.

7.2.2   PMAs in space exploration

Space exploration presents a unique set of challenges that require innovative solutions in order to ensure the success of missions. From navigating the harsh and unforgiving environment of space to ensuring the functionality of spacecraft and robotic systems, the design and application of advanced actuators play a critical role. PMAs are emerging as a crucial technology in space exploration due to their high efficiency, reliability, and ability to perform under extreme conditions. PMAs offer distinct advantages in spacecraft propulsion, robotic manipulation, and the operation of various spacecraft systems, making them indispensable for a variety of space exploration applications. This section delves into the applications of PMAs in space exploration, their benefits, challenges, and the future prospects of these actuators in advancing space technologies.

PMAs, as the name suggests, use permanent magnets to generate the necessary force for actuation, often without the need for an external power source for magnetization. These actuators convert electrical energy into mechanical motion through the interaction between permanent magnets and coils or other electromagnetic elements. Due to their design and construction, PMAs are highly efficient, compact, and capable of providing high torque at low speeds, making them ideal candidates for a variety of space-related tasks.

In space exploration, where mission success is often determined by the ability to operate in remote or harsh environments, PMAs are particularly valuable due to their simple yet robust design. PMAs are resistant to radiation and extreme temperatures, require minimal maintenance, and can operate in vacuum environments, making them well-suited for the challenges of space. Furthermore, their precise control capabilities and high torque-to-weight ratio make them useful for both spacecraft systems and robotic systems on planetary bodies.

The main areas where PMAs are used in space exploration include propulsion systems, robotic manipulation, attitude control, and scientific instrumentation.

PMAs play a key role in spacecraft propulsion systems, particularly in systems where precise control and high reliability are required. One area where PMAs are applied is in the reaction control systems (RCS) of spacecraft, which are responsible for fine-tuning the attitude and orientation of a spacecraft. RCS are used during orbital maneuvers, docking, and even planetary landings. These systems often rely on small thrusters or actuators that can provide the necessary torque to adjust the spacecraft's position.

PMAs are a key component in electric propulsion systems, which are becoming increasingly popular in modern spacecraft due to their high efficiency. Hall-effect thrusters, ion propulsion systems, and electrospray thrusters are some examples of electric propulsion systems where PMAs are integrated into the mechanism for generating the magnetic fields necessary for thrust generation. These systems are particularly advantageous for long-duration missions such as interplanetary exploration, where high efficiency and precision are paramount. PMAs can also be found in gimbal actuators, which adjust the orientation of spacecraft antennas, solar panels, and other critical components. In these applications, PMAs enable smooth and precise movements, ensuring that spacecraft subsystems maintain their alignment and maximize performance during the mission.

PMAs are extensively used in robotic systems for both spacecraft and planetary exploration. These actuators enable precise control of robotic arms and manipulators used in various tasks, such as spacecraft assembly, maintenance, and scientific research. For example, the Mars rovers and other planetary exploration robots require actuators to move and manipulate instruments, collect samples, and interact with the environment. PMAs are particularly advantageous in these scenarios because of their reliability, compactness, and ability to function in the extreme conditions of space and planetary surfaces.

The Canadarm system, for instance, a robotic arm used aboard the Space Shuttle and the International Space Station (ISS), has been integrated with PMAs to provide fine motion control for payload handling and space station assembly. These arms are subjected to zero-gravity and high-radiation environments, where typical electric motors may face challenges. PMAs, with their solid-state design and low maintenance, are well-suited for these environments.

In addition to planetary rovers, PMAs are also integral to space robotics used for satellite servicing, such as the manipulation of delicate satellite components during repairs or upgrades. Their precision and durability allow robots to perform high-risk tasks remotely, minimizing the need for human intervention and maximizing safety during complex operations.

The attitude control of spacecraft refers to the process of controlling the orientation of a spacecraft relative to a fixed point or direction in space. PMAs are used in reaction wheels and control moment gyroscopes (CMGs), which are crucial for stabilizing spacecraft during their missions. Reaction wheels use the principle of angular momentum to control spacecraft orientation, and PMAs are integrated into these systems to help generate the required torque for precise attitude adjustments.

CMGs, which are used for more advanced and larger spacecraft, are a highly efficient method of attitude control, especially for large satellites or space stations that require fine angular positioning. By using PMAs in the control systems, spacecraft can adjust their orientation with high accuracy, ensuring that the spacecraft's instruments, antennas, and other systems maintain proper alignment with the mission goals. For example, the James Webb Space Telescope (JWST), currently in orbit around the second Lagrange point (L2), relies on reaction wheels integrated with PMAs to maintain its orientation while capturing high-resolution images and collecting data. Similarly, satellites used for Earth observation or communication rely on precise attitude control to ensure the continued proper operation of their instruments.

PMAs are also applied in scientific instruments that require precise mechanical actuation for experimentation and data collection. These instruments, which may include telescopes, spectrometers, and particle detectors, often need to move to specific angles or positions to capture data, requiring actuators that can provide smooth and controlled movement. For example, telescopes aboard space observatories use PMAs in their mirror positioning mechanisms. The Hubble Space Telescope, which has been operational for over three decades, uses PMAs for its fine guidance sensors, allowing it to adjust its position to achieve precise measurements of distant astronomical objects. By using PMAs, these instruments benefit from high reliability and minimal risk of mechanical failure in the harsh environment of space.

The use of PMAs in space exploration offers a number of key advantages that make them especially well-suited for the demands of space missions.

First, PMAs are highly efficient because they require little to no external energy to maintain their magnetic field. Unlike conventional electric motors, which rely on electromagnetic induction to create motion, PMAs use the inherent magnetic properties of permanent magnets, reducing the need for continuous energy input. This makes them ideal for long-duration space missions, where energy conservation is crucial.

Second, the compact nature of PMAs allows for a high torque-to-weight ratio, which is a critical consideration in space exploration where weight is a major concern. The reduced size and weight of PMAs allow spacecraft designers to integrate these actuators into compact systems, enabling more efficient use of available space and weight capacity on spacecraft.

Third, space missions often span many years, with spacecraft operating in extreme environments. PMAs are highly reliable and durable due to their simple design, which eliminates the need for moving parts that can wear down over time. The absence of brushes and bearings means that PMAs can operate for extended periods without significant maintenance, making them ideal for missions where human intervention is not feasible.

Fourth, PMAs are particularly effective in space because they can operate in extreme temperature ranges and the vacuum of space. Unlike conventional actuators that may degrade over time due to environmental factors such as temperature extremes, radiation, or vacuum, PMAs remain stable and operational in these conditions. This makes them well-suited for use in space environments, where the challenges of temperature fluctuations, cosmic radiation, and lack of atmosphere require highly reliable and rugged systems.

Finally, PMAs offer high precision and fine control, which is essential in space exploration tasks like spacecraft attitude control, robotic manipulation, and scientific measurements. The precision provided by PMAs enables spacecraft to perform critical tasks with great accuracy, such as docking with other spacecraft, adjusting solar panels for optimal energy generation, and capturing data from distant astronomical objects.

7.3 Disruptive trends in computing and communication

7.3.1 Edge computing for real-time actuation

In recent years, edge computing has emerged as a transformative technology for addressing the growing demand for real-time data processing and decision-making in various industrial, healthcare, automotive, and robotics applications. By enabling the processing of data closer to the source—often at the edge of the network, near the sensors and actuators—edge computing can significantly reduce latency, enhance reliability, and improve the responsiveness of systems that rely on real-time actuation.

The integration of edge computing with actuation systems is particularly beneficial for applications that require fast, accurate, and efficient control over physical processes. In these applications, such as robotic systems, autonomous vehicles, smart manufacturing, and healthcare, the actuation components must respond instantaneously to changing conditions. By distributing processing power closer to the actuators, edge computing enables faster decision-making, reduced reliance on cloud infrastructures, and better overall performance of real-time actuation systems.

This section explores the role of edge computing in facilitating real-time actuation, its benefits, challenges, and its impact on various industries.

Edge computing involves the deployment of computational resources at the edge of a network, closer to the data source and the actuation devices themselves. Unlike traditional cloud-based computing models, where data is sent to centralized data centers for processing, edge computing enables data to be processed locally, often on devices such as gateways, local servers, or even directly on actuators or sensors. This significantly reduces the amount of data that needs to be transmitted over the network and minimizes the time required for data processing and response generation.

In real-time actuation systems, the primary goal is to achieve timely and accurate control over physical processes. This can involve controlling robotic arms, vehicle steering mechanisms, industrial machinery, or medical devices, all of which require rapid feedback loops. For these systems to perform optimally, they must be able to process sensor data, make decisions based on that data, and immediately trigger actuators to perform the required actions. Edge computing plays a key role in achieving this by enabling the processing of sensor data and the decision-making process to occur at the local level, rather than relying on a distant cloud server. This minimizes communication delays and ensures that actuators can respond to changes in real-time, without being hindered by the bandwidth limitations or latency issues that can arise in traditional cloud-based systems.

One of the most significant advantages of edge computing is its ability to drastically reduce latency, which is crucial for real-time actuation. In applications where even small delays can lead to performance degradation or failure—such as in robotic surgery, autonomous driving, or industrial automation—the speed at which decisions are made and acted upon is paramount. By processing data locally, edge computing ensures that decisions can be made and transmitted to actuators within milliseconds, without having to send data to a cloud server and wait for a response.

When critical systems depend on constant communication with cloud services, any disruption in the network can lead to performance degradation or system failure. Edge computing mitigates this risk by allowing actuators to operate autonomously or with reduced reliance on external systems. This is particularly important in environments where connectivity may be unreliable or where remote operation is required, such as in space exploration, disaster recovery, or underwater robotics.

Edge computing allows sensitive data to be processed locally rather than transmitted to centralized cloud servers. This can help address concerns related to data security and privacy, especially in industries like healthcare, financial services, and military applications, where data confidentiality is critical. By minimizing the amount of data sent over the network, edge computing can help protect sensitive information from being intercepted or compromised during transmission.

In many real-time actuation systems, large volumes of data are generated by sensors, which need to be processed to trigger the appropriate responses from actuators. Transmitting this data to the cloud for analysis can be costly and inefficient, especially in environments with limited bandwidth. By processing data locally, edge computing helps reduce the need for extensive data transmission, optimizing bandwidth usage, and reducing operational costs. This is especially important in IoT-based applications, where many sensors and devices are involved, and data traffic can quickly become overwhelming.

Edge computing allows for greater flexibility and scalability in real-time actuation systems. It enables the distribution of processing tasks across various edge devices, such as gateways, edge servers, or embedded processors, rather than relying on a single centralized server. This decentralization makes it easier to scale the system as the number of sensors, actuators, or devices increases, without overwhelming the network or cloud infrastructure. Moreover, edge computing can be adapted to various hardware configurations, from low-power microcontrollers to more powerful edge computing platforms, depending on the application requirements.

Edge computing is revolutionizing Industry 4.0 by enabling faster, more responsive control of manufacturing processes. In smart factories, actuators are used to control everything from robotic arms and conveyor belts to CNC machines and 3D printers. By deploying edge computing at the local level, manufacturers can process sensor data—such as temperature, pressure, or position—directly at the machine level, and immediately adjust the operation of the actuators to optimize performance.

For example, in a smart factory, an actuator controlling the speed of a conveyor belt could be adjusted in real-time based on feedback from sensors monitoring production quality. The local edge device would analyze the data from the sensors, make a decision about the required adjustment, and send a command to the actuator to implement the change. This results in faster processing times, less downtime, and higher operational efficiency. Moreover, predictive maintenance in industrial settings benefits from edge computing by continuously monitoring actuator performance and detecting anomalies or wear and tear. When irregularities are identified, the actuator can be automatically adjusted or shut down, reducing the likelihood of failure and optimizing the maintenance schedule.

In the field of autonomous vehicles, the ability to make real-time decisions and actuate physical systems is essential for safe and efficient operation. Autonomous vehicles rely on a combination of sensors (such as cameras, LIDAR, and radar) to gather data about their environment. Edge computing enables these sensors to process data locally, immediately determining how the vehicle should respond. For instance, if the system detects an obstacle ahead, edge computing processes the data and triggers the appropriate actuator to adjust the vehicle's speed, steering, or braking. The reduced latency afforded by edge computing ensures that decisions are made and executed within milliseconds, which is crucial for ensuring the safety and reliability of autonomous driving systems.

Furthermore, edge computing enhances vehicle autonomy in situations where connectivity to the cloud may be intermittent or unreliable, such as in remote areas or during poor weather conditions. By relying on local processing, the vehicle can continue to function autonomously without needing to connect to a central server for every decision.

In healthcare, particularly in medical robotics and surgical assistance, real-time actuation is critical for ensuring the safety and precision of procedures. Medical robots rely on actuators to control surgical tools with extreme accuracy, making rapid adjustments based on feedback from sensors. By using edge computing, the robot can process this sensor data in real time, making decisions and sending commands to actuators without delay. For example, in robot-assisted surgery, an actuator might adjust the position of a surgical instrument based on feedback from force sensors that monitor the pressure being applied to tissue. With edge computing, the decision to adjust the pressure or change the tool's orientation is made almost instantaneously, ensuring that the procedure is as precise as possible. Moreover, edge computing enhances the safety of these systems by enabling real-time monitoring of patient vitals and system diagnostics, providing immediate alerts in the event of abnormal readings or mechanical failures. This enhances the overall reliability and responsiveness of medical robots, leading to better patient outcomes.

In smart homes and IoT applications, actuators control a variety of systems, from HVAC (heating, ventilation, and air conditioning) to smart lighting and security systems. Edge computing enables these systems to process sensor data locally and trigger immediate actions based on user preferences or environmental conditions.

For example, in a smart thermostat system, edge computing processes data from temperature sensors in real time, adjusting the thermostat settings and controlling the HVAC system to maintain a comfortable environment. This reduces the need for cloud-based processing, lowers the risk of latency, and enhances the overall user experience by providing immediate responses to environmental changes.

7.3.2   Quantum computing in motion control

Quantum computing, with its ability to process and store information in ways that classical computers cannot, has the potential to revolutionize various fields, including motion control systems. Traditional motion control relies on algorithms executed by classical processors, which operate using binary states (0s and 1s). However, quantum computing leverages quantum bits or qubits, which can represent and process multiple states simultaneously through the principles of superposition and entanglement. This unique property allows quantum computers to perform certain calculations exponentially faster than classical computers, opening new avenues for optimization, simulation, and control of complex dynamic systems.

In motion control, the primary goal is to manage the movement of machines, robots, or vehicles with high precision, accuracy, and responsiveness. These systems are typically based on feedback loops, where sensors monitor the system's state and actuators make adjustments to maintain desired performance. As these systems become increasingly complex and require more dynamic responses, the need for faster and more efficient computation grows. Quantum computing, with its potential to solve optimization problems more effectively, can significantly enhance the performance of motion control systems, especially in applications that demand real-time, highly adaptive control, such as robotics, autonomous vehicles, and industrial automation.

This section explores the intersection of quantum computing and motion control, highlighting how quantum algorithms can improve the precision, efficiency, and adaptability of motion control systems across various industries. Quantum computing's influence on motion control is based on its ability to tackle computational problems that are intractable for classical computers, especially when dealing with large, complex systems. Some of the ways quantum computing can impact motion control include:

- Optimization: Motion control often involves the optimization of variables such as speed, position, torque, and energy consumption. For example, optimizing the trajectory of a robotic arm in a manufacturing process or determining the best path for an autonomous vehicle requires solving complex optimization problems, which may involve millions of variables. Classical optimization techniques, such as gradient descent or linear programming, can become computationally expensive and time-consuming as the system's complexity increases. Quantum algorithms, such as the Quantum Approximate Optimization Algorithm (QAOA), offer the potential to solve these optimization problems exponentially faster than classical algorithms, reducing the time required to compute optimal control strategies.
- Simulations and Predictive Modeling: Quantum computers excel at simulating complex systems, especially those that involve large datasets or intricate interactions between multiple variables. For motion control systems, this means that quantum computing could improve the accuracy and speed of simulations used in system design and real-time decision-making. Quantum

computers can simulate the behavior of mechanical systems with greater precision, taking into account complex dynamics and uncertainties that might otherwise require significant computational resources on classical computers. This could lead to better predictive models for system performance, enabling more adaptive and responsive motion control.

- Real-Time Decision-Making: Motion control systems, particularly in robotics and autonomous vehicles, require real-time decision-making based on feedback from sensors and external environments. Quantum computing's potential to process large amounts of data in parallel makes it well-suited for real-time analysis and decision-making. For example, quantum algorithms could improve the efficiency of state estimation and trajectory planning in robotic arms or drones, reducing the latency between sensing and actuation and allowing for faster, more responsive motion control.

Several quantum algorithms show particular promise in improving motion control systems. Some of the most relevant algorithms and methods include:

- Quantum Approximate Optimization Algorithm (QAOA): QAOA is a hybrid quantum-classical algorithm designed for solving combinatorial optimization problems, such as path planning and trajectory optimization. In motion control systems, QAOA can be used to optimize the movements of robots, drones, or vehicles by finding the most efficient paths or motion sequences. Unlike classical methods, which may take too long to compute optimal solutions for large-scale problems, QAOA can potentially provide faster and more accurate solutions, especially as the system's complexity grows.
- Quantum Machine Learning (QML): QML combines quantum computing with ML techniques to enhance the learning and decision-making processes. QML can be used to improve control algorithms by enabling the system to learn more efficiently from large datasets and make more accurate predictions about the system's behavior. For example, in reinforcement learning (RL) algorithms, quantum computing could be used to accelerate the exploration and exploitation phases, enabling faster adaptation to dynamic environments. This could improve the performance of robotic systems, autonomous vehicles, and industrial robots by making their motion control more adaptive and context-aware.
- Quantum Fourier Transform (QFT): The QFT is a key algorithm in quantum computing that can perform Fourier analysis exponentially faster than its classical counterparts. In motion control systems, Fourier transforms are often used to analyze and process signals, particularly in systems where periodic behavior needs to be understood, such as in vibration analysis, signal filtering, or trajectory planning. QFT could significantly speed up these processes, enabling faster signal processing and control adjustments in real-time.
- Quantum Search Algorithms: Algorithms like Grover's Search Algorithm can be used to search through large, unstructured datasets more efficiently than classical search algorithms. In motion control, this could be applied to problems such as fault detection or parameter tuning, where the system needs to

explore large solution spaces to find the optimal configuration or detect anomalies. Grover's algorithm could improve the speed and accuracy of these tasks, making motion control systems more reliable and adaptive.

Quantum computing can enhance motion control systems in several key industries. Below are some of the most promising applications of quantum-enhanced motion control.

In robotics, motion control is critical for ensuring that robots can execute tasks such as object manipulation, assembly, or navigation with high precision. Quantum computing can optimize the trajectories of robotic arms or mobile robots by solving complex path-planning problems faster and more accurately. Additionally, quantum computing can enhance sensor fusion techniques, enabling robots to combine data from multiple sensors (e.g., vision, tactile, and auditory) to make real-time adjustments to their movements. For instance, in a manufacturing environment, a robot equipped with quantum-enhanced motion control could optimize its movements to minimize energy consumption while maximizing throughput. Quantum algorithms could also be used to dynamically adjust the robot's behavior based on feedback from the environment, allowing for more flexible and adaptable systems.

Autonomous vehicles rely heavily on motion control systems to navigate, avoid obstacles, and optimize routes. Quantum computing could enhance the efficiency of trajectory planning in real-time, allowing autonomous vehicles to make better decisions in complex, dynamic environments. By using quantum optimization algorithms, an autonomous vehicle could quickly calculate the most efficient path while considering factors such as road conditions, traffic, and pedestrian movement.

Furthermore, quantum computing could be applied to vehicle-to-vehicle (V2V) communication, enabling faster and more reliable communication between autonomous vehicles to coordinate movements in real-time, thus improving traffic flow and safety.

In aerospace and space exploration, precision in motion control is paramount. Quantum computing could improve the efficiency of spacecraft navigation, robotic arms on space stations, and satellite positioning systems by optimizing trajectories and minimizing fuel consumption. For instance, quantum-enhanced algorithms could be used to determine the optimal fuel usage and flight path for spacecraft, leading to more efficient missions with lower operational costs. Additionally, quantum computing could play a significant role in the autonomous operation of rovers on distant planets, where real-time decision-making and motion control are essential for exploration and data collection. Quantum algorithms could enhance the rover's ability to adapt to unforeseen obstacles or environmental changes, enabling it to perform tasks more autonomously and efficiently.

In smart factories, motion control systems are responsible for the precise operation of robots, conveyors, and other machinery. Quantum computing could optimize the control of robotic arms, production lines, and automated inspection systems by solving complex optimization problems related to speed, timing, and resource allocation. Quantum-enhanced motion control could also improve predictive maintenance by analyzing sensor data more efficiently, allowing for proactive adjustments and reducing the likelihood of system failures. By optimizing manufacturing processes in real-time, quantum computing could help manufacturers reduce energy consumption, increase throughput, and improve the overall efficiency of their operations.

In robot-assisted surgery, motion control must be incredibly precise to ensure that surgical instruments operate accurately within the human body. Quantum computing could be used to enhance the precision of surgical robots by optimizing their movements and reducing latency in feedback loops. Quantum algorithms could also help with the real-time processing of data from multiple sensors (e.g., imaging, force sensors) to guide the robot's actions. Moreover, quantum-enhanced motion control could enable medical robots to better adapt to changes in the environment, such as variations in tissue stiffness or patient movement, improving the safety and effectiveness of surgeries.

7.3.3   6G and beyond

The emergence of 6G technology marks a transformative leap in connectivity, offering ultra-high-speed communication, extremely low latency, and ubiquitous connectivity that can significantly enhance the performance and coordination of motion control systems across various applications. Moving beyond the capabilities of 5G, 6G promises to deliver speeds exceeding 100 Gbps and latencies of less than a millisecond, which is crucial for the real-time coordination of complex systems, such as robotic fleets, autonomous vehicles, and smart factories. As we transition into this new era of communication, the potential for motion control systems to function more seamlessly, efficiently, and responsively will be greatly amplified, unlocking new possibilities for actuator coordination, collaborative robotics, and industrial automation. This section explores how 6G and beyond will impact actuator coordination, specifically in terms of real-time global synchronization of robotic fleets and the broader implications of ultra-high-speed communication on motion control systems.

One of the most significant implications of 6G is its ability to provide ultra-high-speed communication, which will dramatically enhance actuator coordination in distributed motion control systems. In industrial automation, for example, multiple robotic arms, conveyor belts, or drones often need to operate in tandem, adjusting their movements based on shared data from sensors or other actuators. In current systems, even with the relatively fast speeds of 5G, communication delays can hinder the responsiveness and synchronization of these systems. With 6G, however, the low latency and immense bandwidth will allow for near-instantaneous data exchange between actuators, enabling them to work together with a level of precision and speed that was previously unachievable.

In this new era, the ability to communicate in real-time between actuators is a game-changer. For example, in smart factories, robotic arms that are assembling parts on a production line will be able to adjust their movements almost instantaneously based on sensor feedback from their environment or neighboring robots.

This real-time communication is crucial for avoiding errors, optimizing workflow, and maintaining the smooth operation of highly dynamic systems. By utilizing the ultra-low latency of 6G, feedback from one actuator can trigger immediate adjustments in others, allowing systems to adapt quickly to changes, whether they involve unforeseen obstacles or variations in the task being performed.

Moreover, distributed control systems in which multiple actuators work in tandem can be significantly improved by 6G. For instance, in multi-agent systems (MAS), which involve several robots or drones that must coordinate to complete tasks such as material handling, assembly, or exploration, the communication speed and bandwidth of 6G will allow these agents to synchronize their actions more efficiently. These systems can share data about their position, task status, and surrounding environment, enabling them to adjust their actions collaboratively. In a warehouse setting, for example, robots could coordinate the movement of goods across large areas, updating each other in real-time about obstacles, inventory levels, and storage space availability. The result is a much more agile, efficient, and intelligent system than what could be achieved with current communication technologies.

In addition to coordination, 6G will facilitate real-time adjustments and feedback loops, improving system responsiveness. For example, in autonomous vehicles, real-time feedback from the environment—such as sudden changes in road conditions or new obstacles—requires instantaneous adjustments to the vehicle's movements. With 6G, the time delay between sensor detection, decision-making, and actuation will be drastically reduced, ensuring that the vehicle can respond to its surroundings with the highest level of safety and precision. Similarly, in robotics, actuators will be able to adjust their motion in response to real-time data from both internal sensors and external sources, improving their ability to perform delicate operations such as surgery, assembly, or inspection.

Another crucial area where 6G will have a profound impact is in the scalability of large-scale motion control systems. As the number of interconnected actuators grows, traditional communication systems can struggle with the increased data load and coordination demands. However, 6G's ability to support massive IoT networks means that it will be possible to manage vast numbers of interconnected devices and actuators in real-time. In a smart city, for instance, the sensors embedded in infrastructure—such as traffic lights, public transportation, and streetlights—must constantly communicate with one another to adjust their behavior based on changing conditions. 6G will enable global synchronization of these systems, ensuring that they operate in harmony without delays or disruptions.

Beyond industrial applications, autonomous fleets of vehicles or robots, whether drones, delivery trucks, or vehicles, will also benefit greatly from 6G. The ability to synchronize movements of large fleets of autonomous robots or vehicles in real-time is critical for improving efficiency, safety, and overall performance. For example, in autonomous transportation networks, vehicles must coordinate their movements to avoid accidents, optimize traffic flow, and complete tasks such as goods delivery or ride-hailing services. Using 6G's ultra-low latency communication, vehicles can instantly share information about their location, speed, and environmental conditions, enabling them to adjust their trajectories and avoid collisions. This level of real-time coordination is vital for the deployment of autonomous fleets in urban environments, where high-density traffic and dynamic conditions pose significant challenges.

In robotic fleets, synchronization and coordination are also critical, especially in fields such as warehouse automation or agricultural robots. Robotic systems, such as drones used for crop monitoring or autonomous vehicles used in warehouse logistics, often require real-time adjustments based on shared data from other members of the fleet. With 6G, these robots can communicate and coordinate their actions almost instantaneously, improving the overall efficiency and reducing the likelihood of errors or collisions. In warehouse automation, for example, robotic arms, automated guided vehicles (AGVs), and conveyor systems must work together seamlessly to transport goods, optimize storage, and manage inventory. 6G will ensure that these systems can operate in sync, minimizing downtime and improving throughput.

Another area where 6G will have a major impact is in autonomous drone swarms. Drones are increasingly used for a wide variety of applications, including agriculture, environmental monitoring, disaster relief, and logistics. When operating in a swarm, drones must be able to communicate with one another to maintain formation, avoid collisions, and optimize their flight paths. 6G will enable these drones to share data and synchronize their movements in real-time, even when operating in large fleets or across vast distances. This will enable more efficient operations, allowing drone fleets to accomplish complex tasks—such as large-scale agricultural surveys or search and rescue missions—with greater precision and effectiveness.

One of the most exciting possibilities for 6G in motion control is in the realm of autonomous fleets that operate across vast networks or globally. The ability to synchronize these fleets in real-time, with near-zero latency, will make it possible to deploy fleets of robots or vehicles in environments ranging from factories to cities to remote locations like outer space or the deep ocean. This can help reduce operational costs, improve mission efficiency, and enable scalable solutions that were previously unimaginable.

Despite the many advantages 6G brings to actuator coordination and fleet synchronization, there are several challenges to overcome. First, the infrastructure needed to support 6G, including advanced network architecture and hardware, will need to be developed and deployed on a global scale. Additionally, the sheer volume of data generated by massive fleets of robots or autonomous vehicles requires advanced data processing and cloud computing capabilities, along with strategies to ensure data security and privacy. Furthermore, standardization of communication protocols and synchronization methods will be essential to ensure that diverse systems can interoperate seamlessly.

Looking ahead, 6G and beyond will likely continue to push the boundaries of motion control, providing new opportunities to enhance actuator coordination and global synchronization of robotic fleets. As these technologies evolve, the possibilities for autonomous systems and distributed motion control will expand, enabling faster, more efficient, and more intelligent automation systems across industries.

7.4   Societal and ethical implications

7.4.1   *Ethical challenges in advanced robotic systems*

As advanced robotic systems continue to evolve, especially with the integration of PMAs, ethical considerations become increasingly important. These systems, which are already used across a broad spectrum of applications—ranging from industrial automation to autonomous vehicles and surgical robots—pose significant challenges related to accountability, safety, and fairness. As robots become more autonomous, it is essential to address these challenges to ensure that their deployment is beneficial, responsible, and aligned with societal values. This section explores two critical ethical concerns: accountability in autonomous systems using PMAs and ensuring safety and fairness in AI-driven actuation.

One of the most pressing ethical issues surrounding advanced robotic systems that rely on PMAs is the question of accountability. As robots become more autonomous, particularly those with decision-making capabilities powered by artificial intelligence (AI), the issue of who is responsible for their actions becomes increasingly complex. In systems where PMAs are used to control movement with precision, the autonomy of these systems can lead to scenarios where robots make decisions or take actions without direct human intervention. This autonomy is especially prevalent in applications like autonomous vehicles, robotic surgery, or drones used for search and rescue missions.

If something goes wrong, such as a malfunction, an accident, or an unintended consequence of the robot's actions, who is to blame? Is it the manufacturer of the robotic system, the developer of the AI algorithms, the human operators who designed the parameters of the system, or the machine itself? The legal and moral responsibility for a robot's actions is still a gray area, and existing laws have yet to fully address the implications of autonomous technologies. As PMAs enable more precise and independent movement in these systems, the difficulty of determining liability and accountability grows. For example, in the case of autonomous vehicles using PMAs to control their motors and actuators, if an accident occurs, questions arise about who should be held accountable. Is it the car manufacturer, the developer of the software algorithms, or the AI system that decided to perform a specific action that led to the accident? This becomes even more complicated when considering that AI-driven systems may make decisions based on data inputs that are impossible for human operators to predict or understand completely. Transparency in decision-making processes and clear regulations will be crucial to resolving these accountability issues.

The challenge is further compounded by the fact that autonomous systems are constantly evolving. ML algorithms can adapt and optimize themselves over time, leading to actions that may not have been explicitly programmed or anticipated by their creators. In such cases, accountability becomes murky because the system's behavior is a result of an ongoing learning process that may diverge from its initial programming. To address this, traceability mechanisms and explainable AI will become key components of future systems to ensure that when a robot makes a decision, its reasoning is transparent, and its actions are auditable.

In response to these challenges, several approaches are being explored to ensure responsibility is clearly defined. One potential solution is the creation of regulatory frameworks that require manufacturers and developers to implement mechanisms for monitoring, logging, and controlling autonomous systems, ensuring they can be audited if necessary. Additionally, the accountability gap could be addressed by involving multiple stakeholders—including engineers, ethicists, policymakers, and consumers—in the design and governance of autonomous systems, ensuring that the ethical implications of these technologies are considered from the outset.

Another crucial ethical consideration in the deployment of PMA-based robotic systems is ensuring safety and fairness, particularly in AI-driven actuation. The integration of AI with actuation systems powered by PMAs opens up new possibilities in automation, but it also introduces risks, especially when it comes to ensuring that these systems act safely and equitably in a variety of contexts.

When dealing with advanced robotics that rely on actuators to perform tasks autonomously, safety must be a primary concern. For instance, in robot-assisted surgery, robots equipped with PMAs can make extremely precise movements, enabling minimally invasive procedures. However, the potential for error—whether due to a malfunctioning actuator, an incorrect input, or an unintended consequence of the AI algorithm—can lead to serious consequences, including harm to patients or workers. The safety protocols surrounding these systems must be carefully designed and regularly updated to prevent malfunctions, ensure redundancy, and allow for quick intervention in case of failure.

Moreover, as AI algorithms drive more autonomous action in robots, the question of how to ensure these systems are safe under all conditions becomes more complicated. AI-based systems that control PMAs may face situations that they have never encountered before, requiring them to make decisions based on incomplete or conflicting data. For example, an autonomous delivery drone may face unexpected weather conditions or obstacles, leading to a situation where the AI must decide how to proceed. While human operators could intervene in real-time, this may not always be possible in the context of highly autonomous systems. As such, the development of AI systems for PMA-driven robots must include rigorous testing, continuous monitoring, and the inclusion of safety overrides and emergency protocols to prevent accidents or harmful actions.

Additionally, safety extends beyond just the machine and the people it interacts with. In applications like autonomous vehicles or drones, the impact of these technologies on public safety must also be considered. Robotic systems equipped with PMAs that interact with public spaces or other autonomous systems must be designed with strict safety standards to minimize risks. This includes ensuring that systems can communicate effectively with each other to avoid accidents and that there are systems in place to deal with failures safely, whether that involves returning to a base for repairs or safely halting operations.

The use of AI in actuators also raises the question of fairness. AI-driven systems are increasingly responsible for decision-making in various fields, including healthcare, finance, law enforcement, and employment. However, AI algorithms can often reflect biases inherent in the data they are trained on, leading to decisions that may be unfair or discriminatory. In the context of robotics, this becomes particularly important when the robots are tasked with interacting with humans, making decisions that affect their lives. For example, consider a robot using PMAs in a warehouse automation system. If the robot's AI algorithms were trained on biased data—perhaps favoring one type of worker or operational process over others—this could result in unfair outcomes, such as unequal task distribution, unfair treatment of workers, or even the prioritization of certain operations at the expense of others. Ensuring fairness in the development and deployment of AI-driven actuation systems is crucial for promoting equity in increasingly automated industries.

Similarly, in healthcare settings where robotic surgery or diagnostic systems using PMAs are employed, AI must be designed to ensure that it does not exhibit biases based on race, gender, socioeconomic status, or other factors that could lead to unfair treatment. For instance, a biased AI system could influence the choice of treatment or surgical procedures in ways that disproportionately affect certain populations. The fairness of these systems must be carefully scrutinized, ensuring that AI-driven robots operate in ways that are equitable, just, and respectful of human rights.

Addressing fairness requires not only diverse and representative training data but also the implementation of transparent algorithms that can be audited to identify and mitigate bias. Furthermore, systems should be designed to incorporate feedback loops where the performance of AI-driven actuators is monitored for fairness and accuracy, and adjustments are made as necessary. In the case of PMA-based systems, this might involve ensuring that AI-driven motion control adheres to ethical principles and does not inadvertently cause harm or discrimination.

7.4.2   Actuator systems and human augmentation

The integration of PMAs into advanced human augmentation systems presents a significant leap forward in enhancing human capabilities. From prosthetics and exoskeletons to brain-machine interfaces (BMIs), PMA-based technologies are helping to restore lost functions, improve physical performance, and extend human potential. However, as with any technology that directly impacts the human body and mind, the use of actuators in human augmentation raises profound ethical considerations, particularly concerning the boundaries of innovation and the potential for unintended consequences. This section explores both the implications of PMA technologies in enhancing human capabilities and the challenges of balancing innovation with ethical boundaries.

PMAs are already a critical component in the development of assistive devices that help people overcome physical limitations. In the field of prosthetics, for instance, PMAs are used in artificial limbs to provide more fluid and natural movement, mimicking the action of muscles and joints. These actuators enable the precise control of movement and force, allowing users to interact with their environment in ways that were previously impossible. For people who have lost limbs due to injury, disease, or congenital conditions, PMA-powered prosthetics offer a way to regain independence and improve their quality of life.

In the case of exoskeletons, which are wearable robotic systems designed to assist or enhance human movement, PMAs are instrumental in providing the necessary actuation force to support and augment physical capabilities. Exoskeletons powered by PMAs have been developed to aid individuals with mobility impairments, enabling them to stand, walk, or even climb stairs. These systems are also being used in rehabilitation to help people recover lost physical abilities after injuries such as spinal cord damage or stroke. For example, PMAs can assist the wearer in maintaining balance, generating enough force to mimic natural walking, and even enabling the performance of heavy-lifting tasks, potentially allowing users to regain mobility or improve their functional capacity.

In human augmentation for performance enhancement, PMAs are also being integrated into wearable technologies for athletes, soldiers, and workers in high-demand professions. These systems provide increased strength, endurance, and precision, extending the physical and cognitive boundaries of the human body. For instance, a military exoskeleton might be designed to enhance a soldier's ability to carry heavy loads without fatigue, improve mobility in difficult terrain, or increase agility during combat operations. Similarly, advanced industrial exoskeletons are being developed to help factory workers lift heavy materials or maintain a steady, repetitive motion without risk of injury or fatigue.

In addition to physically augmenting human capabilities, PMA-driven systems are being explored for neural interfaces that connect directly to the human brain, allowing individuals to control robotic limbs or devices using thought alone. These BMIs have the potential to not only restore lost functions, such as enabling paralyzed individuals to control prosthetics or wheelchairs, but also to open new frontiers in direct brain-to-computer communication. By combining PMA technology with BMIs, future systems could facilitate new forms of human-computer interaction that are more intuitive and natural than current input methods like keyboards, touchscreens, or voice commands.

These innovations promise profound benefits, enabling individuals to regain physical functions or surpass natural human limitations. However, as PMA-based systems move from assistive technologies to enhancement technologies, there are important ethical concerns about the implications of this progress. The ability to enhance human capabilities raises critical questions about what it means to be human and where we should draw the line between supporting individuals with disabilities and altering human nature itself.

While the potential benefits of PMA-based human augmentation technologies are clear, the development and application of these systems must be approached with caution to ensure that ethical boundaries are not crossed. As these technologies push the limits of human performance and capability, they also bring with them the possibility of profound societal and moral dilemmas.

One of the primary ethical concerns surrounding human augmentation is the issue of inequality. As PMA-driven augmentation technologies become more advanced and widely available, there is the potential for societal divides to form between those who have access to these technologies and those who do not. Access to exoskeletons, prosthetics, and enhancement technologies could become a privilege reserved for the wealthy or particular sectors of society, creating disparities in health, performance, and opportunity. This could deepen existing social inequalities and create a divide between those with access to enhanced capabilities and those without.

Furthermore, the issue of coercion comes into play in industries or environments where human augmentation is used to enhance performance. For example, if PMA-driven exoskeletons or augmentation devices become standard in high-performance environments—such as professional sports, military operations, or heavy industry—there may be implicit pressure on individuals to adopt these technologies to stay competitive. In such cases, the line between voluntary enhancement and coerced augmentation may blur, leading to concerns about personal autonomy and consent. The desire for enhanced performance could become a form of social pressure, compelling individuals to adopt these technologies even if they are uncomfortable or unwilling to do so.

Moreover, there is the risk that human enhancement technologies could be used in ways that go beyond helping individuals overcome disabilities or increase their physical capabilities and into the realm of human modification. With the rapid advancement of PMA-driven augmentation technologies, there may be increasing pressure to use these systems for non-therapeutic enhancements, such as boosting intelligence, strength, or endurance beyond natural human limits. While these enhancements may initially seem to offer individual benefits, they could lead to unintended consequences, including altering fundamental aspects of human identity and experience. The introduction of superhuman capabilities could fundamentally change the human condition, leading to concerns about human nature, freedom, and autonomy.

The ethical challenge, therefore, lies in ensuring that innovations in PMA-driven human augmentation are developed and applied in a way that is responsible, equitable, and respectful of human rights. This involves considering informed consent, where individuals must have the right to make autonomous decisions about their participation in augmentation technologies. It also requires ensuring that these technologies are available to all, rather than just a privileged few, and that they are not used to perpetuate existing inequalities or create new forms of discrimination.

Furthermore, there must be an ongoing dialogue about the moral implications of human enhancement. Just because a technology is possible does not necessarily mean it should be pursued. Ethical frameworks must be established that prioritize human dignity, well-being, and autonomy over the pursuit of technological advancement. Ethical oversight and regulatory bodies must play a role in guiding the development of PMA-based technologies for human augmentation, ensuring that these systems are used to enhance human life in a way that aligns with broader societal values and principles.

Finally, it is crucial to recognize that human augmentation is not merely about increasing physical strength or cognitive ability—it is about improving quality of life and empowering individuals. PMA-driven technologies should be seen as tools for empowerment rather than as means of creating inequality or human modification. The focus should remain on using these technologies to help people who face challenges due to physical disabilities, enabling them to regain function, independence, and autonomy. The ultimate goal of human augmentation technologies should be to enhance human dignity and freedom, not to exploit or alter the essence of what it means to be human.

7.5   Sustainability and global trends

As the global focus on sustainability intensifies, the role of technology in promoting environmental responsibility and resource efficiency becomes increasingly critical. PMAs, which are integral to various industries ranging from automotive and robotics to renewable energy and manufacturing, present both opportunities and challenges in the context of sustainability. The adoption and development of PMA technologies must be aligned with global sustainability goals, and this includes efforts to minimize environmental impact, promote circular economy practices, and contribute positively to climate action. In this section, we examine two key aspects of the sustainability of PMAs: the integration of PMAs into the circular economy and their role in climate action.

7.5.1   Circular economy for PMAs

A key consideration in advancing PMAs sustainably is their integration into the circular economy, which emphasizes the importance of reusing, recycling, and regenerating materials to extend the lifecycle of products and minimize waste. As PMAs increasingly find their way into high-demand applications—such as in electric vehicles, wind turbines, and robotic systems—ensuring that these actuators are designed with sustainability in mind is crucial. A major environmental concern lies in the use of rare earth metals in permanent magnets, particularly materials like neodymium, dysprosium, and praseodymium, which are essential for PMA performance. These materials are not only critical to actuator functionality but are also associated with mining practices that can have significant ecological and geopolitical implications. Therefore, integrating PMAs into the circular economy involves addressing both end-of-life recycling and economic models that support sustainable manufacturing.

One of the fundamental challenges in the lifecycle of PMAs is the end-of-life disposal of actuators. Given that permanent magnets, particularly neodymium-iron-boron (NdFeB) magnets, are integral to PMA systems, ensuring that these magnets can be recovered and reused at the end of their service life is a priority for reducing the environmental impact of PMA technologies. Recycling rare earth materials presents several challenges, as these materials are often integrated into complex systems and are difficult to extract without damaging other components. However, technological advancements are being made in magnet recycling technologies, which focus on recovering valuable rare earth materials from obsolete or discarded PMAs.

Recycling PMAs involves extracting the rare earth metals from used magnets and remanufacturing them into new magnets for use in future PMA systems. This process helps reduce the dependence on primary mining, which has environmental and ethical implications, such as habitat destruction, water pollution, and labor concerns in mining regions. Advanced hydrometallurgical and mechanical processing methods are being developed to increase the efficiency of rare earth recovery, and new techniques are being explored to remanufacture magnets with high performance using recycled materials. The reuse of rare earth elements in this manner could not only cut down on the consumption of virgin resources but also reduce the carbon footprint associated with producing new magnets from raw materials.

Furthermore, incorporating the principles of design for disassembly into PMA manufacturing can improve the ease of recovering materials for recycling. This includes designing PMA-based systems in such a way that magnets and other key components can be easily separated at the end of the actuator's lifecycle. Establishing a more efficient circular economy for PMAs involves the development of global collection and recycling infrastructure, where PMAs can be returned to manufacturers or specialized recyclers to reclaim valuable materials. Such systems will need to be supported by policies and incentives that promote recycling and discourage the disposal of PMA-powered devices in landfills.

In addition to recycling and reuse efforts, the economic models that govern the manufacturing of PMAs need to shift toward more sustainable practices. Traditional manufacturing methods often rely on energy-intensive processes that can result in significant carbon emissions and resource depletion. To address these challenges, manufacturers are exploring new, more sustainable production methods, including the use of cleaner energy sources, such as solar, wind, or hydropower, in manufacturing facilities. Incorporating energy-efficient techniques in the production of PMAs can significantly reduce the environmental impact of their production.

Moreover, the economies of scale play an important role in the sustainability of PMA technologies. As the demand for PMAs grows across industries, innovations in materials and production technologies will be necessary to meet this demand sustainably. The development of low-cost, high-performance permanent magnets, such as those made from alternative materials or with reduced rare earth content, could contribute to reducing environmental and economic costs. Additionally, advanced manufacturing techniques, such as additive manufacturing (3D printing), may offer ways to produce PMAs in more efficient, localized, and sustainable ways, minimizing waste and energy consumption.

Lastly, policymakers can support the transition to sustainable manufacturing models for PMAs by introducing tax incentives, subsidies for green technologies, and investment in R&D for cleaner production processes. Collaboration between industry leaders, environmental organizations, and government bodies will be essential to drive the adoption of sustainable practices in PMA manufacturing.

7.5.2   PMA technologies in climate action

PMAs are poised to play a significant role in climate action through their use in renewable energy systems and their potential to contribute to the reduction of carbon emissions. As the world increasingly turns to sustainable energy solutions, PMAs are becoming essential components of systems that help reduce global carbon footprints and mitigate the effects of climate change. Whether in wind turbines, solar tracking systems, or electric vehicles, PMAs are at the heart of technologies that support green energy initiatives and enhance the efficiency of energy systems.

In the context of renewable energy, PMAs are most commonly found in wind turbines, where they serve as the core components in the electric generators that convert the mechanical energy of wind into electrical energy. The use of PMAs in wind turbines offers significant advantages in terms of efficiency and reliability. PMAs are known for their high torque density and low maintenance requirements, making them ideal for systems that must operate continuously in harsh outdoor environments. By increasing the efficiency of wind energy production, PMAs help to reduce the cost of renewable energy and improve the economic viability of wind power as a major energy source.

In addition to wind turbines, PMAs are also employed in solar tracking systems, which adjust the orientation of solar panels to follow the movement of the sun throughout the day. These systems maximize the energy output of solar panels by ensuring that they are always positioned at the optimal angle to capture sunlight. PMAs provide the necessary precision actuation to position the panels accurately and quickly, enhancing the efficiency of solar power systems and reducing the need for additional energy consumption in panel positioning. By improving the efficiency of solar energy systems, PMAs contribute to the global push for carbon-neutral and sustainable energy solutions.

Furthermore, PMAs play a crucial role in electric vehicles (EVs), where they are used in motors and actuation systems to control movement and energy distribution. As the adoption of electric vehicles grows, the demand for efficient, high-performance actuators increases. The integration of PMAs into EV powertrains improves vehicle performance, energy efficiency, and driving range, contributing to the reduction of greenhouse gas emissions from the transportation sector, one of the largest contributors to global carbon emissions.

PMAs are also crucial in the development of high-efficiency systems that reduce the overall carbon footprint of various industries. Their ability to provide precise and efficient actuation leads to improvements in the energy consumption of systems in a variety of sectors, from manufacturing and logistics to consumer electronics and industrial automation. For instance, in automated manufacturing systems, PMA-driven robotics can operate with high precision and minimal energy consumption, reducing the environmental impact of industrial processes. Additionally, PMAs enable more efficient transportation systems, such as electric and hybrid vehicles, which use actuators to control power delivery, reduce drag, and increase overall fuel efficiency.

By replacing older, less efficient mechanical systems with PMA-powered solutions, industries can reduce energy waste, lower carbon emissions, and support the global effort to mitigate climate change. As PMAs continue to evolve and become more energy-efficient, their contribution to carbon footprint reduction will become even more significant. The ongoing research into next-generation PMAs, including those that use sustainable materials or that offer even higher levels of energy efficiency, will further accelerate their role in global climate action.

7.6   New opportunities

7.6.1   Vision for autonomous microbots

The future of robotics is evolving rapidly, and one of the most exciting frontiers in this evolution lies in the development of autonomous microbots. These small, highly efficient robots, powered by PMAs, hold the potential to revolutionize a variety of industries, including medicine, agriculture, and environmental monitoring. The integration of PMAs into these tiny systems offers numerous benefits, such as precise control, low energy consumption, and high maneuverability. As these technologies progress, autonomous microbots could lead to groundbreaking advancements, offering solutions that are not only innovative but also highly impactful across several fields.

7.6.1.1   PMA—powered microbots in medicine, agriculture, and environmental monitoring

In the medical field, the use of PMA-powered microbots presents an exciting opportunity to improve precision medicine and minimally invasive procedures. Microbots, especially when equipped with PMAs, can perform complex tasks inside the human body, such as targeted drug delivery, biopsy sampling, and even tissue repair at the cellular level. Their small size and the precision of PMA actuators would enable doctors to perform delicate procedures with reduced risk to patients. Magnetic manipulation techniques can control these microbots externally, allowing for precise navigation through blood vessels or even specific organs, thus opening new avenues in non-invasive surgeries and diagnostics. For example, in the treatment of cancer, PMA-powered microbots could be used to deliver chemotherapy drugs directly to the tumor site, minimizing systemic side effects and maximizing the therapeutic effect. Additionally, PMAs can help power magnetic resonance imaging (MRI) contrast agents that are used to guide the bots through the body in real-time, offering a new layer of precision in diagnostic procedures.

In agriculture, PMA-powered microbots could assist in improving crop production, monitoring soil health, and facilitating precision agriculture. These robots could be used to monitor plant health at the individual level, identifying signs of pests, diseases, or nutrient deficiencies. Powered by small PMAs, these microbots can navigate dense crops with ease, detect issues early, and even deliver targeted treatments directly to plants, reducing the need for broad pesticide applications. The autonomous nature of these microbots would make them ideal for large-scale monitoring tasks, as they can operate continuously and provide real-time data to farmers.

For environmental monitoring, the use of PMA-powered microbots can significantly enhance our ability to monitor ecosystems, air and water quality, and pollution levels. These microbots, designed to be environmentally friendly, could be deployed in challenging and hazardous environments, such as deep-sea exploration or hazardous waste sites. In addition, they could be employed in monitoring the effects of climate change, assessing air and water pollution in real-time, and identifying environmental hazards before they become severe. Their small size allows them to access spaces that are too dangerous or inaccessible for human intervention, making them powerful tools in the ongoing efforts to monitor and protect our planet.

While the vision of autonomous PMA-powered microbots holds great promise, several technological and practical challenges must be overcome before these microbots can become a widespread reality. The long-term feasibility of these systems hinges on addressing issues related to miniaturization, power supply, autonomy, and material durability, among others.

One of the challenges in realizing PMA-powered microbots is the miniaturization of the actuators and the overall system. Although PMAs are already used in a range of robotics applications, their application in microscale robots presents new demands. The size and power requirements of PMAs must be significantly reduced to fit into microbots without compromising performance. This requires advances in microfabrication techniques that allow for the precise design and integration of PMA systems into small robotic structures.

The miniaturization of actuators also raises questions about precision control. At the micro scale, even small forces can have a significant impact on the behavior of the robot. Ensuring that PMA-powered microbots can maintain high levels of precision in highly constrained environments will require sophisticated control algorithms and feedback systems that can respond to real-time conditions.

Another significant barrier to the widespread deployment of PMA-powered microbots is the issue of power supply. In medical, agricultural, and environmental applications, these microbots would need to operate autonomously for extended periods, often in environments where access to external power sources is limited. Battery technology, particularly for small-scale applications, has not yet reached a level where it can support the long-lasting, efficient operation of these microbots.

The development of energy-harvesting systems could provide a solution to this challenge. For instance, magnetic energy harvesting techniques could be explored, allowing the microbots to generate power from their own movement or from surrounding environmental conditions. Alternatively, small wireless power transfer technologies may allow for recharging the microbots during operation, ensuring that they can remain functional for longer durations.

Moreover, PMAs are known for their energy efficiency in many applications, but even more advances are needed to optimize the energy consumption of these tiny actuators. Power-efficient designs, such as low-power microcontrollers or adaptive power systems, will be essential to ensure that these robots can operate without exhausting their energy supply too quickly.

The long-term success of autonomous microbots will also depend on the advancements in AI and ML. These systems must be capable of navigating complex environments, making decisions in real time, and learning from past experiences to improve their performance. The development of AI-powered control systems will be crucial for ensuring that microbots can operate with high levels of autonomy, reducing the need for human intervention. For instance, in medical applications, microbots could autonomously detect and navigate to tumors, making real-time decisions about treatment without requiring external commands. In agriculture, they could detect early signs of plant diseases and autonomously apply treatments to specific plants. However, ensuring that these bots are both safe and reliable in their autonomous decision-making is a significant challenge. Any malfunction or unintended behavior in such sensitive environments could have serious consequences, particularly in medical or environmental contexts.

Another critical factor in the long-term feasibility of PMA-powered microbots is their durability. Given their small size and the harsh environments they will need to navigate—whether in the human body, in soil, or underwater—these robots must be resilient to physical wear and tear, extreme temperatures, and other environmental stresses. Ensuring the longevity of these robots requires the development of new materials and protective coatings that can withstand challenging conditions without deteriorating over time.

In medical applications, for example, microbots must be biocompatible and able to function for extended periods within the body without causing adverse effects. Similarly, in environmental monitoring, they must be designed to endure exposure to corrosive substances, extreme temperatures, and pressure changes without losing functionality.

7.6.2 Self-healing actuator systems

Self-healing actuator systems represent an exciting frontier in the design of advanced robotics and automation technologies. These systems allow actuators to autonomously recover from faults, enhancing their resilience and reducing the need for external intervention or maintenance. This ability to self-repair is especially valuable in environments where repairs are difficult or impossible to carry out manually, such as in deep-sea exploration, space missions, or remote industrial settings. The core idea behind self-healing actuators is that the actuators, by incorporating specific materials and design principles, are able to detect damage, initiate a healing process, and restore their functionality. This chapter explores the materials and designs that enable such autonomous fault recovery, the theoretical underpinnings of these systems, and the practical challenges involved in their realization.

At the heart of self-healing actuators are smart materials—materials that have the intrinsic ability to detect and repair damage. These materials react to specific stimuli, such as mechanical stress, temperature changes, or electrical currents, to restore their functional properties. In actuator systems, these materials play a crucial role in enabling the automatic recovery of the actuator's mechanical or electrical performance. Some of the most promising smart materials for self-healing actuators include SMAs, conductive polymers, hydrogels, and CNT composites.

SMAs, for example, exhibit a unique property in which they return to their original shape when exposed to a specific temperature range. This property allows SMAs to be used in actuator systems that require self-repair. For instance, when an actuator experiences deformation, the SMA can be activated by heat to return to its original configuration, effectively repairing the damage. This capability is particularly useful in applications where actuators need to maintain their structural integrity under constant mechanical strain.

Conductive polymers, which combine the properties of both polymers and conductors, are also gaining attention for use in self-healing actuators. These materials can be designed to repair themselves when subjected to mechanical damage. In actuators that rely on electrical conductivity, such as those used in robotics or flexible electronics, conductive polymers can restore electrical pathways when disrupted by damage. When the material undergoes mechanical deformation, the polymer's molecular structure can reorganize itself, reestablishing conductivity and restoring actuator functionality.

Hydrogels, known for their high water content, offer another avenue for self-healing actuator systems. These materials can swell and absorb moisture when exposed to environmental stimuli, which can initiate a self-repair process. The flexibility and adaptability of hydrogels make them ideal for use in soft actuators or actuators that require flexibility in addition to strength. When damaged, hydrogels can reform their structure and regain their original properties, making them suitable for use in bioinspired actuators or medical robotics.

CNT composites are another material gaining prominence in the development of self-healing actuators. CNTs are known for their remarkable mechanical strength and conductivity, making them an ideal candidate for the reinforcement of actuator materials. When incorporated into composites, CNTs create a network that can be reorganized or reconnected after damage. This self-healing mechanism works by restoring the physical and electrical properties of the material, enabling the actuator to maintain its performance despite damage. CNT-based composites offer significant advantages, including high strength-to-weight ratios and excellent electrical conductivity, both of which are essential for actuator systems in demanding environments.

Another key approach in the design of self-healing actuators is the use of microcapsule-based systems. These systems involve embedding tiny capsules filled with a healing agent within the actuator material. When the actuator experiences damage, such as a crack or fracture, the capsules break open, releasing the healing agent into the damaged area. The healing agent then reacts with the material, repairing the crack and restoring the actuator's functionality. This mechanism is inspired by the natural process of wound healing in living organisms, where the body automatically responds to injury by releasing healing factors to repair the damage.

The healing agents used in microcapsule systems can vary depending on the type of actuator and the nature of the damage. For instance, in mechanical actuators, the healing agent might be a polymer that hardens upon exposure to air, filling the crack or void and restoring the material's structural integrity. In electrical actuators, the healing agent might be a conductive material, such as silver nanoparticles, that can restore electrical conductivity when the actuator experiences a short circuit or other electrical failure. The use of microcapsules in self-healing actuator systems offers a simple yet effective way to restore actuator functionality without the need for external intervention.

Vascular networks, inspired by biological systems, offer another innovative approach to self-healing actuators. In this design, actuator components are equipped with a network of channels, much like the vascular system in animals, which transport healing agents throughout the actuator. When damage occurs, the vascular network delivers the healing agent directly to the affected area, where it can begin the repair process. These vascular networks can be filled with a variety of healing agents, including liquid or gel-based materials, that flow through the channels and actively repair the damage.

In robotic systems, vascular networks could be used to heal damage in components such as actuators, sensors, or gears. The channels within the actuator would deliver the healing agent precisely where it is needed, enabling the actuator to recover from structural damage and continue functioning. The design of these vascular systems presents several challenges, including ensuring that the channels are large enough to deliver an adequate supply of healing agents but small enough to avoid compromising the actuator's overall performance.

The theoretical foundation of self-healing actuators is rooted in several key principles of materials science, mechanics, and control theory. The first step in designing a self-healing actuator system is to establish a mechanism for damage detection. This typically involves embedding sensors within the actuator that can monitor parameters such as strain, temperature, voltage, or pressure. When a change in these parameters occurs, indicating damage, the actuator's control system triggers the healing process.

In addition to damage detection, the healing process itself must be carefully modeled. For example, in microcapsule-based systems, the healing process is governed by fracture mechanics, which describe how cracks propagate and how materials respond to stress. By understanding these principles, researchers can design healing agents and materials that can effectively fill cracks or repair damaged components. In vascular systems, the fluid dynamics of the healing agent must be modeled to ensure that it flows effectively through the channels and reaches the damaged area in a timely manner.

The feedback control systems that monitor actuator performance and trigger the healing process are another critical component of self-healing actuators. These systems must be able to detect damage, initiate the healing process, and monitor the effectiveness of the repair. Advanced ML techniques can be applied to optimize this process, enabling the actuator to learn from past failures and improve its response to damage over time.

Despite the promise of self-healing actuators, there are several practical challenges that must be addressed before these systems can be widely deployed. One of the main challenges is the selection of materials that can provide the right balance of strength, flexibility, durability, and self-healing ability. For example, while SMAs are effective at restoring shape, they may not offer the same level of strength as other materials. Similarly, conductive polymers may be well-suited for electrical applications but may not be as effective in high-stress mechanical environments.

Another significant challenge is the complexity of integration. Self-healing materials must be incorporated into actuator designs without compromising their performance or adding excessive weight or complexity. For example, embedding microcapsules or vascular networks within actuator components requires careful consideration of the actuator's design and material composition. Moreover, these systems must be able to operate autonomously, which requires sophisticated control systems capable of detecting damage, initiating the healing process, and monitoring the effectiveness of the repair.

Finally, the long-term reliability of self-healing actuators is an ongoing concern. Many self-healing materials are still in the experimental stages, and their ability to repeatedly recover from damage over extended periods of time remains uncertain. As actuators are subjected to repeated stresses and strains, the self-healing materials must demonstrate the ability to maintain their properties and continue functioning without degradation. Ensuring the durability of self-healing systems is crucial for applications in critical industries such as aerospace, robotics, and energy.

7.6.3 Open questions for researchers

The rapidly evolving field of PMAs and their applications in robotics and automation has brought about numerous technological advancements. However, as the technology matures, there are still many open questions and unexplored opportunities that remain for researchers to tackle. These opportunities not only relate to advancing the fundamental principles of actuator design but also extend to interdisciplinary challenges that span across materials science, control theory, ML, robotics, and other fields. Identifying and addressing these open questions will be critical for unlocking the next generation of PMA technologies and their potential applications.

7.6.3.1 Integration of emerging materials for PMAs

One of the key challenges in the development of PMAs is the exploration of new materials that can enhance performance while maintaining efficiency, durability, and affordability. Many of the current materials used for PMAs, such as permanent magnets and specialized alloys, have limitations in terms of power density, mechanical fatigue, and energy efficiency. Researchers have begun to explore a variety of novel materials, including advanced composites, SMAs, and materials inspired by biology, but significant gaps remain in understanding how to fully integrate these materials into PMA systems.

One important open question for researchers is how to effectively combine these advanced materials with existing PMA technologies to create hybrid systems that maximize the strengths of both. For example, can a material like graphene be combined with rare-earth magnets to enhance the overall power output and thermal stability of actuators? How can piezoelectric materials be utilized to improve the precision of PMAs in applications such as soft robotics or micro-manipulation tasks?

Additionally, bio-inspired materials—such as those that mimic the properties of biological muscles or tissues—hold significant promise for creating more adaptive and flexible actuators. However, the challenge lies in scaling these materials for real-world applications and integrating them into PMA systems that must operate in harsh environments. Researchers will need to bridge the gap between fundamental material properties and practical engineering constraints to develop actuators that can be deployed reliably in commercial products.

7.6.3.2   Energy efficiency and power scaling

As the demand for more powerful, efficient, and lightweight actuator systems continues to grow, energy efficiency remains one of the most pressing challenges for researchers. While PMAs are known for their high torque density and precision, energy consumption and heat dissipation are significant concerns in high-power applications. Researchers are actively seeking ways to improve the energy efficiency of PMAs without compromising performance, but several questions remain open. For example, how can advanced cooling systems or heat-dissipation techniques be incorporated into PMA systems to ensure their reliability and longevity in high-power environments? What role can AI play in optimizing power consumption dynamically, adjusting actuator performance based on workload and environmental factors?

In addition, energy storage is another critical factor in PMA design. Most current systems rely on external power supplies or batteries, which can add bulk and reduce the overall portability of the actuator system. How can wireless energy transfer or advanced battery technologies, such as solid-state or lithium-sulfur batteries, be integrated to create fully autonomous, energy-efficient PMA systems? Further research into energy harvesting techniques, such as those using vibrational energy or solar power, could also open new opportunities for making PMAs more self-sufficient in remote or off-grid applications.

7.6.3.3   Autonomous control and learning algorithms

The application of AI and ML in PMA control systems is an exciting and rapidly evolving area of research. Advanced control algorithms, such as adaptive control, model predictive control, and neural network-based controllers, have already demonstrated their potential in improving the performance of PMAs in complex and dynamic environments. However, many challenges remain in developing control systems that are both adaptable and efficient enough to handle the diversity of tasks and environments that PMAs are likely to encounter in real-world applications.

One open question is how to create self-learning control systems that allow PMAs to autonomously adapt to new conditions or unexpected failures. For example, in the context of robotics, a robotic arm using PMAs may be required to operate in different terrains or environments, each with its unique challenges. How can ML algorithms help the actuator system "learn" the best operating parameters for each specific task? Can we design control algorithms that allow PMAs to learn from their own mistakes and optimize their performance over time?

Moreover, real-time coordination of multiple PMAs, especially in multi-robot systems or collaborative robotics, presents its own set of challenges. How can distributed control systems enable multiple actuators to communicate and coordinate in real time, especially in the face of communication delays or unreliable networks? The development of edge computing and distributed AI could offer solutions to these challenges, but further research is needed to develop efficient, scalable algorithms that can handle the complexity of multi-actuator coordination.

7.6.3.4 Fault detection and self-healing actuators

The integration of self-healing capabilities in actuator systems is an emerging area of interest, with the potential to revolutionize the reliability of PMAs. However, several open questions remain regarding how to effectively design actuators that can not only detect faults but also recover from them autonomously. While materials such as SMAs and conductive polymers show promise in providing self-healing properties, integrating these materials into PMAs in a way that guarantees long-term reliability is still an unsolved challenge.

Researchers will need to investigate new approaches to damage detection, monitoring and repair mechanisms. For example, can nano-sensors be embedded within the actuator material to monitor stress, strain, and other indicators of damage in real time? How can ML algorithms be used to predict when a failure is likely to occur, and how can this information be used to trigger self-healing processes before a full failure takes place? Additionally, how can researchers design actuators that can heal themselves multiple times without degradation in performance over time?

7.6.3.5 Interdisciplinary collaboration and cross-domain innovation

As PMAs continue to evolve, the need for interdisciplinary research will only increase. Researchers from a wide range of fields—such as robotics, materials science, mechanical engineering, electrical engineering, and AI—must collaborate closely to address the numerous challenges that PMAs face. A holistic approach, where expertise from different domains is integrated, will be crucial to advancing actuator technology.

One unexplored opportunity lies in the intersection between biomimicry and PMA design. Researchers could look to biological systems, such as muscles, tendons, and joints, for inspiration in creating actuators that can replicate the flexibility, strength, and self-healing capabilities of biological organisms. Collaboration between biologists, material scientists, and robotic engineers could pave the way for actuators that not only function more effectively but also demonstrate a greater degree of resilience and adaptability in changing environments.

Moreover, the rise of quantum technologies, nano-engineering, and bioelectronics could offer new materials and mechanisms that can significantly improve PMA performance. Interdisciplinary research could uncover ways to combine quantum computing with motion control algorithms to create super-efficient actuator systems with near-perfect precision.

7.6.3.6   Ethical and social implications

Finally, the development of self-healing, autonomous, and intelligent PMA systems will raise important ethical and social questions. As PMAs become integral to critical sectors like healthcare, autonomous vehicles, and military applications, researchers must consider how to ensure these systems are safe, fair, and transparent. How can researchers ensure that PMA-driven robots and automation systems operate in a way that aligns with societal values and does not unintentionally cause harm?

Furthermore, as PMAs enable more advanced and autonomous robots, it will be important to establish clear ethical guidelines and regulatory frameworks to govern their use. How can we balance innovation with responsibility, ensuring that these technologies are developed in a way that benefits society as a whole while mitigating risks such as job displacement, privacy concerns, and the potential for misuse?

References

[1] C. Luo and J. Sun, "Semi-interior permanent-magnet actuators for high-magnet-utilisation and low-cost applications," *IET Electric Power Applications*, vol. 13, no. 2, pp. 215–221, 2019.

[2] M. Abbes, K. Belharet, H. Mekki and G. Poisson, "Permanent magnets based actuator for microrobots navigation," *2019 IEEE/RSJ International Conference on Intelligent Robots and Systems (IROS)*, Macau, China, 2019, pp. 7062–7067.

[3] C. Gong, *Crash Safety of High-Voltage Powertrain Based Electric Vehicles: Electric Shock Risk Prevention*, Springer Nature, Cham, 2021.

[4] J. Liu, C. Gong and Z. Han, "Tuning method for digital PI controllers of PMSM closed-loop driving system," *Electric Machines and Control*, vol. 22, no. 4, pp. 26–34, 2018.

[5] J. Liu, C. Gong, Z. Han and E. Zhang, "An improved adaptive fuzzy PID controller for PMSM and a novel stability analysis method," *2017 IEEE 3rd International Future Energy Electronics Conference and ECCE Asia (IFEEC 2017 – ECCE Asia)*, Kaohsiung, 2017, pp. 2161–2164.

[6] F. Salem, Mohamed I. Mossad and M. Awadallah, "A comparative study of MPC and optimised PID control," *International Journal of Industrial Electronics and Drives*, vol. 2, no. 4, pp. 242–250, 2015.

[7] J. Gao, C. Gong, W. Li and J. Liu, "Novel compensation strategy for calculation delay of finite control set model predictive current control in PMSM," *IEEE Transactions on Industrial Electronics*, vol. 67, no. 7, pp. 5816–5819, 2020.

[8] W. Li and J. Liu, "Improved high-frequency square-wave voltage signal injection sensorless strategy for interior permanent magnet synchronous motors," *IECON 2019 – 45th Annual Conference of the IEEE Industrial Electronics Society*, Lisbon, Portugal, 2019, pp. 3205–3209.

[9] W. Li, J. Liu, J. Gao and C. Gong, "High frequency response current self-demodulation method for sensorless control of interior permanent magnet synchronous motor," *IEEE Access*, vol. 9, pp. 157093–157105, 2021.

[10] C. Gong, Y. Hu, J. Gao, Y. Wang and L. Yan, "An improved delay-suppressed sliding-mode observer for sensorless vector-controlled PMSM," *IEEE Transactions on Industrial Electronics*, vol. 67, no. 7, pp. 5913–5923, 2020.

[11] C. Gong, X. Ding, Y. Liu, L. Ge and J. Liu, "Sensorless control method for surface-mounted permanent magnet servo motors over low-speed range in

robotics," *2024 IEEE 7th International Electrical and Energy Conference (CIEEC)*, Harbin, China, 2024, pp. 4259–4262.

[12] C. Gong, Y. R. Li and N. R. Zargari, "An overview of advancements in multimotor drives: structural diversity, advanced control, specific technical challenges, and solutions," *Proceedings of the IEEE*, vol. 112, no. 3, pp. 184–209, 2024.

[13] W. Gui, J. Gao, C. Yang, T. Peng, C. Yang and Y. Han, "Optimized FCS-MPCC based on disturbance feedback rejection for IPMSMs under demagnetization fault in high-speed trains," *Control Engineering Practice*, vol. 141, p. 105670, 2023.

Index

www.ingramcontent.com/pod-product-compliance
Lightning Source LLC
Chambersburg PA
CBHW050511190326
41458CB00005B/1498